The Age of Epistemology

Also available from Bloomsbury

A History and Philosophy of Expertise, by Jamie Carlin Watson
Beyond Hellenistic Epistemology, by Charles E. Snyder
Kant on Spontaneity, by Marco Sgarbi
Knowledge, Number and Reality, edited by Nils Kürbis,
Bahram Assadian and Jonathan Nassim
Locke on Knowledge, Politics and Religion, edited by
Kiyoshi Shimokawa and Peter R. Anstey
The Aftermath of Syllogism, edited by Marco Sgarbi and Matteo Cosci

The Age of Epistemology

Aristotelian Logic in Early Modern Philosophy 1500–1700

Marco Sgarbi

BLOOMSBURY ACADEMIC
LONDON • NEW YORK • OXFORD • NEW DELHI • SYDNEY

BLOOMSBURY ACADEMIC
Bloomsbury Publishing Plc
50 Bedford Square, London, WC1B 3DP, UK
1385 Broadway, New York, NY 10018, USA
29 Earlsfort Terrace, Dublin 2, Ireland

BLOOMSBURY, BLOOMSBURY ACADEMIC and the Diana logo are
trademarks of Bloomsbury Publishing Plc

First published in Great Britain 2023
This paperback edition published 2024

Copyright © Marco Sgarbi, 2023

Marco Sgarbi has asserted his right under the Copyright, Designs and
Patents Act, 1988, to be identified as Author of this work.

Cover design: Narcissus Marsh's Table of *Regressus* in his *Institutio logicae* (1679)
Cover image: Frontispiece to Niccolo Tartaglia, "Nova Scientia" (*New Science*), 1537.
Niccolo Fontana Tartaglia (1499 – 1557) (© Science History Images / Alamy)

All rights reserved. No part of this publication may be reproduced or transmitted in any
form or by any means, electronic or mechanical, including photocopying,
recording, or any information storage or retrieval system, without prior
permission in writing from the publishers.

Bloomsbury Publishing Plc does not have any control over, or responsibility for, any
third-party websites referred to or in this book. All internet addresses given in this
book were correct at the time of going to press. The author and publisher regret any
inconvenience caused if addresses have changed or sites have ceased to exist,
but can accept no responsibility for any such changes.

A catalogue record for this book is available from the British Library.

A catalog record for this book is available from the Library of Congress.

ISBN: HB: 978-1-3503-2654-5
PB: 978-1-3503-2669-9
ePDF: 978-1-3503-2655-2
eBook: 978-1-3503-2656-9

Typeset by Newgen KnowledgeWorks Pvt. Ltd., Chennai, India

To find out more about our authors and books visit www.bloomsbury.com
and sign up for our newsletters.

Amicus Plato, Amicus Aristoteles, magis amica Veritas
— Isaac Newton

Contents

Sources and abbreviations — viii

Introduction — 1

1 Averroists — 13
2 Alexandrists — 31
3 Francis Bacon — 45
4 Galileo Galilei — 65
5 Thomas Hobbes — 93
6 René Descartes — 107
7 Isaac Barrow — 133
8 Robert Hooke — 149
9 John Locke — 167
10 Gottfried Wilhelm Leibniz — 181
11 Isaac Newton — 195

Conclusion — 223

Notes — 239
References — 275
Index — 295

Sources and abbreviations

All Greek and Roman authors are cited in their most familiar single-name form, both in the text and in the bibliography, for example, Cicero (not Marcus Tullius Cicero) and Quintilian (not Marcus Fabius Quintilianus). My general rule has been to preserve original spelling and punctuation, even when erroneous, except where there are critical editions. Sometimes, when fitting quotations around the text, I have silently changed a lowercase initial letter to an upper, or vice versa, as the sentence requires. I have provided all titles in English, even when originally written in another language (mainly Latin, Italian, French and German). The only exceptions are Francis Bacon's *Valerius Terminus*, where the scholarly convention is not to translate, and Newton's *Optice* in order to make clear its distinction from the *Opticks*.

All quotes from Aristotle are from Aristotle, *Complete Works of Aristotle*, edited by Jonathan Barnes (Princeton: Princeton University Press, 1984), according to Immanuel Bekker's edition. All the translations from Zabarella are from Jacopo Zabarella, *On Methods, On Regressus*, edited by J. P. McCaskey (Cambridge, MA: Harvard University Press, 2013).

A   *Leibniz: Sämtliche Schriften und Briefe* (Darmstadt-Berlin: Akademie-Verlag, 1923–).

AT   *Oeuvres complètes de René Descartes*, edited by Charles Adam and Paul Tannery (Paris: Cerf, 1897–1900). All English translations have been taken from *The Philosophical Writings of Descartes* (Cambridge: Cambridge University Press, 1985).

CL   *Opuscules et fragments inédits de Leibniz*, edited by Louis Couturat (Paris: Félix Alcan, 1903).

EW   *The English Works of Thomas Hobbes* (London: Bohn, 1839–1845).

GP   *Die philosophischen Schriften von Gottfried Wilhelm Leibniz*, edited by Carl I. Gerhardt (Berlin: Weidmann, 1875–90).

KGS   *Kants gesammelte Schriften*, edited by the Royal Prussian (later German, then Berlin-Brandenburg) Academy of Sciences (Berlin: Georg Reimer, later Walter de Gruyter, 1900–). The one exception to this rule is the *Critique of Pure Reason*, where passages are referenced by numbers from 'A', the first edition of 1781, or 'B', the second edition of 1787.

OFB   *The Oxford Francis Bacon* (Oxford: Oxford University Press, 1996–).

OG   *Le opere di Galileo Galilei*, edited by Antonio Favaro (Florence: Barbera, 1890–1907).

OL   *Opera philosophica*, edited by William Molesworth (London: Bohn, 1839–45).

SEH   *The Works of Francis Bacon*, edited by James Spedding, Robert Leslie Ellis and Douglas Denon Heath (London: Longman, 1857–74).

Introduction

This book is about the history of epistemology from the Renaissance to Newton. It traces the development of a number of epistemological ideas and doctrines which concern the interaction between the empirical and intellectual moments of knowledge. The historians of philosophy have traditionally been tempted to divide early modern philosophy into two major approaches: (1) empiricism and (2) rationalism. In this chapter, I show that these two categories are largely misleading and should now be abandoned. Labels like 'empiric', 'empiricist', 'experimental', 'constructivist', 'dogmatic', 'intellectual' and 'rationalist' are used in this book reservedly.[1] Indeed, as will become clear in the unfolding of my argument, even the most rationalist thinker appeals to experience, and the most faithful empiricist feels the urge to rationalize his findings. To afford a better picture of early modern epistemology, it is tempting to substitute for the distinction between empiricism and rationalism other categories such as experimental natural philosophy and speculative natural philosophy. However, as Peter R. Anstey has pointed out, any such attempt is destined to fail, because clear and consistent epistemological boundaries are hard to establish in a period of methodological novelties.[2]

Since my main interest is the history of epistemology and logic, I will select those aspects which most clearly illuminate issues debated by philosophers of that time such as induction, deduction, negotiation, meditation, regressus, syllogism, the role of experience, experiment and observation, the certainty of mathematics, and the foundations of scientific knowledge. In general, these are the issues that concern the interaction between the empirical and the rationalist stages of knowledge, or how knowledge can be transformed by the intellect from unclear and confused cognition into clear and evident knowledge: in other words, how ordinary experience with all its cognitive defects of being limited, provisional, accidental and conjectural can become science, that is universal, certain and necessary knowledge. Larry Laudan puts it very well when he writes that 'the chief role of epistemology in the 17th century was that of redefining the character of the relation between theoretical knowledge and sensory experience'.[3]

These epistemological redefinitions, developments and transformations span a period of three centuries, in which philosophy underwent substantial changes. The historical contexts and intellectual frameworks vary quite significantly. Nonetheless, all the authors of this volume shared a common university education which was mainly

Aristotelian. Aristotelianism persisted at least until the end of the seventeenth century as the university philosophy par excellence. However, even Aristotelianism changed over time, absorbing doctrines coming from other philosophical traditions and becoming even more eclectic in its nature. Albeit the teachings in the universities were based on Aristotelian texts, *what was taught* in Italy, France, England or Germany and *how it was taught* could differ, generating a variety of interpretations of Aristotle's doctrines. These interpretations were also affected by interactions with new philosophical and scientific tendencies, producing thus original perspectives and different responses to Aristotle. The Aristotle read in Padua was different from that taught by Jesuits at the College of La Flèche or that discussed among Oxford physiologists. In this book, therefore, I adopt the standpoint of the history of epistemology, contextualizing in their specific historical contexts the diverse approaches and reactions to Aristotelian philosophy.

It comes as no surprise that I start the book with Aristotle and various sects of Aristotelians. Indeed, the history of early modern philosophy is usually depicted as a long farewell to the Aristotelian tradition. For many early modern philosophers, Aristotle enjoyed the fortune – or misfortune – of being considered either a champion of empiricism against the intellectualism of his mentor Plato, or the founder of a cathedral of truths based on empty syllogistic reasoning lacking the yardstick of experience. This book does not escape this narrative but aims to revisit this leave-taking in a fresh and singular manner, elucidating the specific interactions between the empirical and rational stages of knowledge, which during the Renaissance was epitomized by the Aristotelian theory of regressus.

In no way, however, does the book support the idea that all early modern philosophers and scientists were Aristotelian. This would be anachronistic for a period of raging anti-Aristotelianism.[4] Neither is it my goal to suggest that Galileo Galilei was influenced by Johannes Lorinus, Ludovico Carbo or Paulus Vallius, that René Descartes picked up important ideas from Jesuit thinkers, that Thomas Hobbes based his method on Jacopo Zabarella, that Samuel Smith or Robert Sanderson had an impact on Isaac Newton. My research is not a genealogy. The question of the sources is scarcely relevant because my aim is to look at how in *different* knowledge-making situations philosophers reacted in *diverse* ways to *similar* problems, creating *multiple* epistemological grammars.[5] In stating that some philosophers tackle the issues of regressus or the negotiation of the intellect – even if they do not employ these terms – my intention is not to apply misleading or anachronistic categories or labels, but just to show the similarities and analogies between epistemological problems and cognitive processes, being fully conscious that terminologies vary across time and that different historical and geographical contexts generate different questions.

To take a case in point, the problem of 'clear and distinct' knowledge has two different epistemological references before and after Descartes. The epistemic context in which Locke, Leibniz and Newton use the expression ('clear and distinct') is Cartesian – or at least it can be scarcely understood without reference to this conceptuality. Other authors like Galileo and Descartes himself had Renaissance Aristotelian theory of method as their epistemic framework. This does not prevent us from tracing the developments from one conception to another, revealing differences and similarities.

Something similar may be said of concepts like 'analysis' and 'synthesis' or 'resolution' and 'composition', which changed their meanings and functions following the massive growth of mathematical thought and chemistry in natural philosophy. Again, over these three centuries, notions like 'principles' and 'causes' lose their metaphysical and ontological connotations as constituent elements of reality, becoming the epistemic grounds for founding scientific knowledge. The same relation between subject and predicate in a proposition of scientific knowledge moves away from designating an essential connection that links cause and effect, substance and attribute, and becomes an epistemic functional link of pertinence without any ontological commitment.

In looking at epistemological problems from this perspective, however, it is fruitful to better understand the theories of these authors. We should avoid the idea that they were isolated geniuses capable of completely subverting a tradition, or that the history of philosophy is written by a progressive uncovering of previous failures and errors. On the contrary, the main tenet of my investigation is that starting from a pretty secure Aristotelian environment, or one in which Aristotelian concepts were still very much alive – albeit starting to crumble – these authors took up the devices of Peripatetic philosophy and transformed them into something of their own.

The basic idea in what follows is that the farewell to Aristotle was not a refusal or denial of his philosophy. Rather, it was an attempt to solve some unresolved issues arising from his doctrines through the most advanced understandings and ideas of the time. I call these open issues 'trading zones' between old and new epistemological doctrines,[6] in which the foundations of the theory of knowledge are left unexplored or presupposed, opening up the problem of underdetermination,[7] or what in other cases has been called opacity.

I use the term 'opaque' following Gerd Buchdahl's seminal interpretation of early modern epistemology from René Descartes to Immanuel Kant. Buchdahl's basic idea is that in the foundation of an epistemology, the entire system is or can be coherent, only if the epistemological core is based on extra-epistemological and extra-logical, and even sometimes extra-philosophical, convictions and presuppositions, which are not justified within the system itself. Such characteristics define these epistemological centres as eminently opaque. To make more explicit what characterizes an opaque epistemological element, I will refer directly to Buchdahl's example of the Cartesian theory of knowledge, which I do not discuss in my reconstruction of Cartesian epistemology. Descartes attaches great importance to the ideal of mathematical knowledge, which is the field of clear and distinct ideas, and to the fact that mathematics is a powerful instrument for understanding reality. In order to support this thinking, he introduced a number of opaque conceptions, among which the most notable and well-known is that 'God is no deceiver', which are not justifiable within the epistemological system itself, but which are essential for its existence. For instance, from the conception that 'God is no deceiver' it follows that, although 'our perception of the external world does not reach intrinsically the standard of mathematical knowledge, it is nonetheless perfectly dependable'.[8] This leaves the problem of the foundations of scientific knowledge completely underdetermined from the epistemological standpoint. Similarly, Buchdahl finds analogous arguments in Locke for the reliability of sensory knowledge, in Berkeley where God serves as the foundation of inductive uniformity, in Leibniz where God guarantees the uniformity of knowledge, and in Kant with the Aristotelian categories, which provide the certainty of the scientific knowledge of reality.

In this book, I will explore some of these opaque and underdetermined zones in major authors like Francis Bacon, Galileo Galilei, Thomas Hobbes, René Descartes, John Locke, Gottfried Wilhelm Leibniz and Isaac Newton, for whom Aristotle's doctrine of the scientific foundations of knowledge constituted, directly or indirectly, an enduring problem to be faced in the specific historical contexts in which they lived. The various positions for and against Aristotle generated answers, which are only misleadingly cast as a rejection of his philosophy, when in truth they are a new appropriation and transformation of the Aristotelian tradition, an echo of which we can find even in contemporary epistemology and philosophy of mind.[9]

I propose to read the transition from a widely accepted Aristotelian theory of knowledge or epistemology, straddling more than 1,500 years in different and even at times contradictory versions, to multiple and concurring epistemologies. The evolution of these epistemological ideas is not a progressive denouncement and suppression of errors by superior epistemologies. The idea of rejection, refusal and denial of Aristotelianism as underpinning the rise of early modern philosophy is old in origin and based on the self-proclaimed novelty and claims of anti-Aristotelianism by the so-called moderns, known in fact to historiography as *novatores*. Already in the titles of their works one senses the wind of change that appears to characterize a radical programmatic turning away from the past, and from the Aristotelian tradition especially. It is a widespread European phenomenon which concerns both Latin and the vernacular languages (English, French, Italian, Spanish, Dutch and German), and all disciplines, from astronomy to zoology, from botany to mechanics, passing through chemistry, geology, geography, logic, mathematics and medicine. Among various titles, we may recall Niccolò Tartaglia's *Scientia nova* (1537), William Gilbert's *De magnete … physiologia nova* (1600), Johannes Kepler's *Astronomia nova* (1609), Francis Bacon's *Novum Organum* (1620), Galileo Galilei's *Discorsi e dimostrazioni intorno a due nuove scienze* (1638), Blaise Pascal's *Expériences nouvelles touchant le vide* (1647), and Robert Boyle's *New Experiments Physico-Mechanicall* (1660), all of which emphasize their own original and innovative character in opposition to the Aristotelian tradition and a sterile manner of thinking which is incapable of producing and discovering new knowledge.[10]

How far these self-declarations of originality correspond to the truth, or rather constitute a case of collective blatant self-deception and the same kind of cultural propaganda as Erwin Panofsky recognized in humanists and Renaissance thinkers who were convinced that they lived in an epoch of Renascence, is an open debate.[11] Often in the history of thought, philosophers and scientists have declared themselves to be revolutionaries or subverters of the status quo in order to establish their ideas, exalt their own genius and disseminate their thought: witness Kant and his Copernican revolution.

Is it possible to trust these claims of novelty and anti-Aristotelianism, or are we in fact misinformed by the moderns? Clearly, one cannot question the very existence of these novelties and discoveries over time: they were real and had a huge impact on the histories of philosophy, science and technology. What one must ask is whether the perception of these findings and their impact were truly as revolutionary and innovative as was suggested. Reading the historiographical reconstructions of the period raises an element of doubt. Generally, authors like Bacon, Galileo and Boyle were seen as *novatores*, but this label did not necessarily have the positive connotations it enjoys today. It could simply mean that they were perceived to be living in a new epoch. Polymath Daniel Georg Morhof, for instance, wrote in his *Polyhistor* (1688) that most of the moderns

> attempted a rebellion and strove to conquer the tyranny of the Peripatetics … In addition these innovators do not share the same genius; some of them are mad in their reasoning, some are entirely inept and have introduced nothing new into philosophy apart from various terminologies … The quest for a little fame obsessed many, but only a few were led by the study of truth.[12]

In Morhof's view, desire for fame and glory, more than the truth, moved these *novatores* to criticize Aristotelianism. He believed that emphasizing the original character of their research was not instrumental to the quest for the truth opposed to the supposed falsities of the ancients, rather it was a symptom of the sense of rivalry and competitiveness among colleagues racing to unearth new discoveries that would be remembered in the annals of the history of science and thought.

We may be tempted to accuse Morhof of conservatism, or that he did not understand the implications of the novelties, but it would be a charge that would fall short of the target: Morhof is well aware of the extent of the new findings, and in *Polyhistor* he lists them precisely for each field of science and philosophy. Despite this, these discoveries were not considered revolutionary, but as products of the development of the knowledge-seeking activities of the time. Of course, while Morhof's testimony cannot be taken as gospel, nevertheless it is evidence of a critical feeling towards developments of science in those years.

The moderns' own self-depiction as anti-Aristotelians led twentieth-century historiographical research not only to consider early modern philosophy as a reaction against Aristotelianism but also to neglect the study of Aristotelianism within the intellectual framework that facilitated and conditioned the innovations and discoveries of the 'new philosophy'. Past studies have presented Aristotelian culture as effectively stagnant, a context within which philosophers were obliged to teach the Peripatetic doctrine according to specific canons and standards: these Aristotelians could hardly be expected to bring radical novelties in their comments and explanations of the Aristotelian texts. The situation has radically changed with the new millennium. Scholars have increasingly appreciated the contribution of Aristotelian and Scholastic doctrines to the formation of early modern philosophical and scientific culture.[13] More recently, Daniel Garber has struck the definitive blow against the idea that the philosophy of the *novatores* took over from Aristotelian philosophy, and that 'modern science' simply replaced Aristotelian *scientia*.[14] Indeed, neither Bacon's alleged experimentalism nor Galilei's mathematicism, neither Descartes's supposed rationalism nor even Hobbes's so-called constructivism were capable of establishing themselves globally and supplanting Aristotelianism: 'The diversity of alternative anti-Aristotelian programs that blossomed in the late sixteenth and early seventeenth centuries never completely sorted itself out in a single alternative to the Aristotelian program.'[15]

This is particularly true if we look at epistemology in the early modern era, the broad topic of the present book. Although prominent scholars like Julia Annas and Jonathan Barnes have referred to this period as the 'age of epistemology',[16] one might find it anachronistic to deal with epistemology in a time when the term itself did not exist, let alone the discipline.[17] Indeed, 'there is no sign of anything that corresponds very closely to epistemology as we now think of it'.[18] However, this does not mean, as Robert Pasnau has correctly argued, 'that premodern philosophy neglected epistemology' or that the concerns of epistemology found no place in the older divisions of philosophical disciplines.[19] Perhaps the essence of the problem is to understand with due regard to the history and context what might be thought of as epistemology in early modern philosophy without imposing on the past our ideas of what epistemology has become or is in the present.[20] Current understandings of epistemology no doubt differ from the epistemology of the past. Indeed, many of the topics discussed in this book were in ancient and in medieval times – and especially in the Aristotelian tradition – subject to investigation within a single discipline, that is, logic. However, logic itself underwent radical change at the outset of the early modern period, abandoning the traditional formal apparatus of syllogistic and incorporating psychological and metaphysical themes, which were alien to the earlier tradition.[21] The term epistemology – made up of the Greek words ἐπιστήμη (scientific knowledge) and λόγος (reasoning, doctrine) – seems to capture the idea of a broader doctrine, embracing not only the method for acquiring scientific knowledge but also its subjects, foundations and limits. In working with the appropriation and transformation of the Aristotelian tradition in various historical and intellectual contexts, this book focuses in particular on a specific epistemological problem which the Aristotelians left open and opaque, one that they did not fully convincingly resolve, that of the certainty of the foundations or principles of scientific knowledge and of their acquisition.[22]

*

The entire Aristotelian tradition struggled to reconcile – on the one side – the empirical strand that Peripatetic philosophy assumed with Theophrastus with – on the other – the typical idealistic perspective developed by Aristotle's mentor Plato. This dynamic led Aristotelians to develop multiple epistemologies to tackle problems related to the theory of knowledge and the foundations of science. The struggle between these two positions reached its culmination during the Renaissance with the attempt to develop a perspective based on the theory of regressus – an attempt destined to influence the rise of early modern epistemology.[23] The theory of regressus arises for solving a crucial problem of Aristotelian epistemology.

Aristotle's epistemology is based on two very well-known assumptions: (1) scientific knowledge is always knowledge of first causes and principles and (2) scientific knowledge is always demonstrative. These two assumptions entail also (1) a form of (direct or indirect) realism according to which the world corresponds to mental conceptions and

(2) the possibility of knowing the essences of things existing in the world.[24] All other kinds of cognition lead only to probable and not scientific knowledge. This does not mean that they have no epistemic value but that their conclusions are not necessary, certain and universal. The idea that knowledge is first of all knowledge of causes was commonly shared with different degrees of nuance by early modern philosophers, according to how they conceived of a 'cause' or a 'principle'. In contrast, the fact that scientific knowledge should be demonstrative was much more controversial, meaning that knowledge is never intuitive, but always discursive, and most of the time based on a particular kind of syllogism called demonstration. Nonetheless, most authors from the Renaissance to Newton shared the view of the importance of demonstration for constructing a system of knowledge. Before Descartes, however, demonstrative knowledge of causes and principles gave access to the real essences of things, as they are. After Cartesian philosophy, this possibility is undermined by authors such as John Locke, Isaac Newton and Immanuel Kant.[25]

Aristotelian epistemology is based on the idea that a cause is the reason why a thing exists, that it is the cause of this, and that this cannot be otherwise.[26] This means that from the standpoint of the content, knowledge of causes is always necessary to explain how a thing is and cannot be in any other way. However, for Aristotle knowledge is necessary not only by reason of its content but also by reason of its formal structure being demonstrative. A demonstration is a particular kind of reasoning from which necessary and certain conclusions follow if and only if they are inferred from true, primary, immediate (or self-evident), better-known and prior premises. The Aristotelian problem is how to know with necessity and certainty these premises, avoiding the dilemma of infinite regress. Indeed, if premises are to be known scientifically, then they should be demonstrated, but the premises from which each premise needs to be demonstrated must in turn be scientifically known, and either this process continues forever, generating an infinite regress, without producing knowledge, or it comes to an end at some point.

Aristotle tries to solve the problem by arguing that knowledge of first principles, that is of the foundations of scientific knowledge, is of a different kind and not in itself strictly scientific or demonstrative. He makes clear his solution in *Posterior Analytics* II.19. Aristotle explains that the first objects of knowledge are always sensibles; indeed without sensation no knowledge would be possible. Sensibles then rest in the mind of some animals. If they rest in the mind, then following various sensations, a kind of scientific knowledge is made possible through an extended and laborious effort of the mind. Indeed, memory originates in sensation, shaping experience of the many particulars. From experience a 'generic concept' (καθόλου) of these particulars is formed that remains in the mind. In this way, it is possible to acquire a disposition for scientific knowledge. The mental process, which infers from the various particulars what is the same in all of them, is a kind of inductive process (ἐπαγωγή). This form of induction is a process of notification of knowledge from matter to form, from particulars to a universal. At the end of the cognitive process, what was initially particular in sensation and in the external world becomes a universal concept that reflects the object of experience. The object remains the same, while the cognitive representation changes.

By contrast, Aristotle terms intellection (νοεῖν) the mental process that produces knowledge of the product of this induction. This is performed by the understanding (νοῦς), and its knowledge is true and certain and exclusively intellectual. The process of acquiring and knowing general concepts and principles is therefore twofold. On the one hand, we have the formation of preliminary and rough and imprecise knowledge, which relies on induction (ἐπᾰγωγή), and on the other hand, we have the actual cognition of the intelligibles (νοεῖν), which is a kind of intuitive and immediate act of grasping what is given and generated by experience.

This kind of actual cognition of immediate first principles, with which every scientific demonstration begins, is qualitatively different from the cognition that follows the conclusions of the demonstration, since there is a passage from a generic concept to a true universal concept. In fact, for Aristotle the formation and intellection of general concepts and principles produces only temporary knowledge, which must be proven discursively by means of demonstration before becoming scientific knowledge.

What is left unclear according to the early modern interpreters is the nature of the transition from induction to intellection, from generic knowledge coming from experience to universal knowledge generated by the understanding. The Aristotelian problem can be stated as follows: if according to Aristotle, *nihil est in intellectu quod prius non fuerit in sensu*, knowledge of the principles comes from experience and induction, that is, it relies on something accidental and provisory, which cannot constitute the premise of a scientific demonstration, which, according to Aristotle, should be true and certain. Therefore, how is it possible to acquire scientific knowledge if the demonstration is not based on valid, certain and truthful premises? How is the transition from experience to science possible? How is objective knowledge possible? In other words, as the early modern thinkers understood the problem, how might confused and obscure cognition be transformed into clear and distinct knowledge? This is the big question which early modern epistemology addresses confronted by Aristotle's theory of scientific knowledge.[27] As we shall see, early modern philosophers answered this challenge in different ways, making a fundamental contribution to the reshaping of the Aristotelian tradition and the making of early modern epistemology.

*

This book deals with the reaction of the so-called *novatores* to the Aristotelian theory of knowledge in their attempt to develop their own epistemological systems. These can hardly be categorized under a single label for the reason that they manifest their own individual originality. Many important authors like Spinoza, Berkeley and Hume are left out of the account because their philosophical projects came from a different direction, both in terms of background and perspective. Equally, some were not perceived in their time as contributing to epistemological discussion. Indeed, both Spinoza and Hume were famous for their ethical, political and theological ideas, rather than for their epistemologies.

Furthermore, the focus here is solely on those philosophers and scientists whose epistemologies confronted the Aristotelian approach either directly or indirectly: many names are well-known and have been the subject of detailed investigations for their contribution to the theory of knowledge, and in this sense the book would seem quite traditional. However, the perspective adopted in this research is unique in looking at them for their transformations of Aristotelian epistemology. This is the reason why the book does not deal with women philosophers, though not – as far as we know – engaging with Aristotelian epistemology as a direct stimulus. The most interesting reflections in this sense are made by Margaret Cavendish in her *Observations upon Experimental Philosophy* (1666), but her focus is exclusively on Aristotelian natural philosophy, rather than logic, methodology and epistemology as they are understood in the present investigation.[28]

Finally, in dealing with the evolution of epistemological ideas within the same philosopher, we will encounter many different positions, some of them conflicting. The idea is not to make their epistemology consistent and coherent, but to give a picture of their thought at the moment of its conception.

After the present introduction, Chapters 2 and 3 focus on the various Aristotelian epistemologies originating from the reading of different commentators on Aristotle from the fifteenth century to Galileo Galilei and beyond. They constitute the first full-fledged investigation of the intermediate stage of regressus, the pivotal logical tool for the discovery of the foundations of scientific knowledge. The intermediate stage of the regressus was considered to be the process for transforming empirical and experiential knowledge into intellectual knowledge by providing certain and true premises for demonstration. Called variously mental consideration, negotiation of the intellect, intermediate discourse of the intellect, mental examination, meditation, application and intention of the mind, and work of the mind, this process has been considered the crucial link between Renaissance Aristotelian logic and the early modern experimental method, even if no one has as yet focused on its origin and history. Chapter 2 explains the reason for its introduction among Averroists in order to solve a crucial problem for the elaboration of Aristotelian epistemology, and also points out its main weaknesses.

The Averroist position was rapidly superseded by those logicians, who endorsed an Alexandristic position – that is, an instrumental conception of the relation of logic to science.[29] Chapter 3, therefore, examines Jacopo Zabarella's elaboration of the intermediate stage and its impact on early modern Aristotelianism. The example of Zabarella and his influence shows that the intermediate stage of regressus is mainly, with very few exceptions, conceived as a logical and mental operation of the mind acting as a support for processing experience and the empirical part of the logical process, which by itself cannot lead to science. The intermediate stage becomes responsible for the transition from confused and empirical cognition to distinct and scientific knowledge. In this transition, experiential knowledge achieves the same certainty of mathematical knowledge, becoming science.

Chapter 4 analyses how and by what means Bacon's epistemology digests experience. Usually considered the herald of the new inductive sciences and of experimentalism,[30] Bacon – this chapter shows – through his original conception of induction, tried to elaborate a complex process for finding the axioms of scientific knowledge, the route to discovering the forms of nature. His epistemology offers a perfect balance between the gatherings of empirical data and the rational processing of that information. Specifically, in this chapter, I show that for Bacon experiments play a role mainly, if not exclusively, in the literate experience, but not in induction, which is based on the

Platonic conception of division – rejected by the Aristotelian as an invalid form of method for discovery and recast as the intermediate stage of regressus – and mainly an intellectual activity, a process of clarification of what has already been acquired by experience.

Chapters 5 and 6 deal with the epistemologies of Galileo and Hobbes. Chapter 5 discusses Galileo's original appropriation and re-elaboration of the Aristotelian logical process of the negotiation of the intellect, allowing him to reconcile his experimental activity and his mathematical approach. It focuses on Galileo's understanding of the Aristotelian process and its weaknesses, which need systematization for it to function as the logic of scientific discovery. It deals with the problem in Galileo's early logical writings, while the last part is devoted to the Galilean solution and the transformation of the Aristotelian process into a powerful new investigative tool through the application of mathematics. Chapter 6 reconstructs Thomas Hobbes's idea of scientific method, contextualizing it within the framework of the Aristotelian tradition. Contrary to the orthodox interpretation, I suggest reading Hobbes's method in the *De corpore* as constituting three different stages: (1) analysis, (2) an intermediate step comprising analysis and synthesis, and (3) synthesis. The intermediate step in particular has similarities with the processes of solution and combination typical of the Aristotelian negotiation of the intellect. The chapter shows how Hobbes transforms the epistemology based on regressus, by considering efficient causes and not formal causes as the primary means of proceeding from analysis to synthesis.

Chapter 7 contests the exclusive role that some scholarship has granted to mind in René Descartes's epistemology. In the first part of the chapter, I demonstrate how Descartes's early epistemology and understanding of scientific knowledge in the *Rules for the Direction of the Mind* was influenced by the Renaissance debate on method, specifically in relation to his solving the Aristotelian epistemological problem of the certainty of the foundations of scientific knowledge. Furthermore, it examines how, in addressing this difficulty, Descartes gives precious clues regarding the method for establishing the only criterion for clearness and distinctness that Gottfried Wilhelm Leibniz was looking for in Cartesian writings before the publication of *Posthumous Works on Physics and Mathematics* (1701). The second part of the chapter provides a new account of Descartes's conception of *inspectio mentis*, usually translated as mental scrutiny or mental inspection, the only means by which scientific knowledge is possible. The chapter shows that this notion in the *Second Meditation* (1) is not a kind of intuition but requires hard work of the mind, (2) plays the same role as the negotiation of the intellect, (3) concerns all kinds of ideas and (4) demands a consistent contribution to the experience through enumeration. Reading this mature conception of *inspectio mentis* in light of his earlier methodological attempts provides a new understanding of his epistemology, which seeks to mediate between the active role of the mind and the indispensable function of experience.

Chapters 8 and 9 examine two main exponents for the Royal Society, Isaac Barrow and Robert Hooke. Chapter 8 focuses on Isaac Barrow's ideas on the certainty and causality of mathematics, on the theory of demonstration, and on the origin of the first principles of science. Chapter 9, in turn, considers Robert Hooke's Baconian epistemology in the *Micrographia*, his conception of philosophical algebra, his theory of the reduction of sensations to mathematical standards and finally his original perspective on analysis and synthesis.

Chapter 10 deals with John Locke's epistemological attitude towards experience. Although Locke for a long time was considered the founder of empiricism, I demonstrate – following the path of recent research – the importance of the rational part of his theory of knowledge and his re-elaboration of the Aristotelian tradition. Nonetheless, his exclusive reliance on sensation as the only means of acquiring knowledge and his denial of the appraisal of substance mark a clear break with Aristotelian epistemology, making superfluous any kind of mental or intellectual process in the extension of knowledge. The fruitful interaction between experience and reason *seems* destined to collapse, opening the way to epistemological pessimism.

In Chapter 11, I show how the traditional image of Leibniz as a rationalist, which would argue in favour of the complete obliteration of experience, is largely false. Rather, I show the importance of experience within Leibniz's epistemology as a means of acquiring scientific knowledge and truth, and how he elaborates new forms of analysis to be applied to both mathematics and natural science, which benefit from the discussion of the regressus. His epistemology therefore is not based only on the pure workings of the intellect but requires a fundamental contribution of experience.

Chapter 12, finally, focuses on Isaac Newton's logic of discovery in both his optical writings and in the *Mathematical Principles of Natural Philosophy*. In the first part, I emphasize how Newton conceives of the scientific method as being composed of two moments, analysis and synthesis, leading to a substantial revision of the Aristotelian theory of regressus. In the second part, I examine in what sense Newton's *Rules for Philosophizing* in the 1713 edition of the *Principles* relate to the intermediate stage of regressus, as some recent interpretations have advanced.[31]

The conclusion in Chapter 13 outlines the fate of the latest developments in epistemology as the end point of early modern philosophy, focusing on Immanuel Kant's transcendental idealism from the early pre-critical attempts to his most mature works. His programmatic intent is to overcome dogmatism and empiricism. In overcoming this epistemology, Kant aspires to a scientific knowledge with a level of certainty that is equal to mathematics, but in order to do this he denies the possibility of having scientific knowledge of what Aristotelians called essence. Kant resolves the conflict that Bacon had felt at the end of the Renaissance, originating in the contrast between empiricism and rationalism, though an epistemological tension remains.

1

Averroists

Padua. 1493. Within the university walls, professors were confronted by their different interpretations of Aristotle, following either Thomas Aquinas or Duns Scotus. Bishop Pietro Barozzi had just issued an edict against the Averroistic idea of the uniqueness of the intellect. In these tumultuous years, the first news of the discovery of the West Indies was spreading rapidly. With clear evidence of human life in the torrid zones, the certainties of Aristotelian philosophy were being shaken by experience. Aristotelian wisdom was under attack and held in check.[1] The Aristotelians' embarrassment was evident. There was a revealing episode on 18 March 1523 during Pietro Pomponazzi's lectures on the impossibility of living in the equatorial region according to Aristotle's *Meteorology*. When the class was over, he declared:

> This – my dear students – is what Aristotle and Averroes say. But what should we think today? I believe that when sensation and reason are in contradiction, sensation should prevail ... therefore I tell you that I have received a letter from a friend, who has written to me that the king of Spain sent three ships and ... they passed through the torrid zone, and they discovered that it was inhabited. ... Therefore all the things demonstrated by Aristotle are false.[2]

Aristotelians had to elaborate a new strategy capable of making sense of experience – of justifying and rationalizing what was hitherto inconceivable for their Peripatetic minds, like the habitation of the equatorial zones.[3]

Renaissance Aristotelians felt the urgency to fix their epistemological problems, and they did so by reflecting on the logical theory of regressus. In general, regressus theory comprises two main stages and has its origins in Averroes's interpretation of Aristotelian logic, for which real science was made possible only by a twofold process.[4] The first stage is a kind of resolution or analysis and usually employs the tools of induction and of an argument from effects to cause (also called demonstration τοῦ ὅτι, *quia, quod, ab effectu, ab signo, prius nobis, a posteriori*). The second stage is characterized by composition or synthesis and usually employs the tool of arguing from cause to effects (also called demonstration τοῦ διότι or *propter quid, prius naturae, a priori*).[5]

Scholarship has devoted a lot of attention to analysis and synthesis,[6] but notwithstanding the number of studies and the copious streams of words written on the topic, no one has so far focused on the origin and history of the intermediate process between these two stages. This intermediate process is variously named mental consideration (*mentalis consideratio*), intermediate consideration of the intellect (*intermediam intellectus considerationem*), negotiation of the intellect (*negotiatio intellectus*), intermediate negotiation of the intellect (*negotiatio intermedia intellectus*), intermediate discourse of the intellect (*discursus intellectus intermedio*), examination (*examen*), mental examination (*mentale examen*), meditation (*meditatione*), application and intention of the mind (*applicationem et intentionem mentis*) and work of the mind (*labor mentis*). But what exactly is this intermediate stage? Jacopo Zabarella, one of the most celebrated logicians of his time, asked the same question:

> What sort of thing this mental consideration is and how it is done, I have not seen made clear by anyone. For even though some say that this intermediate negotiation of the intellect is interposed, they nevertheless have not shown how we are led by means of it into distinct knowledge of the cause and what the power of this negotiation is.[7]

The history of the intermediate stage of regressus starts in Padua with Francesco Securo Neritonensis. Neritonensis was a pivotal figure at the University of Padua at the end of the fifteenth century. Dominican and professor *in via Thomae*, Neritonensis was the teacher of Pietro Pomponazzi, Tommaso de Vio (also known as Cajetanus) and Gasparo Contarini. Despite his nickname 'greatly verbose' (*multum verbosus*),[8] he has not left any writing, and we can find traces of his thought only in the works of his pupils and opponents such as Antonio Trombetta, Nicoletto Vernia, Pietro Pomponazzi, Agostino Nifo and Girolamo Balduino.

Neritonensis's main thesis is that intellectual knowledge of the cause cannot be superior and more intense than sensible knowledge. If science is knowledge of causes that proceeds from the known to the unknown, the process from sensation to the intellect cannot be from something unknown in the sensation to something known in the understanding; otherwise, the order of cognition established by Aristotle in *De anima* and in the *Organon* would be reversed. Therefore, sensible knowledge – which comes from the demonstration *quia* (or induction) for Neritonensis – was sufficient for knowledge of the cause. Demonstration *propter quid* – which proceeds from the cause to the effects – added no new content to knowledge, making regressus theory totally unfruitful. Demonstration *propter quid* becomes in this way mainly a dialectical or topical argument[9] – concerning the arrangement and the order of knowledge, rather than a truly demonstrative argument, which discovers new knowledge. Only demonstration *quia* follows the order of cognition and provides firm knowledge of the cause. Indeed, the discovery of the cause from the effect is certain because the effect is, without doubt, cause of the discovered cause, and this is the highest certainty that one can achieve, even higher than knowledge coming from analysis of a cause. In the latter case, there is lacking a necessary and essential connection between cause and effect since multiple causes can produce that same effect, while there is only one discovered cause of the effect. Furthermore, from the existence of the effect we can necessarily infer the existence of the cause, while, vice versa, from a cause we cannot argue for the existence of a particular effect. Therefore, working forward from the cause to discover the effect is a mere *supposition* – that is, a conjecture – whose epistemological evidence and level of certainty fall short of any scientific demonstration.[10]

The criticism against demonstration *propter quid* is even harsher. Neritonensis would have denounced the usefulness of the demonstration *propter quid* for discovery, especially in the field of natural philosophy. Its validity can at most be supported for the mathematical sciences, where the premises – that is, the causes of the demonstration – are axioms conceived artificially by the mind. This validity cannot be assumed in natural philosophy because otherwise we should also have to assume that understanding comprehended something more than sensation, against the Aristotelian and Thomistic principle, 'there is nothing in the intellect that was not earlier in the senses' (*nihil est in intellectu quod prius non fuerit in sensu*) and what was said previously regarding the order of cognition. Having dismissed the contribution of demonstration *propter quid*, Neritonensis also diminishes its role within regressus theory. Indeed, the conclusion of the demonstration *quia* – the place where the demonstration *propter quid* started according to the supporters of regressus theory – does not provide the definition or the essence of the cause, rather merely knowledge of its existence, and thus the final knowledge achieved by the demonstration *propter quid* does not differ, qualitatively and quantitatively speaking, from that acquired by the demonstration *quia*. In other words, in assuming the twofold movement of regressus theory, one would be endorsing circular reasoning explicitly denied by Aristotle in *Posterior Analytics* 72b 25–9. Neritonensis's condemnation of regressus theory is total: his perspective emphasized resolution based on the priority of sensible knowledge and of experience, and therefore of induction or demonstration from the effect. The strong emphasis on the analytical and experiential stage of regressus led Neritonensis to downplay the contribution of the intellect in the discovery of the cause: experience and analysis were adequate for this purpose. Demonstration *quia* was sufficient in its own right for the cognition of the cause, and the demonstration *propter quid* was therefore superfluous to its discovery. Here we find an important preliminary separation being drawn between analysis as a method of invention and synthesis as a method for the arrangement or order of knowledge. Neritonensis is stretching Aristotle, proposing a new interpretation of his epistemology, which represents a distancing from the Aristotelian tradition, a tradition that puzzled his contemporaries. It was indeed a rejection of Averroism too, at least as Ugo Benzi conceived it: he maintained that the first part of regressus – that is, the demonstration *quia* – was sufficient for scientific knowledge because it discovered the cause from the effects, and since science is knowledge of cause (*scientia scire per causam est*), this was enough.[11]

Paduan logicians, who defended regressus theory, in turn conceived an intermediate step between the two kinds of demonstration and called it negotiation of the intellect. Pietro Pomponazzi reveals this episode in his manuscript *quaestio* titled *If There Is Regressus*, written in 1503. Countering Neritonensis's objections, the Scotist Antonio Trombetta was the first to introduce this kind of logical device:

> Our teacher Antonio Trombetta responds to this argument [Neritonensis's argument] in his *Questions* and in his reply to Burley at [his commentary to] the first [book] of *Physica*. For Master Antonius says that although when it is first

apprehended our knowledge of the cause is not so intense as our knowledge of the effect, nevertheless it is pondered through the negotiation of the intellect (*per negotiationem intellectum*), (not, however, strictly speaking) and the appropriate knowledge of the cause is made more intense through certain principles evident to the intellect.[12]

In the third *quaestio* of his *Metaphysical Questions* (written in 1493, but first printed in 1502), Trombetta answers Neritonensis generically. Neritonensis had maintained that the understanding could not know anything without a corresponding image coming from sensation. He had assumed, moreover, that the knowledge of the image is more intense than that of the understanding and, finally, he had argued that this intensity derives not from the understanding but from sensation – otherwise the demonstration would proceed from the unknown to the known, in opposition to the Aristotelian principle mentioned above. According to Trombetta, the process of negotiation justified the validity of regressus, giving as his reason the qualitative gap between, on the one hand, the knowledge of cause from effect and, on the other, the higher knowledge of the cause employed as universal principle in the ensuing demonstration *propter quid* – set in such contrast by Neritonensis. Trombetta never explains in detail the functioning of the negotiation: his position is merely sketched in outline, and the process remains mostly obscure. Pomponazzi, however, provides a clear example of Trombetta's thinking:

> For example, it is supposed that if from the eclipse I come to its cause, which is some obstruction (for the eclipse is deprivation of light from the Moon), the intellect then tracks down this cause using certain self-evident principles. Thus: there is an obstruction between the Sun and the Moon. It is either a body or a non-body. Not a non-body, because then it could not be an obstruction. It is either a transparent or an opaque body. Not a transparent body, because then it could not obstruct … so it is Earth. And thus having acquired a more intense knowledge of that cause, I return to the effect.[13]

This process of division, coming from Plato, however, had already been criticized by Aristotle as a weak, hypothetical and conjectural type of argument, similar to induction.[14] It would be unable to establish the true definition, which should be the premise of a scientific demonstration. Aristotle is clear on this point, and the long passage deserves quoting in full:

> For division is, so to speak, a weak deduction; for what it ought to prove, it begs, and it always deduces something more general than the attribute in question. … In demonstrations, when there is a need to deduce that something belongs, the middle term through which the deduction is formed must always be inferior to and not comprehend the first of the extremes. But division has a contrary intention; for it takes the universal as middle. Let animal be the term signified by *A*, mortal by *B*, and immortal by *C*, and let man, whose definition is to be got, be signified by *D*. The man who divides assumes that every animal is either mortal or immortal: i.e.

whatever is *A* is all either *B* or *C*. Again, always dividing, he lays it down that man is an animal, so he assumes *A* of *D* as belonging to it. Now the deduction is that every *D* is either *B* or *C*, consequently man must be either mortal or immortal, but it is not necessary that man should be a mortal animal – this is begged: and this is what ought to have been deduced.[15]

Aristotle denies that division could really discover anything, but he does not deny that it could make knowledge clearer and more evident, even if by nature conjectural and hypothetical.[16] That was enough for the supporters of regressus to state that the knowledge of the cause discovered through the demonstration *quia* became clearer and more intense after the negotiation, and thus that this cause could be employed as the premise of the demonstration *propter quid*.

Neritonensis countered this reasoning that higher intellectual knowledge should start from more intense sensible knowledge, claiming to base this conclusion on the reading of an unspecified passage in the seventh book of *Physics*. He had this argument and 'nothing more', Pomponazzi wrote,[17] and in response to his personal request to indicate the passage, Neritonensis acknowledged that though absent from the Aristotelian corpus, the position was completely consistent with Aristotelian ideas and their framework.[18] Neritonensis's answer – notwithstanding Pomponazzi's disdain – is rather illuminating because it emphasizes the fact that in order to attain clearer and more evident knowledge of the cause, one should not turn to the generic and abstract working of the intellect but rather to the more accurate propensities of sensation and experience. The negotiation of the intellect should be an empirical or experimental part of regressus, which in Neritonensis's eyes had to do with analysis or resolution and not with synthesis or composition. However, Neritonensis's position was not completely understood, and no further opposition to negotiation remained.

*

In the same years in which Trombetta published his ideas in defence of regressus against Neritonensis, Nicoletto Vernia gave a series of lectures on *Posterior Analytics*, now preserved in the manuscript *Canon. Lat. Misc. 506* at the Bodleian Library. One of these lectures claimed that the passage on circular reasoning in *Posterior Analytics* is a defence of the Averroistic theory of regressus as being the only tool for acquiring scientific knowledge. Vernia manifestly declares that his objective is to confute Neritonensis's thesis, which denied the scientific validity of 'this circle and of this circular demonstration'.[19] The issue at stake is to understand whether demonstration *quia* is sufficient or lacking in providing scientific knowledge. Vernia clearly states that 'the philosopher who proceeds from the process *quia* is very imperfect'. What is more known to the mind is the knowledge of the effect, but this knowledge cannot be limited to the content acquired by sensation. Having identified the cause of the effect, the understanding should 'regress' in order to acquire perfect science, otherwise the analytical process, 'which acquires universal knowledge through individuals', would provide only a hypothetical kind of knowledge. Instead, true scientific knowledge of the cause, according to Vernia, shows that the cause is the cause of the effect *in essendo* and not only *in cognoscendo*. In other words, we have scientific knowledge when we know that the cause is the cause of the being of the effect, which means in Aristotelian terms that it is a formal cause. In the demonstration *quia*, in contrast, for Vernia we know the cause by means of the effect, and therefore we know the effect better than the cause because the effect is the cause of the knowledge of the cause, but not of its being. We conceive of the relationship between the cause and the effect only *in cognoscendo*. In the demonstration *propter quid*, the cause is not only the cause of cognition but also of the being of the effect. Only in this sense is the demonstration *propter quid* a true scientific demonstration.

Demonstration *quia* cannot lead to science since it is reducible to an induction, leading to hasty generalizations that can be disproved in particular circumstances or cases. Resolution, Vernia states,

> is not discursive, but proceeding from senses, memory and experiments (*ex sensibus, memoriis et experimentis*) generates a universal knowledge and, after having acquired the universal, the understanding knows the individual in a reflexive manner and from the universal concludes, and by means of it the understanding makes the induction.[20]

The process outlined by Vernia comes directly from Aristotle's *Posterior Analytics* II.19. The result of the inductive process, herein called *notitia universalis*, is not a true universal concept, but a generic concept (καθόλου) that does not clearly characterize the essence of the various individuals, and for this reason the understanding may yield hasty generalization and fallacious conclusions. Like Trombetta, Vernia uses the example of the lunar eclipse:[21]

> When I say 'the Moon eclipses; something is placed before what eclipses; therefore something is placed before the Moon,' I have an imperfect and confused knowledge, that is, that something is interposed between the Sun and the Moon.[22]

Resolution, therefore, does not provide a proper knowledge of the effect from the cause, because that knowledge of the cause is weak and imperfect. Inductive knowledge is merely accidental, not necessary, and therefore without any scientific value. Vernia thus denounces the approximate character of sensible knowledge. He usually refers to the repetition of sensible knowledge as 'experiments' (*experimentis*). The use of this word does not imply that Vernia is suggesting a kind of experiment performed by early modern science: experiments are only repeated experiences.[23] And even if it were not so – and we took the word in the modern sense – Vernia is attacking this kind of method as inadequate and insufficient for scientific knowledge. We should supplement experiments with a form of rational validation and verification, to the extent that it is not so very important to know the effect or the cause as it is to understand whether they provide an adequate middle term for the demonstration and that the middle term refers to both in the conclusion. For this reason, it is necessary to introduce a new process: in precise terms, the negotiation. Negotiation is, therefore, couched as a support to experiential knowledge, so that this may become science. It is not an experience or an experiment in itself, in so far as that, at least for Vernia, negotiation is limited to intellectual activity. As with Trombetta, for Vernia negotiation works like a disjunctive syllogism (i.e. through division) by means of which we should acquire a clearer knowledge of the cause, because the knowledge that was previously based on two terms or attributes, will now be based on four or eight.[24]

The divisive process of negotiation, as outlined by Vernia, is a process of mental clarification of experience acquired in the stage of resolution. By means of negotiation, the demonstration would be based on necessary principles and not on probable opinions, and for Vernia this was enough to answer Neritonensis's objections to regressus.

Nicoletto Vernia's pupil Agostino Nifo took the concept of the negotiation of the intellect a stage further. In his juvenile work, by his own testimony, he had supported the necessity of the negotiation of the intellect as an intermediate process between the demonstration *quia* and the demonstration *propter quid*:

> In the *Collections* and in the *Commentaries* to the books of *Physics* we believed that beyond the discovery of the cause there was needed a particular negotiation of the intellect, primarily by means of the cause ... And for negotiation of the intellect we meant composition and division, by means of which the understanding discovered the cause, that is, the definition. We called the first process the discovery of the cause, of its existence (*quod est*), but negotiation discovered the definition of the effect. Subsequently in the second process, the effect was known according to its essence (*propter quid*) by means of a causal definition.[25]

Indeed, in the *Exposition of Physics* (1508), as John H. Randall emphasized, Nifo presented an articulated regressus theory consisting of four stages. However, it is hard to say whether Nifo really supported the validity of the negotiation as he suggested years later, because regressus theory is attributed to 'recent thinkers', perhaps Vernia, rather than to himself:

> Recent thinkers maintain that there are four kinds of knowledge. The first kind is of the effect through the senses, or observation; the second is the discovery of the cause through the effect, which is the demonstration by sign; the third is knowledge of the same cause through a negotiation of the intellect, from which there first comes such an increased knowledge of the cause that it is fit to serve as the middle term of a demonstration *simpliciter*; the fourth is a knowledge of that same effect *propter quid*, through that cause known so certainly as to be a middle term.[26]

Nifo makes clear that the most important aspect of regressus theory is not the discovery of the cause, but rather to ascertain whether the cause could become the middle term of the demonstration. For Nifo, as with Vernia, negotiation functions as a reinforcement of the rational epistemological apparatus, which is fully developed only in the demonstration *propter quid*. Unlike his predecessors, Nifo gives a precise characterization of what negotiation is:

> This negotiation is composition (*compositio*) and division (*divisio*). For when the cause itself has been discovered, the intellect composes and divides until it knows

the cause in the form of a middle term. For though cause and middle term be the same thing, they differ in their aspect. For it is called the cause in as much as the effect proceeds from it, whether it be better known than the effect or not. But it is a middle term in as much as it is a definition. From effect to cause is thus the procedure of discovering the cause; negotiation is directed toward the cause as a middle term and a definition. But since a definition is discovered only through composition and division, it is through them that the cause is discovered in the form of a middle term, from which we can then proceed to the effect.[27]

Negotiation for Nifo would be precisely the process outlined by Vernia, were it not for the fact that it is made up of two further processes: composition and division. This twofold process establishes a qualitative difference in cognition between resolution (demonstration τοῦ ὅτι) and composition (demonstration τοῦ διότι). In resolution we merely identify the cause of the effect, while in composition we grasp the actual connection between the cause and that particular effect from which the process of knowledge started. For understanding further developments, for instance, in Thomas Hobbes, it is important to emphasize the distinction between the resolution and analysis of the first part of the regressus from the process of division in the negotiation, and the composition and synthesis of the second part of regressus from the process of composition in the negotiation.

In the subsequent edition of the *Commentaries*, however, Nifo adds a consideration (*recognitio*) in which he would deny the validity of negotiation:

It is customary to deal at length with the regressus in physical demonstrations; I say 'physical,' because there is no regress in mathematics. In this difficulty the most recent thinkers conceive three kinds of knowledge in the demonstrative regress. The first is knowledge that the effect is, that is, that the proposition signifying the effect is true; and this knowledge comes from the senses. For instance, that man has the capacity for science is known by sense. The second kind of knowledge is of the reason why what is observed by sense is so. Thus we consider the reason why man has the capacity for science, and not the brute; and we say, because he has a rational soul. Therefore of the effect, or of the proposition signifying the effect, there are two kinds of knowledge: the one, that it is true, and this is clear to the sense; the other, why it is so, and this is known to us through the discovery of the cause. Of the cause, or of the propositions signifying the cause, there is but one kind of knowledge, and this is discovery, which is nothing else than that it is the cause, or that the propositions that signify the cause are true. Hence these recent thinkers conceive that through this knowledge, which is the discovery of the cause, or that the propositions, which signify the cause are true, there is learned the reason why the effect is so, or why the conclusion which signifies the effect is true. Thus in the regressus in physical demonstration there are three kinds of knowledge, of which two are of the effect while the third is the discovery of the cause. When the last is related to the effect, it is the reason why the effect is so; but when it is related to the cause, it is the fact that it is the cause. And this discovery is made through the effect... From this it is clear that there is no need of any negotiation to render greater our knowledge of the cause, as we formerly held; for the mere knowledge that it is the cause is the reason why the effect is so.[28]

So after being initially drawn to Vernia's position, Nifo is persuaded by recent thinkers and by a closer reading of the Aristotelian Greek commentators that negotiation is useless since the knowledge of the resolutive moment is insufficient to discover the cause.

Among these recent thinkers, we can certainly find Pietro Pomponazzi, who in his 1503–4 lectures on *On the Soul* criticized Nifo's *On the Three Books of the Soul*, devoting a *quaestio* to regressus theory.[29] In the *If There Is Regressus*, against Trombetta's conception of negotiation as a kind of division, Pomponazzi sided with Neritonensis in establishing the priority of the resolution. The target of his polemic is Girolamo D'Ippolito from Monopoli, a Domenican theologian who taught at Naples and Padua. Girolamo argued that the demonstration *quia* led only to knowledge of the existence of the cause, without establishing a necessary and essential connection between cause and effect, since the initial knowledge of the effect is always accidental in being reliant on experience. Regressus, therefore, was necessary to determine that this specific cause was really the cause of the effect. This argument is invalid according to Pomponazzi 'because I cannot imagine that after proceeding from the effect to the cause I have knowledge of the cause but I do not know that that cause is cause of that effect, because – if it were so – I would never discover the cause of that effect from the effect, but I would always be looking for it'.[30] Such an argument is based on Hugo Benzi's *Exposition on Galen's Art of Medicine*,[31] in which resolution leads to the discovery not only of the cause but also of the fact that a particular cause is cause of a specific effect. Indeed, it would be impossible to proceed from the cause without the knowledge that this cause is the true cause of the effect from which the previous argument started.[32] If there is an essential connection between the cause and the effect, one cannot stand without the other.

Following Pomponazzi's argument, regressus would seem completely superfluous, but it is not so. Indeed, Pomponazzi establishes that even if the knowledge of the two moments has the same content, this differs in terms of clarity, distinctness and precision. Knowledge coming from resolution is confused and disorderly, while that coming from composition is distinct and perfect. The problem is, therefore, to identify the process that leads from confused to distinct knowledge – that is, from knowledge of the existence of the cause to knowledge of the causal dependence between cause and effect by means of the middle term. Pomponazzi's answer to this is the negotiation of the intellect:

> And when they ask: how should one proceed in regressus? I say, as it is usual, by means of a negotiation of the intellect; if that cause has prior causes, it is necessary to resolve those prior causes so that we may find the first and adequate cause; if the cause is first ... by means of the understanding one will convert the effect and the cause and will go back with the demonstration *propter quid*.[33]

Pomponazzi thus keeps alive the process of negotiation, but without specifying its function. The idea is pivotal in his philosophy, but remains rather opaque.

Pomponazzi's conception of the negotiation is likely to have exerted a certain degree of influence on Nifo.[34] Contrary to what he had written in his *On the Three Books of the Soul*, in his later works Nifo establishes a qualitative difference in the cognition between resolution and composition. In resolution, the mind merely identifies the cause of the effect, while in composition it grasps the actual connection between the cause and that particular effect from which the process of establishing knowledge started. The latter kind of knowledge does not provide a better understanding of the cause in itself, but only of its connection with the effect and, for this reason, even the demonstration *propter quid*, especially in the field of natural philosophy, retains a hypothetical character because the knowledge of the cause comes from sensation and experience. So Nifo explicitly writes that

> it seems to me that in the regressus made in physical demonstrations the first process, by which the discovery of the cause is put into syllogistic form, is a mere conjectural syllogism, since through it the discovery of the cause is syllogized in a merely conjectural fashion. But the second process by which is syllogized the reason why the effect is so through the discovered cause, is demonstration *propter quid* – not that it makes us know absolutely (*simpliciter*), but conditionally (*ex conditione*), provided that that really is the cause, or provided that the propositions are true that represent it to be the cause, and that nothing else can be the cause ... But you object that in that case the science of nature is not a science at all. We must say that the science of nature is not a science absolutely (*simpliciter*), like mathematics. Yet it is a science *propter quid*, because the discovery of the cause, gained through a conjectural syllogism, is the reason why the effect is so ... That something is a cause can never be so certain as that an effect exists; for the existence of an effect is known to the senses. That it is the cause remains conjectural, even if that conjectural existence is better known than the effect itself in the order of knowledge *propter quid*. For if the discovery of the cause is assumed, the reason why the effect is so is always known. Hence in the *Meteors* Aristotle grants that he is not setting forth the true causes of natural effects, but only insofar as was possible for him, and in conjectural fashion.[35]

Nifo agrees with Pomponazzi that the Aristotelian texts in support of the regressus theory seem to characterize natural philosophy as conjectural and hypothetical by nature, setting it apart from mathematics. Indeed, this conjectural nature was employed in the Aristotelian tradition by Averroes and Aquinas to explain that, even if Ptolemaic astronomy explained phenomena perfectly,[36] this did not mean that the explanation was 'causally' necessary, since effects could have different causes from those hypothesized by astronomic theory.[37] In mathematics, causes are certain because they are either self-evident or conventional, while in natural philosophy causes must be discovered from sensation and experience. In natural philosophy knowledge can be like mathematical knowledge only if we consider the discovered cause in the first stage as self-evident, but this knowledge will always be *propter quid*, and never *simpliciter* as in mathematics. Nifo acknowledges an intrinsic weakness in resolution for the constitution of natural philosophy as science. Such weakness was clear to Aristotle, who in his *Posterior Analytics* 1.12 (78a 5–13) wrote:

Now sometimes it is not possible to make a deduction from the assumptions; and sometimes it is possible, but it is not seen. If it were impossible to prove truth from falsehood, it would be easy to make an analysis; for they would convert from necessity. For let *A* be something that is the case; and if this is the case, then *these* are the case (things which I know to be the case, call them B). From these, therefore, I shall prove that the former is the case. (In mathematics things convert more because they assume nothing accidental – and in this too they differ from argumentations – but only definitions.)

Resolution provides certain results only when an effect can be produced by a specific cause and a specific cause necessarily produces only that effect – that is, when there is a co-implication of cause and effect. This would be possible only in mathematics where there is total convertibility, but not in natural philosophy where the same effect can be produced by various causes. Co-implication is viable only when propositions express essential connections, as is the case in mathematics. This does not happen when propositions express accidental connections, which is the case in natural philosophy – that is, where relations are established only by experience.

This idea shifted completely with the application of mathematics to natural philosophy, or better the reduction of the latter to the former, as happened in seventeenth-century philosophy. All propositions then became convertible. If in natural philosophy all connections might be considered essential, then mathematics would be the proper tool for describing nature – that is, it would be possible to apply geometrical methods and mathematical resolution to all areas of scientific research. However, at least for Aristotle, in nature – where there is matter and change – there are accidental connections, and for this reason mathematical resolution cannot be applied to all natural investigations. For all these reasons, Nifo excludes mathematics from regressus. Mathematical knowledge is always a priori; its object is always intentional, and it proceeds from cause to effect by a demonstration *propter quid*, which does not require a demonstration *quia*. The resolution of the demonstration *quia* is a *prius nobis* that does not coincide with the *prius naturae*, while the mathematical resolution is a *prius naturae* that coincides with the *prius nobis* because the objects are products of the mind.

Aware of the limitations of resolution as applied to natural philosophy, Renaissance Aristotelians gave to negotiation the task of finding out the essential connections and the co-implication, where existing, between the cause and the effect, in order to determine whether the discovered cause could serve as principle of the following demonstration *propter quid*. This determination, at least for Trombetta, Vernia and the young Nifo, was purely intellectual in scope, and no experiment could help it – not even a thought experiment, which would have been merely hypothetical and conjectural in kind.

*

The opacity of the negotiation process was such that many philosophers and logicians reacted against its introduction, including some among the supporters of regressus theory. According to Marco Antonio Zimara, science comes from the regressus – that is, from the combination of analysis and synthesis – but he takes as evidence for the negotiation's uselessness its absence from the Aristotelian corpus:

If that negotiation is necessary, why has it been omitted by the Philosopher [Aristotle] who teaches the art of demonstration? ... If the negotiation is necessary, it is surprising that Aristotle does not teach or employ it.[38]

Girolamo Balduino, by contrast, closely considers the internal dynamics of regressus. Balduino believed that regressus, constituted by the combination of the demonstration *quia*, which notifies the existence of the cause, and of the demonstration *propter quid*, which discovers the connection between the cause and the effect, was the essential instrument for acquiring science. In regressus, these two stages constituted a single reciprocal demonstration where the cause and the effect are convertible. Convertibility and reciprocity are possible, as we have emphasized previously, if and only where there is an essential connection between cause and effect – that is, co-implication. This co-implication is expressed by a definition – the only logical device capable of guaranteeing the passage between the *demonstratio quia* and the *demonstratio propter quid* because it determines the essence of the cause – which is, in other words, the formal cause. Only in this way can we know not only of the existence of the cause but also that this specific cause is the cause of the effect. If this were not the case, regressus would lead merely to accidental and conjectural knowledge, unable to identify the connection between the cause and the effect.

For all the logicians examined above, negotiation was responsible for this process of clarification of the cause. According to Balduino, conversely, this is not possible. Balduino never mentions negotiation explicitly in his work but criticizes that process of composition and division – that is, the process that Nifo characterizes as negotiation – which is said to lead to the essence and definition of the cause. The process of composition and division, indeed, would lead in his view only to the discovery of the parts in relation to the whole, which constitutes the object under investigation, and nothing more. Definition, instead, is the only tool that can determine the essence of the cause, capable of establishing that the terms of the demonstrations involved in the regressus are universal and necessary, and not merely accidental, contingent and particular. Moreover, a perfect regressus requires convertibility between cause and effect, and for this convertibility to occur the cause cannot be a generic cause. The cause discovered by the demonstration *quia*, whatever it is, must be in some way or other reduced to a formal cause, because this is the only cause that reveals a true essential definition. Only in this way may the cause – which will become the middle term of the argument – be both *causa essendi* and *causa inferendi* of the regressus.[39] This is important because otherwise the conclusion of the regressus would be merely conjectural and not necessary. In other words, like Nifo, Balduino knows that in order to acquire the highest degree of certainty of scientific knowledge in the field of natural philosophy, it is necessary to identify all essential connections between cause and effect. In this way, for Balduino, knowledge of natural philosophy has a level of certainty comparable to mathematical knowledge. In saying this, Balduino distances himself from Aristotle, who in the field of natural philosophy acknowledged demonstrations in which cause and effect were not convertible.

In opposition to Nifo, as already indicated, Balduino denies the validity of the process of composition and division – that is, of negotiation – in the identification of such essential connections. Nevertheless, he introduces another process of the intellect, which corresponds to negotiation, and which is totally independent of regressus. This process reduces the discovered cause to a formal cause, extracting the essence of the cause. To the question 'which is the tool employed to know the essence', Balduino answers that 'according to the first [book] of the *Topics* the tool that notifies the essence of a thing is definition'.[40] In contrast, to the question, 'by means of which instrument does one find the definition', Balduino's answer becomes rather blurred:

> One cannot give a specific rule to this. The reason is, according to the first book of the *Ethics* and the second book of the *Metaphysics*, that the way of knowing the subject-matter and the essence is specific to the [various] arts ... the tool will be that which allows the knowledge of the essence; judging how and which it is pertains to a skilled craftsman (*peritum arteficem*).[41]

Balduino's textual references were obvious for the Aristotelians of his time. In the first passage, which corresponds to *Nicomachean Ethics* 1098 a 25–31, Aristotle writes that a carpenter and a geometer see the same angle from two different perspectives; in the second, he establishes that one should not aspire to the same degree of certainty and rigour in all sciences: mathematics is more certain than natural philosophy because the latter deals with accidental things, namely, matter. Balduino shifts the problem of negotiation outside of the regressus, giving to the specific methods of each discipline the role of finding definitions. It is rather a singular notion, because in this way Balduino undermines the possibility of a general method or logic for scientific discovery and gives to the specificity and expertise of the various sciences the task of determining the essence. Balduino does not reveal whether this intermediate stage would be something purely mental, as was negotiation for Trombetta, Vernia and Nifo, or something more experimental; he makes clear that it is not a universalizable process, even if leading to the definition of the cause, and compares it to the knowledge and skill acquired by a craftsman.

Bernardino Tomitano endorses Balduino's conception of the intermediate stage of regressus. A detailed discussion of the regressus is contained in Padua University Library ms. 810.[42] As with Balduino, for Tomitano the transition from demonstration *quia* to demonstration *propter quid* is made possible not by negotiation, but by definition: indeed, only definition makes explicit the formal cause, to which all the other three aspects (efficient, material and final) of the cause should be reduced. In this way, the convertibility of the middle term, avoiding circularity of process, is guaranteed.

A lack of faith in the intermediate stage of regressus had been felt by Alessandro Piccolomini in his *Commentary on the Certainty of Mathematics* (1547), when against the supporters of regressus theory he posited not only that he had never understood what they meant by negotiation of the intellect but also that they were introducing an opaque element into a logical process, which should have been clear and distinct. According to Paolo Palmieri, 'this is the most damning criticism ever levelled at the intermediate stage of regressus theory ... no regressus theorist has ever been able to respond to Piccolomini's strictures'.[43] It is impossible to know whether the Zabarellean formulation would in the end have satisfied Piccolomini's demand – indeed, Zabarella's

On Regressus came out in the same year as Piccolomini's death (1579). However, the issue at stake is straightforward: for Piccolomini, negotiation would have introduced 'an obscure, ultimately illogical element into the logical process'.[44] Piccolomini solves the problem by reducing negotiation to definition.

In his *The Instrument of Philosophy* (1551) Piccolomini agrees with Balduino and Tomitano. Definition is the only tool that allows for reciprocity between the two kinds of demonstration by providing not only the existence of the cause but also its essence, and thus identifying the convertible middle term for the demonstration *propter quid*. After stating that it is hard and laborious to 'make the definitions of things', and after the exclusion of Platonic division as a method for finding definition because it does not lead to necessary conclusions, Piccolomini proposes the combined use of 'three methods [*vie*], that is, division, composition and then syllogism'.[45] In other words, even if Piccolomini refutes negotiation, he reintroduces it as the combined process of these three methods.

Piccolomini places great emphasis on the process of composition, in particular, of the genus with the differences, establishing two methodical rules. First, it is necessary to combine the differences gradually, without jumping to the conclusion, and according to their degrees of extension, starting with those closest to what needs to be defined. Second, the genus should be divided systematically into differences, which are necessary for reconstructing what is being defined. He calls the work of the mind that takes place according to these two rules in the Italian vernacular *negotio*, which seems to be an allusion to the negotiation of the intellect.[46] However, Piccolomini points out that these two rules are not always sufficient. Indeed, it may happen that there is no easy choice between two contradictory differences to attribute to a genus, and in this specific case one should appeal to sensation and experience. This is particularly clear when he applies his logic to natural philosophy in order to establish whether on a specific night there will be a lunar eclipse.[47]

First of all, the definition of eclipse is required. In order to secure this, one should primarily consider its genus – that is, a lack of lunar light. Then one should divide this genus into two differences – that is, a lack of light either with or without a full Moon. With the help of experience and sensation, the conclusion arrived at is a lack of lunar light together with a full Moon. However, from experience one appreciates that this lack of light can happen on other occasions as well, such as when the Moon is below the horizon or where there are clouds, in which case we cannot say that an eclipse is the lack of lunar light with the full Moon. Piccolomini suggests further dividing and reasoning to establish whether this lack of light is possible also on account of the interposition of the Earth between the Sun and the Moon. In order to test whether this difference is necessary to the definition, one should have recourse either to sensation or reasoning. But it is impossible to establish this through the senses – or else we should live on the Moon. Through reasoning, therefore, one becomes persuaded that the true difference is the interposition of the Earth, and that the definition of the eclipse is thus the lack of lunar light with a full Moon, accompanied by the interposition of the Earth between the Moon and the Sun.

As a consequence of this, sensation and experience can be helpful in affirming or negating which of the two differences is the true one. Nonetheless, Piccolomini is aware of the weakness of sensation, which can only know the accidents of things. Since sensation and experience cannot always be diriment, the intellect exchanges and substitutes common characteristics for essential and proper characteristics, falling into error and not determining the real definition and nature of a thing.[48] The role of sensation remains pivotal, and this is the reason why

> philosophers sweated in discovering with long observation and care, with anatomies and dissections of animals, plants, stones and of any other thing for understanding well which nature, part and condition were attributed and led to these or those accidents, in such a way as to know gradually the proper accidents of things.[49]

Piccolomini closes his investigation by saying that these observations, anatomies and dissections for finding the true differences are essential to the foundation of scientific knowledge, opening up empirical and experimental approaches typical of early modern philosophy.[50]

Girolamo Capivacci, who taught medicine at Padua, similarly, in his *On the Differences of Doctrines* (1562), focuses on the role of definition instead of negotiation, giving special emphasis to the experimental dimension. According to Capivacci, true definition 'not only explains the essence of the thing but makes knowable all the properties that come with the essence of the defined things in such a way that all ambiguities vanish'.[51] This kind of definition is possible through resolution. This resolution is not a form of division through which one enumerates the individuals of a species or the species of a kind – a process criticized by Aristotle, as we have seen; rather, it is a process that proceeds from individuals to species and kinds in order to identify the essential elements of an individual, dismissing the accidentals and establishing thus the true definition.[52] This process determines the presence and absence of peculiar characteristics in individuals, which makes possible the formation of groups of individuals with specific definition. It is a purely rational process, although based on the collection of individuals, which are known by sensation. This working of the intellect generates the basis for scientific knowledge: indeed, through this rational process we are able to discover the formal definition of an individual, a definition that will be sufficiently general to act as the principle for the following demonstration *propter quid*. According to Capivacci, there is an alternative way of finding the premise of the demonstration *propter quid*, which is based on experiment. In general, experiments produce technical knowledge (*peritia*) – expertise rather than science. However, if the experiment is introduced into the subsequent process of demonstration, then its result can provide scientific knowledge. This is what Capivacci calls experimental doctrine (*doctrina experimentalis*).

For Capivacci, this experiment consists of the process outlined by Aristotle in *Posterior Analytics* II.19. Repeated sensations produce images and simulacra in memory, and these – being analogous to sensations and experiences – are activated to produce general conclusions that are not properly speaking universal scientific principles. Indeed, being discovered by sensation in an accidental way, such conclusions cannot be categorized as necessary, and their validity should therefore be demonstrated. Only after this may they – in the same way as a formal definition – be properly considered a principle for a possible demonstration that leads to science. Capivacci's example is quite clear: after the repeated experience of seeing that lettuce cures someone who is feverish, we can conclude generically that lettuce cures fever. This kind of conclusion is merely experimental – that is, it is accidental and not necessary. For it to become necessary, one should explain why lettuce cures fever, that is, why what is refreshing cures and generates heat, first arguing that lettuce has a power to refresh and then that lettuce can cure fever. The result of this process is not a definition but a principle that can be used as a premise in a demonstration. In the particular case of the experiment, convertibility – which is required by Paduan logicians and achievable by means of a true analytic process that leads to a formal definition – is not possible. This is not a genuine regressus, properly speaking, but it allows the passage from resolution to composition. Regressus was expressed in this way:

1. Resolution: every A is B, every C is A, then every C is B.
2. Composition: every C is B, every A is C, then every B is A.

After composition, one can clearly see that the argument is not circular because the minor premise of the two syllogisms is inverted and then the conclusions are different. Experiment, in turn, was expressed through this formula:

Analytic stage
X_1 has experimented A and C happened
X_2 has experimented A and C happened
X_3 has experimented A and C happened
...
X_{100} has experimented A and C happened
From every A follows C

This inductive process simply states that from every A follows C, but it does not explain why. A following analysis of the definition of A and C leads to the identification of a series of characteristics, among which is the characteristic D which is contrary to C.

Every A has D
D is contrary to C
Contrary cancels out contrary ← common principle of medicine
Then from every A follows C

In both regressus and the experiment, additional work is necessary to determine the definition or the presence or absence of specific characteristics that explains the reason of the effect. This can be either totally mental or experimental, or a combination of the two. In some cases, mere investigation of the definition can produce sufficient knowledge to determine the formal definition as principle for a demonstration; in other cases, it is necessary to investigate many individuals by means of sensation to understand whether, and in what way, that characteristic is present and to elaborate a table of absence and presence.

In sum, Capivacci's idea of the intermediate stage of regressus is neither purely an intellectual exercise nor merely empirical. In both cases, he admits the necessity for additional work to support the transition from analysis to synthesis, allowing thus the passage from merely technical expertise to science. Whatever the nature of this kind of work, it confirms the weakness of the resolution, unable by itself to guarantee scientific knowledge. Capivacci, like Piccolomini, is aware that a merely empirical approach cannot lead to scientific knowledge, and that at the same time reasoning without the yardstick of experience would be empty. Influenced by his experience as a physician, he develops an epistemology that foreshadows what Bacon looked for at the beginning of the seventeenth century. However, the continual search for a definition would lead early modern philosophers to criticize this methodology as useless for the improvement of knowledge, it being chiefly focused on the explanation of words rather than things.

2

Alexandrists

Padua. 1562. Fresh off the press, the last volume of the massive Aristotle–Averroes edition printed by the Giunta brothers in Venice had just come out. The book, which circulated somewhat freely among the students of the University of Padua, declared to the world that their professor Bernardino Tomitano was the most outstanding logician and philosopher of the age. This edition represented a significant contribution to the history of logic, offering three different Renaissance versions of Averroes's *Long Commentary on the Posterior Analytics*. It represented the highest peak of interest in Averroism during the Renaissance. Soon afterwards, the new Greek commentators on Aristotle like Simplicius and Themistius replaced Averroes, providing a new angle on Peripatetic thought. Alexander of Aphrodisias, in particular, gained a considerable reputation for his audacious readings of Aristotle's psychology and logic.

One of the most important Alexandrists of the time in the field of logic was Tomitano's pupil Jacopo Zabarella. Zabarella has been estimated to be one of the most intelligent and clearest logicians of all time.[1] Famous for his instrumental conception of logic – which he derived from Alexander – Zabarella established once and for all the theory of regressus, making a fundamental contribution to the solution of Aristotelian epistemological problems. The multiple processes that characterize the obscure intermediate stage of regressus created confusion among logicians of his time, and it was for this reason that Jacopo Zabarella decided to determine the nature of this working of the intellect. Regressus, using the standard definition in his acclaimed work *On Regressus*, which was published for the first time in 1578, is

> some sort of reciprocated demonstration, by which, after we have demonstrated an unknown cause from a known effect, we convert the major premise and demonstrate the same effect by means of the same cause, so that we may know scientifically what it is on account of ... Regressus, therefore, is necessarily composed out of three parts. The first, of course, is demonstration *quod*, by which we are led from confused knowledge of the effect to confused knowledge of the cause. The second is that mental consideration by which we acquire distinct knowledge of the cause from confused knowledge of it. And the third is a demonstration *potissima*, by which we are finally led on from the cause known distinctly to distinct knowledge of the effect.[2]

The discussion comes alive in chapter 4. Here he aims to clarify the nature of the first procedure in a regressus, using the example of a demonstration made by Aristotle in the first book of the *Physics* 189 b 20–191 a 22, where it is demonstrated that matter exists. The demonstration *quia* states that 'where there is generation, there is underlying matter there; in a natural body there is generation; therefore in a natural body there is matter'.[3] Zabarella points out that the minor premise – that is, there is generation in a natural body – is known only in a confused way since it is known only through sensation when we discern empirically that a natural body is generated and passes away. Knowledge of why this is the case – that is, of the cause – is completely obscure, although we do understand that there is a cause. More interesting is his reflection on the major premise. The fact that 'where there is generation, there is underlying matter' is not something known through sensation because we do not have sensible knowledge of the underlying matter of substantial change but only of the matter that underlies the alteration of accidental qualities. He adds, however, that the major premise is easily knowable by means of 'some applied mental consideration (*aliqua adhibita mentali consideratione*)',[4] which is a kind of demonstrative induction:

> We see that change of accidents has a subject; we do not see this of substances; but nevertheless, once this has been considered a little, we know it must be so in every change, just as we see in change of accidents. And so we perform that induction that Averroes ... normally calls demonstrative ... Demonstrative induction is performed in necessary matter and in things that have an essential connection to each other. In it, not all particulars are taken account of, since once some have been *inspected*, our mind at once notes the essential connection, and then, leaving aside the singulars, at once gathers the universal. For it knows that it is necessary that it be so in the rest, for from the fact that some predicate belongs to something per se, it may be inferred that it is, therefore, predicated of all. When our mind considers, therefore, the nature of change or of generation, it knows that this predicate is essential to it and necessary, that is, that there is some underlying matter that is receptive to be so in the generation of accidents. Therefore, from the various changes in accidents it infers this universal, 'Every change has an underlying subject', under which it afterward gets this major, 'where there is generation there is underlying matter'.[5]

Demonstrative induction, unlike dialectical induction, discovers from a few particulars an essential connection between two terms through which it is possible to reach a general conclusion.[6] The minor premise is known in a confused manner because while we know that every change has underlying matter, we do not know that the effect of this change depends on this underlying matter – that is, underlying matter is not known to be the cause. Since every change is always conjoined to underlying matter and is not knowable independently of it, what one can say very simply is that when there is a change then there is underlying matter, but we do not know that this underlying matter is the cause of the change. The existence of something changing, therefore, argues in favour of the fact that there is a cause, but does not provide a clear understanding of the cause because the change is 'not known as an effect of that cause, but only as something never separable from it'.[7] A syllogism with such a confused premise can lead only to confused knowledge, and it is for this reason that regressus is so fundamental. The first part of regressus discovers 'only the inherence of the cause in a given subject, not that it is the cause of that effect, but as some sort of necessary and inseparable predicate'.[8] Following Pomponazzi, Zabarella admits that there is a qualitative difference about clarity and distinction in knowledge between the two moments of the regressus. The first leads to confused knowledge, and for this reason the conclusion must be tested further. The question is how one should test this conclusion so that it can become a premise of the demonstration *propter quid*. In order to do that, Zabarella says,

> It is necessary that there intercede some third intermediate work by which we are led into distinct knowledge of that cause, which was known only confusedly. Some, knowing that this is necessary, called it negotiation of the intellect. We can call it a mental examination of the cause or a mental consideration.[9]

The starting point of mental examination is to determine the existence of the cause; only in this way, indeed, is it possible to discover what it is. The second major aspect, 'without which the first would not suffice, is a comparison of the discovered cause with the effect by means of which it was discovered'.[10] This process of comparison would clarify confused knowledge by establishing the conditions of the cause. But what is this comparison? Zabarella refers to it without giving a precise definition. In the third book *On Methods*, Zabarella characterizes this process as a 'diligent inspection' (*diligente inspectione*) or 'mental inspection' (*mentis inspectione*).

The context in which Zabarella discusses this is a consideration of the distinction between what is said to be a known *per se* and what is said to be an unknown *per se* in the broadest theory of definition. An unknown *per se* is what cannot be known in light of its own self-evidence and which therefore requires something else, a middle term, for mediation to take place. Zabarella gives as an example the triangle in being a figure which has three angles equal in total magnitude to two right angles. This nature, according to Zabarella, is not known *per se*, or by means of extended mental examination or diligent inspection, since it is not sensible, suggesting that the inspection or the mental examination is useful only in the case of objects experienced through sensation. In order to understand the nature of a triangle one needs something additional that is better known – that is, a middle term – in order to demonstrate its meaning. It needs a work of the intellect.

A known *per se*, in contrast, has two meanings, as we saw earlier in this chapter: known *per se* is (1) what is evident to everyone; (2) what – if set out systematically – is known without something else being added, either at once or after some consideration of it. Among this second type, Zabarella includes definitions and suppositions which are not known, for instance, to the student before they are expressed by the teacher – in other words, they are something that is believed *per se* when properly set out. Objects of the second kind 'are not actually known but, nevertheless, if they are set out, they are knowable *per se*'.[11] For example,

> the definition of a circle is not actually known to the peasant but, nevertheless, if it is set out for him, he understands it at once, without any middle. So also the

definition of the human being, although it is not actually known to the peasant, nevertheless, as soon as it is set out to him, it is known by him *per se*, not by means of something else.¹²

So in Zabarella's conception we have three classes of objects: objects which are evident *per se*, objects that are evident only if they are carefully inspected and examined by division and objects that do not require further examination or inspection because they are merely intellectual by nature. The second category of objects is particularly interesting for our purposes because – as Zabarella explicitly says – they constitute definitions or suppositions coming from experience – that is, possible premises for a demonstration, once they are known clearly and distinctly.

This clarification occurs by means of a systematic exposure of the essential characteristics of something already known. Zabarella exemplifies this process by an analogy with what happens in sense perception:

> For some things do not appear on first glance that we see afterward when we exert the eyes' acuity more. And hidden by ashes, a burning coal is not discerned but then afterward, after there has been some moving and handling of the ashes or the burning coal itself, it is beheld and moves the sight by means of that [coal] itself. It is not to be doubted that something similar happens with our mind. For some things that are not apprehended by the very first mental inspection (*mentis inspectione*) are known afterward by means of a diligent consideration of them not by means of another, but by means of themselves. Therefore, that those things that we knew appertain to something essential to it and appertain to it by virtue of what it is; we do not show it by means of another, instead, the thing itself comes to light *per se*, when we *effectually fix the eyes of the mind on it*. For by considering such attributes well and comparing them with the thing itself, we come to know that they are essential to it and we constitute the definition from them.¹³

The parallel between what happens in sensation and in the intellect reveals that the process Zabarella suggests is a form of clarification. In other words, no new knowledge is discovered, rather it is better specified. In particular, the nature of the thing is explained by means of a mental inspection – just as we shall see in Chapter 7 in connection with René Descartes – and thus what was previously known only confusedly and obscurely, after inspection is known distinctly and clearly. Accordingly, these natures or definitions are clarified and examined in the mind and by the mind alone, without the contribution of sensation, because the parts and the whole were already known through the senses. The process of clarification is purely mental and intellectual and seemingly without involving sensation. On fixing the eyes of the mind on it, the thing itself comes to light *per se*. Zabarella, like Descartes will do later, outlines a process that proceeds from obscure and confused to clear and distinct knowledge, and argues that this clarification depends on how carefully the mind fixes on the object.

*

This kind of process of clarification happens naturally, as we have already seen, in the case of the so-called demonstrative induction. Here, not all particulars are considered, 'because when we start to enumerate a few, it is at once apparent that the predicate is essential to them, and so the enumeration of the others being but an aside, we gather the universal', that is, the nature or the 'what it is' of the thing,[14] at least as the mind conceives it. Thus Zabarella can conclude that

> in the *mental* consideration of a thing, the essential connection shines forth by means of the thing itself, not by means of something else. This becomes manifest to us without discursive movement when we consider them ... when we intently consider [the thing].[15]

In this case there is a clarification and identification of the essential elements of the thing, which does not happen either by means of syllogism or discursive reasoning or experience, but only by focusing the mind on the thing. This process where something is known by means of itself – that is, when no inference occurs – 'is a proceeding from imperfect to perfect knowledge of the same thing by means of diligent inspection of it'.[16]

We can go a little further with our investigation of what it means to focus the mind on an object. Zabarella states that this act of fixing the mind is a process of distinguishing one thing from others, by means of which the essence of what comes from sensation and imagination is conceived and understood. It is not a mere abstraction from matter but the identification of essential characteristics, through which what was previously known confusedly and indistinctly through sensation and imagination becomes known distinctly and clearly through the intellect.[17] This transition from sensation to intellect, from the sensible to the intelligible, does not always refer to a specific class of objects. For instance, after seeing Socrates with some specific characteristics, and having identified his essence as of a human being, and after seeing Callias with the same specific characteristics and having identified his essence as of a human being also, at the sight of another 'object' with similar characteristics, the intellect acquires the habit of forming an immediate judgement that that object itself is likewise a human being, without having to consider all the characteristics involved. This is possible for Zabarella because the intellect – in particular, agent intellect – is characterized by the act of distinguishing things from others (*actus distinguendi rem a re*), and this faculty is natural to the intellect but sharpened by the inspection of images (*phantasmatis inspectio*), which come from sensation.[18] In this way, once the mind perceives sensations with similar characteristics to those of Socrates and Callias – even if the sensations refer only to characteristics – the mind (through an idea) will infer the existence of a human being.

In conclusion, Zabarella characterizes the process 'from imperfect to perfect knowledge of the thing by means of a careful inspection of it' as division.[19] This comparison leads to the progressive knowledge of the conditions of the cause: once a condition is known, *by analogy*, many other conditions are knowable until we can determine all those conditions that are fundamental for the existence of the effect, as we will see in the case of the magnet in René Descartes's *Rules for the Direction of the*

Mind. Zabarella proposes to start from the confused knowledge of the cause: that is, of its existence. From this knowledge, bit by bit and progressively, one should identify which specific condition of the cause determines the effect. If – removing a specific condition – the effect still stands, then that condition is not an essential condition for the effect, and therefore it cannot be considered as a characteristic of the cause. With this work of accurate selection of the conditions of the cause, one can uncover a core group of essential conditions that determines the effect in such a way that the knowledge of the cause is no longer confused, but clear and distinct. Once all the conditions are established, the cause can operate as a term for the demonstration *propter quid*. The process does not differ from that proposed by Capivacci, with the construction of a list of these conditions' presence or absence, which will have echoes in Francis Bacon as well. Zabarella's example is not altogether clear, but it is the only one he can find in *Physics* 189 b 30–191 a 22, which seems to explain what mental examination is:

> From generation we discover that in a natural body there is the underlying matter of the substantial form. We do not yet know, of course, that matter is the cause of generation itself; we know only that this change cannot happen without some underlying matter, and accordingly that matter is inseparable from change ... Matter discovered by means of generation, therefore, is not yet known to be the cause of generation, because what matter itself is as yet unknown ... to discover what matter is, he progressed as follows. He taught in the first place how it differs from privation, for the function of matter is to underlie contraries and take them on, and the function of contraries is, in return, to impress themselves upon that same matter. Therefore, matter remains underneath each contrary and never passes away, and, since it ought to be the matter and the starting principle of all natural bodies, it also ought to be such that it is able to take on all forms and all privations. And so, according to its own proper nature, it ought to be committed to no form, no definite nature ... Matter, therefore, according to its own nature, ought to lack all forms and have the potential to take on all forms. Without doubt, this is the nature of matter ... and from that, the very definition of matter is gathered ... and thus we learn, little by little, what place matter itself has in generation and what its function is, and we distinguish that from the jobs of the other starting-principles. Once the function is known, each of the conditions that are necessary for it to be able to execute such a function come to light.[20]

The investigation of matter as the cause of generation leads to the identification of those characteristics which matter possesses in order to be classified as matter – that is, to underlie contraries and to be without a specific form. These are also the conditions that make matter the cause of generation. What is interesting for Zabarella is not the scientific validity of Aristotle's definition of matter, but rather the Stagirite's method – that is, the continuous process of comparison to uncover the essential elements that constitute the cause of the effect, a comparison which characterizes the mental examination. Only after this mental examination has taken place is it possible to proceed with the regressus. Zabarella, therefore, seems to make Capivacci's methodology explicit in a more detailed way.

His debt to earlier logicians at Padua, such as Balduino and Piccolomini, is evident in the second explanation of the mental examination, which is based on Aristotle's *Physics* 259 b 21-31. In this or passage, Aristotle – starting from the existence of an eternal motion, already previously established – aimed to demonstrate the existence of a prime eternal mover, although the knowledge of the latter would be confused. Some philosophers – and Zabarella is probably referring to Neritonensis or Pomponazzi – might object that we cannot know the existence of the prime mover without knowing the prime mover to be a cause – indeed, the concept of mover implies the concept of a cause. However, this knowledge is confused, 'for, of course, a mover is an efficient cause of motion, but the proper cause of eternal motion, which is a species of motion, is not a mover taken in the broad sense, but a mover endowed with some definite characteristics'.[21] It is not sufficient, therefore, to know the cause as cause, but the cause should be reduced to a formal cause. This cause must be known by means of all the characteristics that make it a cause 'and these characteristics of the first mover do not become known by means of that first procedure', that is, the demonstration *quia*. For this reason, 'it was necessary for Aristotle to perform that mental consideration or negotiation of the intellect by which the essential characteristics of the first mover came to light – if not all, at least some, that is, those that it was necessary to know so as to know the cause of eternal motion'.[22] In sum, as for Balduino, mental examination is capable of reducing the efficient cause to a formal cause. In the following passage, Aristotle, according to Zabarella, would provide a more exact explanation, by identifying the specific characteristics of the prime mover – such as being immovable – which relate to it not as an efficient cause but as a formal cause. Aristotle, 'who had not fully satisfied himself with that brief mental examination' at the end of the book would decide 'to perform that fuller mental consideration' and to identify as characteristics that 'the first mover is indivisible and completely separated from matter'.[23]

Against those who maintained as sufficient the knowledge of the existence of the cause for knowledge of causal connection, Zabarella shows the necessity of the demonstration *propter quid*:

> From the inspection of smoke, someone argues and shows that there is fire there that he does not see. By this one demonstration, it appears to be demonstrated both that there is fire there and why smoke comes to be. But this is not true. For in knowing these two things, the speed of the understanding deceives him. For the demonstration of fire from smoke, by its very nature, makes clear nothing except that there is fire. And if we appear also to know what that smoke is on account of, it is not by means of the same demonstration but by means of another, one by which we demonstrate smoke from fire and render the reason why there is smoke there.[24]

With this example Zabarella shows that the conclusions of the demonstration *quia* cannot lead directly to scientific knowledge, but the demonstration *propter quid* is necessary. Zabarella adds that with the intermediate mental consideration sometimes it can happen that 'when we demonstrate that a cause is from an effect, we may at the same time discover what that is',[25] but, like Pomponazzi, he emphasizes the difference between knowing that the cause exists and the connection between the cause and the effect.

It is noteworthy that for Zabarella, like his forerunners, mental examination was important in order to guarantee convertibility between cause and effect. In this way, the method of scientific discovery could aspire to the same certainty as that of mathematics. For Zabarella, philosophy of nature has the same rigour and demonstrative character as all the other mathematical sciences because it is made up of necessary propositions, necessarily concatenated, and in which the cause and the effect are convertible. Both mathematics and philosophy of nature are grounded on the convertibility of cause and effect, and their only difference is as follows:

> Mathematical resolution, through which after all demonstrations we retrace and resolve the posterior theorems in the anterior ones, and the latter in the principles, is a kind of exercise for erudites, in opposition to the resolutive method that we now consider; it is indeed a process from more unknown things to more known things, which would be almost useless to the ignorant and to someone who is preparing to learn science, because it would not produce any knowledge. Instead, we deal with that resolutive method which produces knowledge of more unknown things from more known things, and which happens in other sciences, especially in natural science.[26]

Zabarella's passage – which finds a close parallel, as we shall see, in Hobbes's *De corpore* ch. 6, sect. 19 – establishes that the substantial difference between the two methods is that the method of natural philosophy proceeds from known to unknown, while that of mathematics works from unknown to known. By means of mental examination, however, the two methods have the same level of certainty and rigour. In other words, Zabarella's method is a perfect combination of the Aristotelian discovery of the principles, starting from what is best known to the mind – that is, from sensation and experience – with the manner of geometry proceeding from perfectly convertible propositions. This is not exactly an Aristotelian method.[27] Indeed, in completing this idea of combination started by Nifo and Balduino, Zabarella moves away from the Aristotelian position. Co-implication and convertibility were, according to Aristotle, possible only in the field of mathematics, and not in natural philosophy where the contingent and accidents are dominant, and where a number of causes could lead to the same effect. For this reason, Aristotle introduced dialectical instruments such as induction for the investigation of natural phenomena, to be used in addition to demonstrative devices. Conversely, Zabarella limits induction to demonstrative induction, equating it with *demonstratio quia*, and thus all the cause–effect relations become essential and – as a consequence – the certainty to which natural philosophy can aspire is the same as that of mathematics. To sum up, in Zabarella there is a sort of mathematization of the mental examination, which we will find in very different forms in other early modern philosophers.

Zabarella's exposition of mental examination is clearer than that of other contemporary logicians, in particular because he relates this process to a kind of mental inspection or demonstrative induction and explains the importance of the essential connection between the cause and the effect. In any case, the working of the intellect remains largely a mental process, which has nothing to do with experiments in the modern sense of the term.

*

Zabarella's conception of mental examination had a long legacy in the sixteenth and seventeenth centuries; the Alexandristic interpretation of Aristotle spread not only in Italy but all over Europe.[28] Not all philosophers or logicians referred explicitly to it, and not all provided an exhaustive characterization of its functions, but almost all of them employed it. Most of the time, the treatments are very brief and very limited in information because they were published in short summaries and textbooks, but they nonetheless reveal the strength of the tradition.

Numerous examples illustrate Zabarella's influence in the decades around 1600. The Scotist Philippus Fabri attributes to Duns Scotus the identification of the intermediate stage of regressus as a kind of further 'inquiry' (*inquisitione*),[29] but he employs the Zabarellean example of underlying matter. Samuel Smith deals with negotiation as 'accurate consideration of the discovered cause'.[30] John Flavell, quoting Zabarella, states that the intermediate stage of regressus is 'a diligent mental examination by means of which we compare the cause with its effect and the effect with its cause, acquiring thus a distinct knowledge of the cause itself'.[31] Richard Crakanthorpe states that the second moment of the regressus is 'an accurate and diligent mental examination, through which we know the proximate cause of the effect'.[32] Christopher Airay, instead, speaks simply of an 'applied disquisition' through which the conversion of the terms of the demonstration is made possible.[33] Franco Burgersdijk deals with 'a comment of the mind, which, comparing the cause with the effect, knows that the cause is of that specific effect'.[34] Claudio Clemente Burgundo states that the cause and the effect can be known confusedly or distinctly. In order to acquire clear and distinct knowledge, an intermediate inquiry (*intermedia inquisitio*), 'which many call negotiation or mental examination', is required. This inquiry should identify all those elements that are always conjoined with the effect, and which should characterize the nature of cause through which the effect is.[35] Thomas Carleton simply speaks of a 'better penetration' of the cause by the intellect.[36] Cornelius Martini writes that 'through the mental examination one finds the being of the cause', that is, its existence, and that this mental examination intercedes between the demonstration from effect and the demonstration *propter quid*'.[37] Joachim Jungius, in commenting on Zabarella, emphasizes that after the induction, 'a certain mental consideration and applied reasoning' are necessary in order to know what is essential, necessary and substantial. This 'consideration, or reasoning, or negotiation of the intellect or mental examination', makes certain what is uncertain, that is, makes more known what is unknown.[38]

John Newton, in contrast, states that to achieve a scientific knowledge 'our understanding hath proceeded from a kind of confused and experimental knowledge of the effect, as nearer to sense, to the like confused knowledge of the cause; and hath ripened that knowledge by often meditating on, and comparing of the cause to the effect, till it come from a confused to a distinct knowledge'.[39] Narcissus Marsh, again, considers the intermediate stage 'a reasoning of the understanding (from sensation, by means of likes, contraries, concomitants, etc.) through which it acquires a perfect knowledge'.[40] In particular, the understanding discovers the necessary connection between the cause and the effect.

Throughout the sixteenth and the seventeenth centuries, Jesuit philosophers discussed the problem of regressus at length. Franciscus Toletus devoted an entire *quaestio* on *On Demonstrative Regress*, introducing the intermediate stage of the regressus as 'other speculations of the intellect (*alijs intellectus speculationibus*)', which lead to a more clear and distinct knowledge of the cause that could serve as the basis of the following demonstration *propter quid*.[41] The Jesuit Johannes Brutscher bases the entire discipline of logic on the intermediate stage of regressus. Science, he writes, is knowledge of evident and certain conclusions that derive from logical reasoning. The logical instrument of science is the demonstration *propter quid*, which is preceded by two additionally fundamental kinds of argument. The first is induction based on experience, and the second is mental examination, which is a diligent inspection, consideration and investigation of the conclusions generated by experience so that possible errors can be identified and discovered. Without this mental inspection, it would be impossible to acquire evident and certain conclusions, which are at the basis of any scientific argument. The mental examination cannot work *ad infinitum* in its comparison of the cause and the effect on the basis of experience, or no science would be possible. The inspection, therefore, should proceed until it is sufficiently probable that the conclusions are correct, in such a way that 'the conclusion of the logical demonstration [*quia*], which is in itself unclear, becomes evident through the mental examination'.[42]

Johannes Lorinus and Paulus Vallius provide two interesting accounts of mental examination, most notably because, according to William A. Wallace, they are Galileo Galilei's own sources for Zabarella's conception of the intermediate stage of regressus.[43] In his *Commentaries to Aristotle's Logic*,[44] a transcription of the lectures given in 1584,[45] Lorinus, following Balduino and Zabarella, states that the demonstration *quia* makes possible only knowledge of the existence of the cause, while the cause in itself is known only confusedly. For this reason, an 'examination, that is, negotiation (*negationem* [sic]), or mental consideration', must intercede, through which the cause – previously known imperfectly and indistinctly – is now knowable perfectly and distinctly 'considering its nature and conditions', by means of a continuous comparison between cause and effect.[46] This operation, Lorinus writes, was according to Balduino possible only through definition. Rather, it is possible in at least two ways: either by means of an immediate apprehension of the natural light of the intellect (*apprehensione lumine naturali*), which instantaneously identifies the connection between effect and cause or, alternatively, through many arguments and extended analysis.

Paulus Vallius, once more, devoted two chapters to negotiation in his *Logic*.[47] This work is likely to be the transcription of lectures he gave in 1588, the manuscript of which has unfortunately been lost. If, as seems probable, Ludovico Carbo's *Additions* (1597) blatantly plagiarized Vallius's lectures, then at that time Vallius was not concerned about regressus, since the problem is entirely absent from Carbo's book.[48] According to Vallius, we can distinguish two stages in regressus that follow the process of our mental reasoning. The first, represented by the demonstration *quia*, produces the knowledge of the cause starting from the effect, while the second, represented by the demonstration *propter quid*, explains the necessary connection between cause and effect. Regressus also has two modalities. The first makes no transition between the two demonstrations and does not require a negotiation. This happens only in very rare cases when the cause is self-evident. The second and more frequent mode requires a more articulated work of the intellect to ensure the transition between the two demonstrations, since there is no clear and distinct knowledge of the cause. An intermediate time (*tempore intermedio*) must intervene for the cause to be known formally and distinctly, that is, to reduce the efficient cause to a formal cause:

> This intermediate consideration of the understanding, some call it negotiation of the intellect, others mental examination, others meditation, others application and intention of the mind and others in other ways, but in whatever way it is called, this consideration must intercede.[49]

On the usefulness of this negotiation, Vallius is clear:

> Through this negotiation, we see the effect to be always necessarily connected to the cause and that the cause convenes to all the conditions of the cause and the effect convenes to all the conditions of the effects, and thus comparing the cause with the effect and knowing the essence of the cause, we know that it is cause and that on it the effect depends, and in this way we have the knowledge of the cause by resolution and by the application of logical rules, by means of which we can know what is the cause and the effect.[50]

For Vallius negotiation is a logical process, completely formal and mental, and so supporting the insufficient empirical and experiential aspects of knowledge acquisition. It consists of determining those essential conditions that are prerequisite for such an effect, and these conditions should characterize the nature of the cause in a necessary way. Indeed, Vallius adds, this examination 'or negotiation of the intellect, interceding between two demonstrations, ensures that the cause and the effect, which in the first demonstration were known only materially, are now known formally'.[51] Vallius states that this mental examination is helped 'by a certain induction and by an intermediate reasoning of the intellect, but specifically by that resolution and application of logical rules through which one can know the cause and the effect'.[52] Negotiation of the intellect is therefore an exclusively logical and mental process, which is part of the resolutive and inventive stage of the scientific method and makes the essence of the cause knowable. However, this same mental examination, Vallius emphasizes, cannot act as a substitute for the demonstration *propter quid*, because the aim of science is not simply to know the essence of a cause but to discover the connection between cause and effect. Of indeterminate length, according to the perspicacity of the understanding, this intermediate stage is necessary because it is the only tool that makes possible a clear and distinct knowledge of the cause, which is the basis of scientific demonstration.

In his *Logical Institutions*, published in 1641, Gherardo Baldi provides a detailed explanation of the role of the negotiation. Mental examination provides a different level of engagement with the cause: before the examination, the cause is known only *materialiter*; after, it is known *formaliter*.[53] Before, we know of its existence; after, we understand its essence. This examination by its nature consists of a form of reasoning within the understanding that discovers the properties of the cause and investigates those properties that are essential for the existence of the effect.[54] This process guarantees the transition from confused to distinct knowledge of the cause. The transition is made possible through the negotiation of the intellect, which 'arguing the essence of the cause and knowing distinctly, makes this knowledge of the middle [term] required for the argument *propter quid*'.[55] Indeed, knowledge *propter quid* is not sufficient for knowing the existence of the cause discovered in the first part of the regressus. Knowledge of the existence of the cause in itself does not amount to science merely because it concerns a cause. Only if we know that it is the cause of an effect do we enter the realm of scientific knowledge, for science focuses mainly on the relationship between cause and effect. For this reason, definition is not an adequate substitute for negotiation, because definition simply provides a basic explanation of the essence of the cause, but not of its relation to the effect. But, most of all, definition does not work with the unknown as a starting point, but rather it makes explicit what is already known.[56] Negotiation, instead, is helpful in identifying the inherent nature of cause within effect, and in explaining the unknown essence of the cause. Indeed, mental examination, 'comparing the cause with the effect, finds the properties and the essence of the cause and consequently we know that the effect is effect of the cause'.[57] Finally, Baldi writes that 'mental examination is nothing more than the resolutive method',[58] or, better, it works when the resolution, after a demonstration from effect or a demonstrative induction, is over.

In whichever direction we turn, with the exception of Piccolomini and Capivacci, as we encountered in Chapter 2, the intermediate stage of regressus appears generally to be taken for a mental process that fills the gap of resolution, that is, a passage of knowledge coming from experience, confused and obscure, emerging as clear and distinct. For Renaissance Aristotelians, negotiation or mental examination was the tool for transforming mere experience into scientific knowledge. For most, it was a useful way of guaranteeing that investigations into nature had the same certainty as mathematics, by reason of the convertibility of the terms of demonstration and the reduction of the cause to a formal cause. With this operation, conclusions in the field of natural philosophy acquire the highest degree of certainty, even though the enquiry is still qualitative and strictly related to the discovery of definition, essence and substance. Their epistemology was anything but naïve. It presents a high level of complexity in theorizing about nature and makes possible the interaction between experiential and rational moments.

The intermediate stage of regressus reveals awareness of the importance of experience, but at the same time acknowledges its weakness, because experience provides a merely provisional and accidental type of knowledge entirely at odds with the status of mathematical knowledge. One of the last heirs of the Paduan Aristotelian tradition, Cesare Cremonini, left a clear account of this interaction between experience and mathematics. In the *Treatise on Education* (1596), Cremonini states that experience in natural philosophy differs from that in mathematics since the former is always confused and, for this reason, requires laborious attention (*negotiosa animadversio*):

> The same manner of experience is not found everywhere. In mathematics for the confirming of principles it is sufficient to employ induction based on the observation of what is in the matter whence mathematics is abstracted; in that field the truth of the principles is immediately evident. But in natural philosophy such observation is not so obvious a way of gaining principles, nor is the collection of principles by its employment so easy. There, laborious attention is indeed required, procured from the application to things; and even with this the principles are arrived at not without keen cognition. Moreover, this experience is necessary not only for the natural investigator, if he is to arrive at first principles (*pro primis principiis comparandis*); it is requisite for almost every manner of science. For experience is likewise required in morals in much the same fashion, and even in theology, since we do not ascend to those abstract causes without a manifold and laborious attention (*multa & negotiosa animadversione*) to their effects.[59]

Mathematics understands inductively – that is, intuitively (which is demonstrative induction) – and immediately, its truths, which are evident and clear. To acquire such truths in its turn, natural philosophy requires an application of the intellect that is accurate and precise – in short, nothing other than the negotiation and mental examination of regressus. In order to resolve this opposition between mathematics and natural philosophy, Zabarella introduced the idea of demonstrative induction, which was in fact a form of intuition – that is, it immediately grasps the essential connection between cause and effect. Moreover, negotiation was intended to guarantee for natural philosophy the same degree of certainty as mathematics, which is clear evidence of a concomitant epistemological re-evaluation of mathematics. The idea of a negotiation focusing on the convertibility and essential connection between cause and effect led to a neglect of many natural phenomena that could not be reduced to a simple definition or formal cause. All of this slowed down the growth and development of scientific knowledge in a broader sense. This clearly shows, as Craig Martin has pointed out, that 'while epistemological ideals played a role in the presentation of concepts and guided the direction of natural philosophy, they should not be mistaken for actual methods of research'.[60] The negotiation of the intellect or mental examination was the first attempt by the Aristotelians to respond within their own epistemology to the new discoveries coming from experience, and to make sense of them.

3

Francis Bacon

Cambridge. 1573. At the age of twelve, Francis Bacon walked for the first time through the doors of Trinity College. The course of study was dominated by Cicero; humanist texts were all the rage, in fact. Cambridge was the centre of English Ramism, and the ideas of Paduan Aristotelianism and Jesuits remained in the distance. This education must have seemed dry and futile to the young Bacon, more interested in the promise of a political career. However, this context offered a fertile soil of heated debate on method between William Temple and Everard Digby, influencing the very dynamic of epistemology in late sixteenth century England. Bacon did not yet know that these logical discussions would come to occupy his life as a philosopher, leading him to develop a new epistemology, which since the publication of John Stuart Mill's *A System of Logic* (1843) has been largely considered experimental and aimed at the reform of Aristotelian methodology, being to a large extent rooted in the concepts of experience and induction.[1]

This picture of Bacon's epistemology has, in recent years, undergone substantial revision.[2] It has been shown that what is truly distinctive about Bacon's approach in opposition to the Aristotelian tradition is that induction represents the only method of discovery, since 'it justifies the reduction of the senses, is the active motive of natural history, is the core of demonstration and the only way of establishing first propositions and first notions'.[3] This induction, therefore, is not the simple enumeration usually attributed to the Aristotelians; rather, it is something more complex, which cannot be understood, however, independently of the framework of the Aristotelian tradition.[4]

In this chapter, I propose a new reading of Bacon's conception of induction as a special tool through which he makes clear what Aristotelians left somewhat vague and undetermined – that is, the method for finding, within his epistemology, foundations and principles for science, such as universal notions and axioms.[5] My specific point is to close in on a narrower notion involved in the Aristotelian epistemology, that of mental examination, which Bacon transformed and elaborated into a totally new logical tool for discovery. Indeed, what Bacon called induction reflected more the negotiation of the intellect than Aristotelian induction. It is quite obvious that Bacon departed from the Aristotelian tradition, but the Aristotelian framework is the starting point from which we can fully appreciate his contribution to the history of epistemology.

Bacon's first attack on traditional logic is in *Of the Proficience and Advancement of Learning, Divine and Humane* (1605). He asserts that logicians' use of induction as a tool for the discovery of first principles is 'utterly vitious and incompetent',[6] because it concludes 'uppon an enumeration of particulars, without instance contradictorie'.[7] The main issue at stake is therefore that induction would proceed only in an affirmative way through a listing of particulars without having any further corrective or proof. What is gathered from experience and sensation can lead to the truth but not immediately, rather only 'by comparison, by help of instrument ... and other like assistance'.[8] Bacon demands the production of a logical tool capable of working by comparison, through which sensible knowledge can be amended. However, before any comparison can take place, there must be an accurate gathering of data. This process is called 'experientia literata'. Only after this experience is it possible to proceed to discovery, which means 'to discover that we know not', and not 'to recover or resummon that which wee alreadie knowe'.[9] This literate experience is also the foundation of the most complex kinds of demonstrations, like those of syllogism.

Bacon examines more specifically the contribution of correct induction in *Thoughts and Conclusions* (1607–9). He states that syllogism can be useful, but much of the time it is based on 'generic, imprecise, not well-defined' notions, especially concerning natural things. According to Bacon, the solution would be to employ induction, which would collect in an adequate manner all the information to be provided to the intellect. However, Bacon is highly critical of previous conceptions of induction. Firstly, it was always considered an easy shortcut rather than a laborious working of the mind. Secondly, it was applied only to the discovery of the first principles of knowledge. Finally, its treatment was limited to its elementary and basic form, proceeding by enumeration and resulting, therefore, in doubtful and contingent conclusions that form no basis for demonstration. Bacon, therefore, places an emphasis on the fact that induction is a difficult, exacting and toilsome discursive process, which has nothing to do with a kind of 'magical' intuition capable of immediately grasping first causes and principles. It covers more than the task of finding first principles and notions and can be extended to every kind of knowledge, proceeding not only affirmatively but also negatively.

It is in the *New Organon* (1620), however, that Bacon realizes his project for the demolition of the traditional concept of induction. Yet in the preface Bacon identifies in dialectical tools, and in particular in classical induction, the cause of an obscure and confused type of knowledge, which should be amended:

> Now there never was the slightest hope that the errors which have flourished and will forever flourish would (if the mind were left to itself) put themselves right one after another either by native force of intellect or the help and support of dialectic. And why? Because the primary notions of things (whence all other notions flow) which a slack mind effortlessly absorbs, lays up, and accumulates are corrupt, disordered, and recklessly abstracted from things, and there is no less arbitrariness and instability in the secondary notions and other notions.[10]

Induction, which has the task of abstracting from matter notions regarding physical and sensible things, produces only contingent and uncertain knowledge. The task of Bacon's new method is precisely that of amending this kind of induction and of transforming knowledge that is obscure and confused into something clear and distinct. The entire Baconian method aims to perform a task analogous to mental examination or negotiation of the intellect in the Aristotelian tradition. In doing this, Bacon proposes a new way of interpreting nature, a new logic that is different from the traditional one.

Traditional logic was wrapped around syllogism. Induction did not have a special role, rather its study would have been somewhat superficial. Bacon rejects syllogism because, as he had already affirmed in the *Thoughts and Conclusions*, it is based on notions, which 'are ineptly and recklessly abstracted from things, and vague, insufficiently delimited and circumscribed, and indeed rotten in many ways',[11] thus summarizing what he had written in the 'Preface'. Induction is susceptible to these problems because it leaps 'immediately from sense and particulars to the most general propositions', proceeds 'by simple enumeration' jumping to conclusions, is exposed 'to the danger of instant contradiction' and observes only familiar things, achieving no result.[12] Bacon's attack on induction is concerned primarily with its being reducible to sensible knowledge. Being content with information coming directly from sensation leads to uncertain knowledge, because sensation falls into error and deceives the mind. Furthermore, sensation is in essence narrow and limited:

> The sense fails us in two ways: for it either deserts or deceives us. For first there are many things which escape the sense even when it is properly managed and not obstructed at all ... And even when the sense does get a grip on something. Its hold is not terribly secure. For the testimony and information of the sense is always made to the measure of man and not the universe; it is a very great mistake to say that the sense is the measure of things.[13]

For this reason, an intellectual instrument to rectify sensation is necessary:

> We do this not so much with instruments as with experiments. For the subtlety of experiments is far greater than that of the sense itself, even when it has precise instruments to help it (and here I speak of those experiments designed specifically for the purpose, and thought out and applied with skill).[14]

Bacon, therefore, does not attribute much value to sensation in considering it a source of basic and even sometimes deceptive knowledge, and he looks for a mental device capable of overcoming such deficiencies. Sensible knowledge must be ordered and systematized, otherwise the logical inferences built on it are weak, fragile and with no solid foundations. This cognitive method has the advantage of being the most direct, leaping from the particular to the most general axioms. However, such knowledge is uncertain,[15] because the axioms

> flow from a handful of slender experiences and from a few particulars which crop up often, and are pretty much made and tailored to fit them, so that it is no wonder if they do not lead to new particulars.[16]

What makes the discovery of an axiom from some particulars problematic for Bacon is the fact that if a new instance turns up which was previously unknown, the schools, in particular the Peripatetic philosophers, retained it 'by some silly distinction, when it would have been better to put right the axiom itself'.[17] On this point, he attacks the old Scholastic approach for its dogmatism, incapable of explaining the new geographical and celestial discoveries. In contrast, as we shall see, for Bacon, negative instances should represent a corrective tool in order to identify the true nature of the cause.

Bacon calls this weak, confused, obscure and indistinct knowledge 'anticipation of nature', – very effective in finding agreement among people, but invalid from an epistemic standpoint, leading to a vague experience (*vaga experientia*), or a mere palpation (*mera palpatio*).[18] What is deficient in this knowledge stems from the method: that is, the way of proceeding by enumeration in a merely affirmative way to determine the most general and universal of principles and causes. From this method will arise fancies, 'mere opinions, ill-defined and notional conclusions, and axioms altered daily'.[19] Only God, or perhaps higher intelligences, can proceed by affirmation and immediate knowledge, whereas the human mind can proceed on a secure path only negatively or discursively, by means of rejections and exclusions, which constitute, as we shall see, the core of the Baconian induction.[20]

In greater detail Bacon establishes that this fallacious method is constituted of four steps which lead the mind towards false conclusions. Firstly, 'the impressions of itself are faulty, for both lets us down and betrays us'.[21] A correct method, that of induction, should be used to compensate for this error. Secondly, notions 'are abstracted badly from the impressions of the senses, and are indeterminate and confused, whereas they should be sharp and well defined'.[22] This was the kind of knowledge acquired at the end of the first stage of regressus in the Aristotelian tradition. Thirdly, induction by simple enumeration 'without bringing nature's proper exclusions (*exclusionibus*) and divisions (*solutionibus*), or separations (*separationibus*) to bear' is very insubstantial.[23] The process of induction needed to be rethought if it were to lead to scientific knowledge. Fourth and lastly, the method of demonstration and proof from general principles 'is the very mother of error and blight of all the sciences'.[24] Indeed, it is not a genuine method of discovery, because – for Bacon, as we have seen – only induction can lead to the invention of new knowledge.

Bacon's concern is therefore to make clear and distinct these notions that traditional logic was unable to determine. That Bacon has a bad opinion of traditional induction comes as no surprise. If we look at the most important logical writings that he likely had access to, we can clearly see how induction received poor treatment, with only very rare exceptions. One of the most important Cambridge Aristotelians, Everard Digby, understands induction as that instrument which proceeds from particulars to universals, and without which scientific and universal knowledge is not possible.[25] John Case states that the first principles of knowledge are acquired by means of sensation and induction, and that their precise knowledge relies on the various degrees of induction.[26] Griffith Powell, after showing how the process of acquisition of first principles is based on sensation, memory and experience, establishes that the way to their knowledge is induction, through which the intellect grasps as universal what the sensation knows only as particular.[27] Induction proceeds through the comparison of many particulars encountered by sensation, from which a generic idea is generated.[28] Like Powell, Robert Balfour characterizes the process of induction as a form of comparison without which it is impossible to apprehend first principles.[29]

Samuel Smith states that induction is the appropriate tool for the knowledge of scientific principles. This kind of induction, following Zabarella, is properly speaking called demonstrative. In order to lead to scientific knowledge, induction must have three features: (1) its subject matter must be necessary and not contingent, and this seems to exclude a direct application to natural sciences; (2) there must be a necessary relation between subject and predicate; and (3) these principles must be known by themselves or should require minimal demonstration.[30] Demonstrative induction is different from dialectical induction and also demonstration *ab effectu* for two reasons. The subject matter of dialectical induction is not necessary, but accidental, probable and contingent. Demonstration *ab effectu*, conversely, has the task of demonstrating principles unknown by nature, while induction is concerned with the task of discovering principles which are more known by nature and thus considered fundamental.[31]

Robert Sanderson, author of one of the most widely diffused handbooks of logic of the seventeenth century lists four possible means of discovery, through which the mind gradually acquires new knowledge. The first is sensation, through which we acquire a particular cognition. The second is accurate observation or history, through which we grasp and collect in the mind what comes from sensation. The third is experience, generated by the application of the mind to further observation. The last is induction, which infers a universal conclusion out of many experiences.[32] Induction involves the enumeration of all particulars, but when a correlation in nature is evident, it can introduce the clause *et sic de caeteris* according to which if '*iste magnes trahit ferrum, et ille, et hic, et pariter se habet in reliquis, ergo omnis magnes trahit ferrum*'. However, Sanderson is aware of the fallibility of induction, so much so that he states that the process can be falsified by a single contradictory instance. For this reason, induction in itself does not lead to scientific knowledge.[33]

Following Zabarella, John Flavell maintains that without induction it is impossible to achieve common notions and universal principles.[34] Yet he also points out that induction is not a mode of logical inference through which the unknown becomes known, but rather a notification of something deriving from sensation to the intellect. This notification starts from something that is in the first instance evident to sensation and provides something that is obscure and awaiting clarification by the intellect. Flavell, however, does not explain how this process works.[35] Also relying on Zabarella, Bartholomeus Keckermann held that induction is a process of notification by means of which it is possible to contemplate universals and general propositions applicable to all particular cases by adding the clause *nec potest dari dissimile exemplum*.[36] Christoph Scheibler, on the other hand, states that induction should enumerate all cases, even if the cases are not always enumerated in a distinctive manner. Indeed, in most instances, particular cases are implicit or arranged chaotically. Like Zabarella, Scheibler points out that in demonstrative induction it is not necessary to render a complete list, but only to identify essential characteristics common to all particular cases, and from there it is possible to proceed to the generalization.[37]

*

Bacon aims to amend this supposedly childish concept of induction developed by the Aristotelian tradition popular at this time, and in doing so he reclaims and re-elaborates the function of the negotiation of the intellect, which, as I have shown in the previous chapters, made knowledge coming from sensation and experience clear and distinct. Bacon seeks to elucidate this unresolved aspect of Aristotelian epistemology, which remained a weak point.

An early attempt to explain how to give clarity to these notions generated by sensation and experience is in *A Description of the Intellectual Globe* (1612):

> For the images of individuals are taken up by the sense and fixed in the memory. They pass into the memory as it were whole, in the same form in which they crop up. Then the mind recalls and reflects on them, and, exercising its true function, compounds (*componit*) and divides (*dividit*) their portions. For, single individuals have something in common with each other and, on the other hand, something distinct and manifold. Now this composition (*compositio*) and division (*divisio*) takes place either according to the will (*arbitrio*) of the mind, or according as we find it in things ... if these same portions of individuals are compounded and divided according to the very evidence of things, and according as they truly reveal themselves in nature, or at least as they are seen to reveal themselves to each man according to his capacity, these are the functions of reason and to reason do we assign the whole business of managing such things.[38]

The inductive process for arriving at common notions is that of composition and division – that is, those two operations that, according to Paduan logicians, characterized the negotiation of the intellect. Bacon points out that the task of reason is that of dividing and compounding the things as they are, without introducing fancies or simplifications, in order to know their real nature, in contrast to what happens if the mind is free to use its imagination. A cornerstone in experience is necessary.

For the moment, I will not consider the subject matter of these division and composition, rather I shall focus on the logical process. The starting point, of course, is that Baconian induction is not that of the Aristotelians, even if the *Description of the Intellectual Globe* seems to refer discreetly to the procedure in *Posterior Analytics* II.19. Already in the *Advancement* Bacon stated that induction by simple enumeration, without advancing contradictory instances, was capable only of conjecturing, of formulating hypotheses, and thus was incapable of concluding scientifically.

In contrast to traditional logic, for Bacon induction entails a demonstrative inference that supports sensation and investigates the nature of things. In order to acquire scientific knowledge, a kind of induction is necessary that 'divides (*solvat*) and separates (*separet*)' experience and that is not merely a simple enumeration of particular cases.[39] Induction is thus for Bacon an eminently analytical process. He speaks of this as a kind of 'introspection' of nature (*naturam introspicere*), as a dissection of the real world from which comes the discovery of things themselves.[40] This form of analysis achieves conclusions that are necessary in themselves and that are therefore scientific. According to Bacon, and unlike the Aristotelians, induction by itself is capable of scientific knowledge, which is acquired by means of exclusions (*exclusiones*) and rejections (*rejectiones*).⁴¹

Like traditional induction, Baconian induction starts with sensation and particular cases in order to discover axioms; unlike the former, it proceeds gradually, without interruption:

> But we should hope for better things from the sciences only when we ascend the proper ladder by successive, uninterrupted or unbroken steps, from particulars to lower axioms, then to middle ones, each higher than the last until eventually we come to the most general. For the lowest axioms barely differ from naked experience. The highest and most ones in between are the true, solid and living axioms on which men's fortune and affairs depend, but above them again we come at last to the most general ones themselves, such as are not abstract but properly limited by these middle ones.⁴²

According to Bacon, these axioms, which are discovered through a specific process, are the most adequate tool for finding new particulars, while he warns against looking for new particulars on the basis of other particulars:⁴³

> Our way and method (as we have often said clearly and are happy to say again) is not to draw results from results or experiments from experiments … but (as true Interpreters of Nature) from both results and experiments to draw causes and axioms, and from causes and axioms in turn to draw new results and experiments.⁴⁴

The manifold quality of particulars is without doubt a source of hope for scientific discovery because they are certainly fewer in number than the fancies of imagination – however, they must be well arranged. The orderly collection of these particulars is given a specific name by Bacon: literate experience.

Literate experience, unlike the simpler and vaguer Aristotelian experience, is not merely based on sensation, from which one infers by means of logic, but is grounded in experiments. This aspect seems to be absent in sixteenth-century Aristotelians, except for Piccolomini and Capivacci, who, as we have seen, introduced elements of experimentalism and sensible proofs into the process of the negotiation of the intellect.

Bacon is clear on the importance and centrality of the experiment in the literate experience, whose aim is to offer an orderly basis on which to achieve knowledge that is clear and distinct. Intellect acquires clear and distinct scientific knowledge:

> not so much with instruments as by experiments. For the subtlety of experiments is far greater than that of the sense itself, even when it has precise instruments to help it (and here I speak of those experiments designed specifically for the purpose, and thought out and applied with skill).⁴⁵

Indeed, 'all truer interpretation of nature is accomplished by means of instances, and apt and appropriate experiments'.⁴⁶ Those experiments, which lead to knowledge of causes and axioms, are called illuminating experiments (*lucifera experimenta*).

Literate experience in itself does not lead to scientific knowledge, because it is – as Sanderson would say – only a history, that is, a collection of experiences. Rather, literate experience is the ground on which axioms are discovered. Only from axioms, then, can one descend and discover new particulars. This process of ascension to the axiom and descent to particulars is a re-elaboration of the Aristotelian regressus. On other occasions, Bacon repeats the same concept:

> This route is not laid on the flat but goes up and down – ascending first to the axioms, then descending to the effects.[47]

> The directions concerning the interpretation of nature comprise two general stages: the first is to do with extracting and fetching up axioms from experience; the second with deriving and drawing down new experiments from axioms.[48]

Bacon depicts this twofold approach by way of analogy with the work of ants, spiders and bees. In sect. 95, Bacon writes as follows:

> Those who have dealt with the sciences have either been empirics or dogmatists. The empirics, in the manner of the ant, only store up and use things; the rationalists, in the manner of spiders, spin webs from their own entrails; but the bee takes the middle path: it collects its material from the flowers of field and garden, but its special gift is to convert and digest it. The true job of philosophy is not much different, for it depends not only or mainly on the powers of the mind, nor does it take the material gathered from natural history and mechanical experiments and store it unaltered in the memory but lays it up in the intellect changed and elaborated. Therefore from a closer a purer alliance (not so far achieved) of these two faculties (the experimental and the rational) we should have good hopes.[49]

It was not the first time that Bacon applied the zoological metaphor of ants, spiders, and bees to epistemological models. In the *Thoughts and Conclusions* and in the *Refutation of the Philosophies* (1608) we can find the very same passage.[50] Bacon describes three different epistemological attitudes. The first is that of the *ants*, which gather data without any systematization. They collect histories, but they are incapable of understanding the larger framework and significance of their findings. They proceed from a particular case to infer particular conclusions. Their knowledge is only conjectural and provisional. The second is that of *spiders*, which with their webs construct complex philosophical truths, which goes beyond the solid ground of experience. With their pure thoughts, they aim to know all reality with certainty and necessity, but indeed their cognition is largely fictional and groundless. The third epistemological model is that of *bees*. Bees collect and gather material from experience, but they do not store and use to draw conclusions. Rather they transform and elaborate this raw material into more refined and accurate cognition, which make sense of individuals through a larger and more comprehensive system of knowledge. Their knowledge is generated by the interaction and negotiation between the empirical or experiential and the rational or intellectual stages of the process. According to Bacon, epistemology should follow the work of bees in defining its scope and instruments. It should overcome empiricism and dogmatism (or rationalism) and elaborate a third way.

Bacon takes this distinction from Aulus Cornelius Celsus's *On the Medicine*, a text largely forgotten in mediaeval Europe and which became widely available only with the printing of the *editio princeps* in 1478 in Florence. In this pivotal text for understanding the history of medicine in antiquity, the Roman encyclopaedist classified the ancient medical schools as empirics (or experimentalists) and dogmatics (or rationalists). Lexically, there is a strict correspondence between Celsus and Bacon, which is reflected also in their respective epistemologies. According to Celsus, empirics believe that 'the sole knowledge necessary is derived from experience' and they based the art of medicine 'on practice alone'.[51] They hold that methods of medical practice 'differ according to the nature of localities',[52] opening up the idea of medical relativism, and therefore the impossibility of establishing scientific, universal knowledge. At the opposite end of the scale, dogmatists based the art of medicine upon speculation and they maintain that 'if the causes which produce diseases were everywhere the same, the same remedies should be used everywhere'.[53] They propose 'a reasoned theory of medicine',[54] which has 'as requisites, first, a knowledge of hidden causes involving diseases, next, of evident causes, after these natural actions also, and lastly of the internal parts'.[55] Celsus makes clear that rationalists 'do not deny that experience is also necessary',[56] but argue rather that 'it is impossible to arrive at what should be done unless through some course of reasoning'.[57] In opposition to dogmatists, empirics accept 'evident causes as necessary, but they contend that enquiry about obscure and natural actions is superfluous, because nature is not to be comprehended'.[58] From a range of different opinions and dogmas, a decision cannot arise only through reasoning, but requires an experiment. The validity of dogmatic epistemology depends only on argument and on authority. Empirics believed that 'even philosophers would have become the greatest of medical practitioners, if reasoning from theory could have made them so'.[59] Empirics confronted dogmatics with the assertion that when they heal patients this happens not because they deduce the means of healing from hidden causes concerning which different opinions are held, 'but from experiences of what had previously succeeded'.[60] In other words, medicine is not 'deduced from such questionings, but from experience'.[61]

Celsus's two medical sects mirror Bacon's ants and spiders.[62] Celsus, like Bacon, does not side with one sect or the other, but believes rather that experience and reason should work together for the perfection of the art of medicine.[63] Citing Celsus, Bacon makes his ideas clear in the *Valerius Terminus* (1603?) in very different words, within the context of discussing the Aristotelian epistemology on how to establish causes:

> And this did Celsus note wisely and truly, how that the causes which are in use and whereof the knowledge now received does consist, were in time minors and subsequents to the knowledge of the particulars out of which they were induced and collected; and that it was not the light of those causes which discovered particulars, but only the particulars being first found, did men fall on glossing and discoursing of the causes.[64]

Bacon distinguishes two inseparable steps that lead to scientific knowledge: to proceed from particulars to universals, and then from the universals one may argue for the particulars. These two steps essentially embodied the epistemologies of empirics and rationalists. It is remarkable that what allows the alliance between these two moments is a peculiar labour of the mind, as in the negotiation of the intellect, which digests and makes sense of empirical data. Bacon's main achievement in the history of early modern epistemology is to propose a new way for establishing this alliance, based on solid grounds departing from the Aristotelian tradition.

As Antonio Pérez-Ramos has suggested, the combination of experimental and rational stages has clear formal analogies with the Aristotelian methodology of regressus, although 'the logical content of this Baconian regressus has little to do with, say, Zabarella's'.[65] In spite of the differences, it is important to emphasize, however, that in Bacon there is always a rational and deductive moment after the empirical stage. According to the Lord Chancellor, the twofold movement from experiments to axioms, and then from axioms to experiments, was ineluctable.

For Bacon, the pathway that leads to the discovery of the axioms and first principles – that is, the parallel procedure developed by Aristotle in *Posterior Analytics* II.19 – is composed of 'three kinds of service: service to the senses, service to the memory and service to the mind or reason'.[66] He adds that this process must be based on an 'adequate natural and experimental history'.[67] One should not invent by imagination or reasoning the subject matter of knowledge, but one should investigate and discover what nature does. In the *Description*, Bacon stresses the importance of performing the composition and division 'according to the evidence of things', and not 'according to the will of the mind', because the work of imagination, being not bound by any law and necessity of nature, 'can join together objects which do not at all come together in nature, and tear apart things which are never found apart'.[68] Bacon demands that experience be the cornerstone.

In the *New Organon*, Bacon writes that division and composition can be laid out systematically in series of tables, which was, as we have seen, a technique already used by Paduan Aristotelians – Capivacci being one of them – who applied the logic of regressus to natural investigations, in particular in the field of medicine.[69] Again, not by chance, Bacon in many of his writings aims to amend Paracelsus's and Gerolamo Cardano's analytical methods. Finally, Bacon criticizes the unsystematic and irrational experience of empirics – that is, that ancient medical school that cherished the collection of particulars.

In order to counter this absence of systematicity, Bacon advances the idea of using tables, as a means of bringing together experiments and experiences in an appropriate order. The first table shows essence or presence. It is the result of what traditional induction produced affirmatively – that is, a list of all possible attributes of the essence of a thing is presented to the intellect. The second table, by contrast, presents to the intellect those instances of which a given nature is devoid. It is quite clear for Bacon that this list could potentially be infinite and, for this reason, the attributes should be considered alongside what is lacking in closely related subjects where a specific nature exists and is manifest.[70] Finally, the third table presents the instances 'in which the nature under investigation exists to a greater or lesser degree … by comparing its increase or decrease in the same subject, or its intensity in different subjects compared one with the other'.⁷¹ In this way, it is possible – according to Bacon – to determine all the attributes that essentially inhere in the subject under investigation, and to what extent and to what degree they participate in its nature.

What is striking about these tables is that they are used to make a presentation to the intellect (*comparentia ad intellectum*),⁷² which means that the activity that orders and arranges material deriving from experience is purely intellectual. Equally intellectual is the capacity of true induction 'to discover which nature [that is, attribute] appears constantly with a given nature or not, which grows with it or decreases with it'.⁷³ Indeed, Bacon reaffirms the concept that the 'complete division (*solutio*) and separation (*separatio*) of a nature' is made 'not by fire,⁷⁴ but with the mind (*per mentem*), which is a kind of divine fire'.⁷⁵ In particular, the task

> of true induction is to reject or exclude individual natures which we cannot find in any instance where the nature under investigation is present, but can find in any instance where the given nature is absent, or find growing in any instance when the given nature wanes, or waning when the given nature grows.⁷⁶

There is no room for experiments in this kind of true induction; rather, everything is mental, a working of the intellect based on literate experience constituted from previous experiments.

In a previous passage, Bacon expressed the same concept with almost identical words, stating that

> separation (*separatio*) and division (*solutio*) of bodies is certainly not to be achieved through fire, but by reason (*per rationem*) and true induction, with auxiliary experiments; and by comparing them with other bodies, and reducing them to simple natures and their forms, which in a compound body come together and become intertwined.⁷⁷

What Bacon emphasizes in this passage is that experiments are merely auxiliary and not necessary to this process. It is worth noting Bacon's emphasis on this twofold process of separation (*separatio*) and division (*solutio*). At first glance, it could seem no more than a duplication of words. But it is not so: the terminology is accurately applied. During the sixteenth century, separation and division assumed very well-defined characteristics. Separation concerns the analysis of a real and concrete thing in order to identify its physical or chemical components, whereas division concerns the analysis of a concept in order to determine its attributes.⁷⁸ The fact that Bacon conceives these two processes as conjoined is extremely significant for his philosophy of nature and represents a radical break with the tradition. Only if real separation relating to actual qualities of real phenomena and conceptual division relating to logical distinctions go hand in hand, does the transition from speculation to production – from knowledge of all the attributes that constitute gold to the making of gold itself, starting with these same attributes – become possible, as we shall see.

These two processes are the same thing, but no one, according to Bacon, performed them appropriately, except Plato, 'who does to some extent use this form of induction for scrutinising definitions and ideas'.[79] Indeed, this form of induction should be employed not only to discover axioms but also to determine notions.[80] In the *Valerius Terminus*, Bacon makes clear that this kind of Platonic reasoning is based on composition and division.[81]

In the *Thoughts and Conclusions* as well, Bacon commends Plato because he 'made use of induction, not only for his first principles but throughout'.[82] In *The Refutation of the Philosophies* (1608) Bacon writes that Plato

> made use of induction throughout, not only to establish first principles but also middle propositions ... But he corrupted them and made them fruitless by aiming only at abstract forms and taking the material for his inductions only from superficial and vulgar experience. Instances of this kind, being known to everybody are suited to discussions but not to research. Accordingly, since he did not practice serious study and observation of natural phenomena, which are the only basis of philosophy, it is no matter for wonder that neither his lofty genius nor his happy method accomplished much.[83]

Plato's main problem, therefore, is not the logical structure of his argument, but rather his exclusive focus on conceptual division and not also on real separation. In the *New Organon*, restating what he had previously said in the *Valerius Terminus*,[84] Bacon makes clear that the method Plato had elaborated was that of division (διαίρεσις) and combination (συναγωγή), outlined in the *Phaedrus* for determining definitions.[85]

In order to better understand Bacon's view, it is important to unpack the Platonic conception a little. Plato states that division is the process whereby 'we are able to dissect in accordance with the forms respecting the natural joints',[86] while combination is 'where we bring many dispersed things which we grasp together under a single idea in order to define each thing (that is dispersed), and so to make clear whatever might be selected as the subject of discourse'.[87] According to Michael V. Wendin, combination presupposes division in the sense that a combination 'is always a collection of subkinds of a higher kind' and a proper combination 'will also be a proper division of the higher kind in question'. Furthermore, the single idea figuring in combination 'must be definitionally relevant', which means that the various parts constitute the essence of the thing.[88] In other words, it is the process of division and composition already mentioned by Aristotelians as foundational to one of the modalities of the negotiation of the intellect. We saw in the previous chapters, however, that this form of inductive reasoning based on division had been criticized by Aristotle for its inability to discover, as opposed to providing a more clear and distinct knowledge of the already acquired. Plato himself speaks of this process not as a kind of logic of discovery, but as a logical tool for making knowledge clearer and more precise.[89]

Writing about induction in considering the arts of discourse, in the *Advancement* Bacon specifies that judgement 'hath a coincidence with invention', but this invention is 'no other but out of the knowledge, whereof our minde is alreadie possest, to drawe foorth or call before us that which may bee pertinent to the purpose which wee take into our consideration'.[90] Bacon adds that this is not invention, but 'remembrance, and suggestion',[91] with implicit reference to the Platonic conception of recollection.

It is therefore legitimate to question if, how and why Bacon's induction has the capacity to discover; or, whether, to the contrary, it is merely a mental and intellectual procedure, like the negotiation of the intellect, designed to make clear and distinct knowledge already discovered by experiments and brought together systematically through the literate experience.

In the *New Organon*, experiments are foundational to the experimental history or literate experience, but they appear to be auxiliary for induction. As Bacon outlines experiments in relation to induction, his idea is not to see how nature works in different or unobserved situations; to the contrary, as Thomas S. Khun pointed out,[92] the experiment within induction seems to be the tool that generates clear and distinct knowledge from what is already known: the experiment is helpful in being a means of accepting or rejecting previously acquired knowledge in the majority of cases. Indeed, induction is rooted in the rejection and exclusion of something already known.

In the *Advancement*, Bacon states that induction can be called invention 'so as it be perceyved and discerned, that the scope and end of this invention is readynesse and present use of our knowledge, and not addition or amplification thereof'.[93] One might argue whether Bacon's conception of induction matches this notion. There is no reason to suspect otherwise. Bacon believes that there are two means of procuring this ready use of knowledge: namely, (1) preparation and (2) suggestion. Preparation, however, is scarcely an aspect of knowledge, while one cannot say of suggestion that it 'is deficient'.[94] The process envisaged by the term 'suggestion' assigns and directs us 'to certaine markes or places, which may excite our minde to returne and produce such knowledge as it hath formerly collected, to the end wee may make use thereof'.[95] Bacon's reference seems to be the Renaissance art of memory and Topics, whereby the argument proceeded by means of common topical places.[96] No doubt this allusion is present, but there is something more because he adds that this type of suggestion does not only 'furnish argument to dispute probably with others', but it ministers 'unto our iudgment to conclude aright within ourselves'.[97] This part of invention called suggestion appears to be a rational and logical operation of the mind, which Bacon compares with the 'faculty of wise interrogating'.[98] The interrogating faculty is not associated with the art of memory or Topics, but with Plato's method of inquiry. Referring to the *Meno*, Bacon states that whosoever seeks, knows that which he seeks as a 'general notion' – that is, generically – otherwise he will never know that he has found something when he finds it.[99] He adds that it is for this reason that the larger the 'anticipation' is, the more direct and compendious is the research. This 'general notion' designates a quality of knowledge which is not yet defined, clear and distinct, but which requires further insight, that of the suggestion. If the reference to Plato is not casual, the way to make clear and distinct the knowledge relating to this 'general notion' is through division, as previously mentioned, and this would not imply a need for experiments.

Nonetheless, the 'general notion' rests on preliminary and generic knowledge elsewhere defined as *digestione prima*[100] – that is, a type of knowledge similar to that which the negotiation of the intellect had to make clearer and more distinct. This initial material of knowledge cannot be that previously rejected as unable to produce true knowledge, but rather it must be a kind of cognition which has the capacity to lead to scientific knowledge. For the same reason, I would exclude the possibility that this 'anticipation' outlined in the *Advancement* is the anticipation of the mind or of nature developed in the *New Organon*. Indeed, the anticipation of the mind is considered as a method set in contrast with serious research,[101] and the anticipation of nature is seen as a form of conjectural knowledge at variance with the true interpretation of nature.[102] Thus the anticipation of the *New Organon* corresponds much more readily with the 'preparation' of the *Advancement*.

The kind of knowledge on which the 'general notion' is grounded is for Bacon similar to that of a particular Topics, which has been 'rejected generally as inartificial and variable' in the schools.[103] He does

> receive particular topiques, that is places or directions of invention and inquirie in every particular knowledg, as things of great use, being mixtures of logique with the matter of sciences; for in these it holdeth, 'Ars inveniendi adolescit cum inventis;' for as in going of a way, wee doe not onely gaine that part of the waye which is passed; but wee gaine the better sight of that part of the waye which remayneth: so everie degree of proceeding in a science giveth a light to that which followeth; which light if wee strengthen, by drawing it foorth into questions or places of inquirie, we doe greatly advance our poursuyte.[104]

If the logic of invention improves with the discoveries, this particular Topics that provides preparatory material on which the mind exercises its operation of clarification of the general notion is what Bacon calls literate experience in the *New Organon* and that in the *Advancement*, just a few lines before the passage we have examined, he epitomizes using precisely this name.[105]

If my interpretation is correct – and the logic of invention is based on the discoveries, and on the discoveries the mind advances in knowledge – then the experiment is concerned primarily, even if not exclusively, with the literate experience, while induction busies itself with its operation of clarifying these discoveries – that is, of interpreting nature.

For this reason, experiments are more relevant to literate experience than to induction. There are cases, however, in which induction, operating through exclusions and rejections, requires specific experiences or instances. The instances are aids (*auxilia*) for true and perfect induction. Although they deal with intellectual activity and induction is primarily a logical operation based on division, these instances can resort to experience, sensation and observation. Among the various kinds of instances at least one is particularly significant for explaining the inductive process of exclusion and rejection. This is the famous case of crucial instances. They intervene when the intellect is unable to determine from two alternatives which is the true cause of the nature under investigation. These instances would show that the union of a single form or cause with the nature under investigation is 'constant and indissoluble (*indissolubile*)', while the others are 'variable and separable'.[106] Their objective is to determine which causes and forms are essential to the nature and, therefore, what can be excluded and rejected as not necessary. This is the specific task of crucial instances, and it is also that of induction – to the extent that interpretation of nature frequently ends with them.

These crucial instances can comprise other instances, or be present in other instances, but, according to Bacon, more often they are sought and conceived through considerable effort. Scholars have interpreted these instances as forerunners of Robert Boyle's crucial experiment.[107] The process of exclusion and rejection of induction should thus be based on information gathered by experiment. In other words, there would be a necessary experimental moment in Baconian induction.

The examples Bacon provides are vague on this point, however, and crucial instances do not always involve exclusions by experiment, as Daniel Schwartz has correctly pointed out.[108] In the case of ebbs and flows, Bacon states that in order to determine the true crucial instance, it is simply sufficient to use a plumb line. This is not a real experiment, rather a strategy for more accurate observation. In the case of the spontaneous movement of rotation, Bacon claims that faithful accounts and histories can be trusted. In the investigation of gravity, Bacon suggests trying multiple experiments in different conditions. In sum, in relation to the inductive stage, experiment plays only a supporting and not necessary role for Bacon in the foundation of his new method.

*

We have left aside for a while the subject matter of induction, that is, what undergoes the process of division and composition. Now the time is right for understanding the novelty introduced by Bacon into his natural philosophy with respect to the Aristotelian tradition, and how its conception represents a step forward in solving problems left open with the negotiation of the intellect. In the *New Organon*, as we have seen, Bacon states that composition and division cannot be performed on fancies, but only on nature, considering 'matter, and its structures (*schematismi*), and structural changes (*meta-schematismi*), and pure act, and the law of act or motion'.[109] He categorizes all these various elements under the umbrella of the general term 'form'. In another passage, Bacon defines forms as encompassing the true differences of things or the laws of pure act.[110]

At the beginning of the second book of the *New Organon*, Bacon establishes that 'the work and aim of human knowledge is to discover the form, or true difference, or *natura naturans*, or source from which a given nature arises'.[111] He is aware of a lack of precision in his definition; indeed he writes these are merely the words that 'come closest to describing the thing' he means.[112] All these characterizations make the concept of form far from being clear and univocal.

The best way to understand Bacon's conception of form is to frame his position historically. In the *Advancement*, using Aristotelian terminology, Bacon assigns the task of inquiring formal causes to metaphysics, stating that this inquiry is the worthiest that can be undertaken and not useless as some believe, discrediting the possibility of the human mind to grasp 'essentiall formes or true differences'.[113] Bacon, therefore, explicitly points to his conception of forms as expressing true differences and claims that the human mind is perfectly capable of achieving their knowledge. This idea goes against sceptic and even some Aristotelian movements, according to which rarely and only under specific conditions can the human mind contemplate the essences, because this is the true activity of angelic and divine minds. The human being was better characterized by its active, rather than speculative, intellect. However, Bacon narrows the field of these forms by referring to Plato. He agrees with Plato that 'forms were the true object of knowledge', however, the Greek philosopher 'lost the reall fruite of his opinion by considering of formes, as absolutely abstracted from matter, & not confined and determined by matter'.[114] Thus Plato abandoned natural philosophy in favour of theology, and these particular theological forms were not graspable by the human mind. Bacon's criticism against Plato is similar to Aristotle's, which brought forms back into the natural world. It is legitimate to expect Bacon to deal with forms according to their Aristotelian conception, but he did it only marginally.

At the beginning of the second book of the *New Organon*, Bacon establishes that to know truly is to know by causes, as the Aristotelians assert, though only the knowledge of the formal cause leads to scientific knowledge. Indeed, final causes are useful only in ethics, not in natural philosophy, while material and efficient causes as remote causes are 'perfunctory and superficial'.[115] In Bacon, forms should not be understood as mere essences or universals, because these would only be abstract or fictional notions of the mind – for him, nothing in nature exists except individuals. Rather, they should be considered as a law according to which individuals are generates and which represents 'the foundation both of knowing and doing'.[116] Only the cognition of the form as law can provide complete and perfect knowledge. Indeed, he who knows efficient or material causes 'cannot shift the underlying limits of things' being able to produce effects only on some materials.[117] There is no necessity but only contingency between material and efficient causes and the nature of things. Indeed, the same efficient cause can produce different effects, and the same matter can constitute various objects. This was a problem for the Aristotelians, as we have seen in the previous chapters. This does not happen to the form, according to Bacon, for

> the form of any nature is such that if it be in place the given nature invariably follows. Thus it is constantly present when that nature is present, and universally asserts it, and inheres in the whole of it. The same form is such that if it departs, the given nature infallibly disappears. Thus it is always absent when that nature is absent, and always withholds it, and inheres in it not at all.[118]

Forms – although not forms exclusively – cover the function of the essential attributes sought by the Aristotelians in the negotiation of the intellect. Their presence and absence designate the existence or non-existence of a thing by defining the cause. As with the Aristotelians, for Bacon too the relation is reciprocal: determinate forms correspond to specific natures, and vice versa. Not surprising, therefore, is Bacon's statement according to which 'a true form is such that it draws up the given nature from some source of being which inheres in many other things, and is better known to nature than the form itself'.[119] Indeed, forms are constitutive elements of essence.

Like the Aristotelians, Bacon expresses the reciprocal character of the relation between essence and forms, or nature and forms, through their convertibility. The sum of forms more known by mind corresponds to the nature of the thing more known by nature. True and scientific knowledge of a thing is possible if and only if that sum constitutes that essence. For this reason, Bacon repeats more than once that special attention must be given to the fact that 'a thing's form is present in each and every instance in which the thing itself is present, otherwise it would not be the form'.[120]

Angus Fletcher has pointed out how Bacon's emphasis on this convertibility could be a Ramist legacy,[121] in particular considering the third law of method, also called the law of wisdom in which it is stated that 'a general axiom is one in which the predicate is always true of the subject, not only in all cases and in itself, but also reciprocally'.[122] There is no need to bother with Ramus or the Ramists, however, to attribute the criterion for establishing the necessity between cause and effect or subject and predicate to convertibility. Rather, this is not an unusual position, so much so, in fact, that Ramus himself believed that he was indebted in this respect to Aristotle's *Posterior Analytics* I.4. In the *Valerius Terminus*, Bacon writes that 'this notion Aristotle had in light, though not in use'.[123] Convertible terms, however, are discovered by means of Platonic division, and are 'the form or the formal cause', 'the true difference'.[124] Beyond what he states in the *New Organon*, in the *Valerius Terminus* Bacon adds two 'notes of caution'. The first is that 'the nature discovered be more original than the nature supposed, and not more secondary or of the like degree'; and the second is that 'the nature inquired be collected by division before composition'.[125] I have already pointed out how Bacon employed this twofold procedure of division and composition in *A Description of the Intellectual Globe*, and how this comes from the Aristotelian context of the mental examination through a re-elaboration of Plato's notion of division.

That Bacon is working within the Aristotelian framework and that he is transforming it is evident in his systematic conception of the application of knowledge of these forms as being constitutive of the nature of the things. Bacon categorizes the axiom of the transformation of bodies into two kinds. The first kind considers the body as a combination of forms comprising simple natures. For instance, in gold one can find the characteristics that are the colour yellow, a specific gravity, a peculiar malleability and so on. Therefore, whoever knows these forms and how to introduce them into a body could create gold as a concrete object. The second kind of axiom, in turn, does not proceed from the forms of simple natures but from concrete bodies, identifying the latent process – that is, the process of generation and transformation – and thus anticipating what Hobbes would maintain some years later in the *Elements of Philosophy the First Section, Concerning Body*. In every generation and transformation, one should investigate 'what is lost and given off, what stays behind, what enters; what is dilated, what contracted; what is put together, what separated; what is continued, what cut off; what spurs on, what curbs; what dominates, what yields; and much else besides'.[126] These elements are no less than forms considered in another light, as we have previously seen.

Unlike the Aristotelian model, however, Baconian forms are what makes possible the reproduction and recreation of the effect: once one knows all the essential and constitutive elements of gold, it is possible to create it. It is clear that in this case the distinction between natural and artificial collapses. According to Bacon, what is natural exactly coincides with what is artificial – that is, the human mind can penetrate it in such a way as to know how to produce it.

Renaissance Aristotelians, especially after Pietro Pomponazzi's *De immortalitate animae*, developed a kind of scepticism about the human mind's ability to penetrate nature so deeply. They believed that one can only really know a thing if it is completely under the control of the will, and this is usually the case in ethics, in poetics and in the mechanical arts. This is not so for the natural world, which is for Aristotle simply a subject for contemplation, not manipulation. Human beings do not produce natural things, but only artefacts. Natural things are a subject of contemplation, but even if this operation of the mind is a potential possibility for all human beings, only a few can cultivate it. For this reason, the proper essence of the human being is not speculating and contemplating but acting and producing. The two realms of speculation and action are completely different. This separation collapses with Bacon because there is a continuity between natural philosophy or physics – that is, speculation – and mechanics – that is, making or doing. For Aristotle, there is no continuity between natural philosophy and mechanics: they deal with two different subjects, the former with natural movements and the latter with violent motion. It is only with Renaissance mechanics that this distinction progressively seeps away.

Here lies the major epistemological difference between Bacon and Aristotle (and his followers), namely, for the former knowledge is power, power of doing and making, while for the latter true knowledge excludes making and doing. In the Aristotelian tradition, knowing and doing are two distinct models, represented by the contemplative and the active life. Conversely for Bacon, 'human knowledge and human power come to the same thing, because ignorance of cause frustrates effect'.[127]

Bacon's *Novum organum* is therefore new in relation to the Aristotelian *Organon* – not for the logical processes involved, which aspired to improve and systematize what the Platonic tradition and Renaissance Aristotelians had already conceived with the negotiation of the intellect – but for his ultimate objective. Baconian literate experience and induction are not mere instruments designed to know something, but productive tools for the actual making, and this conception is totally alien from the Aristotelian perspective. Bacon believed he had solved the problem of Aristotelian epistemology, and he was sure of the validity of his new conception of induction and his new method for establishing the grounds for science. However, as is well-known, none of his followers developed his investigation on induction any further, quite possibly because Baconian induction did not represent a genuine turn in the elaboration of a logic of scientific discovery. Its task was mainly limited to making clear and distinct the knowledge acquired through literate experience.

Being based on literate experience, Baconian induction suffered from the same implicit weakness as induction by enumeration or demonstrative induction, that is, underdetermination. In other words, Baconian induction would have been based on the assumption that – in presenting the causes of each nature to the intellect – all logically conceivable possibilities had been found and catalogued through literate experience; otherwise, the induction would have inferred, by the processes of rejection and elimination, invalid knowledge. If it were not based on this assumption, then Baconian induction could not lead to true scientific knowledge, unless its function was merely conjectural in the field of natural philosophy. But mere conjecture does not match with Bacon's conception of knowledge as knowledge of causes and forms. Bacon never thought to apply a mathematical tool to solving the problem; indeed, he was generally averse to the possibility of making claims that related to natural philosophy by mathematical means.[128]

Bacon's idea of induction, therefore, represents a step forward in the elaboration of a new logic of scientific discovery. Nonetheless, on account of its ambiguous and opaque character, it shared the same destiny as the Aristotelian negotiation of the intellect: Bacon's followers transformed it into something different again, a new heuristic tool – that of the crucial experiment (*experimentum crucis*).[129]

4

Galileo Galilei

Venice. 1610. It was cold winter, the sky was limpid and the air was clear when Galileo Galilei pointed his telescope for the first time in human history to the Moon and to Jupiter, discovering admirable mountains, low-lying areas and stars never seen before. Through observation he destroyed a thousand years of Aristotelian theories on the heavens. But bare experience was not enough to dismantle such an established vision of the universe, or reach his disruptive conclusion, in particular in relation to earthshine and lunar mountains. Galileo needed a powerful investigative and demonstrative tool in order to win consensus among his Aristotelian adversaries. In this chapter, I want to show how Galileo elaborates this tool from the ashes of Aristotelian epistemology by introducing new forms of mathematical reasoning.

Galileo's epistemology and logic of scientific discovery have ceased to be the matter of discussion they were thirty years ago. It is as if this were now an exhausted topic: 'the problem of Galileo's method has been solved'.[1] Indeed, 'the method actually used by Galileo in his mature scientific works was precisely the regressive method perfected by Zabarella that Galileo had found in these Jesuit lectures'.[2] The situation no longer seems so settled and recent studies have even begun to highlight the quality of Galileo's originality, his quintessentially problem-solving approach and high degree of eclecticism. What can perhaps be said is that almost all the passages in which Galileo deals with method and epistemology have been commented upon seemingly from all directions, almost exhaustively.[3]

Scholarship has discussed at length the thesis according to which Renaissance Paduan Aristotelianism, and especially the twofold method of regressus, was the source of Galileo's scientific method of experimental discovery. This idea was initially advanced by John H. Randall,[4] but subsequently contested because missing the historical connection between Paduan Aristotelians and Galileo.[5] However, this same connection was subsequently reaffirmed, the missing link being found in the Jesuit lectures at the Collegio Romano.[6] By this account, the key to understanding Galileo's experimental methodology lies in the intermediate stage of regressus, which is characterized by

> the work of the intellect, testing to see if this is a cause convertible with the effect, eliminating other possibilities. This usually requires a period of time, during which the work is that of the mind (*negotiatio intellectus*), not the senses, although

sensible experience plays an important and essential part. Basic to this stage is a *mentale ipsius causae examen* (literally 'a mental examination of the cause itself'), where the Latin *examen* corresponds to the Greek *peira*, a term that is the root for the Latin *periculum* (meaning test) or *experimentum* (meaning experiment or experience). The main task is thus one of testing, for example, investigating and eliminating other possibilities, and so seeing the cause as required wherever the effect is present. At the end of this period, the cause is grasped formally by the mind, that is, precisely as it is the cause, and the unique cause, of the particular effect.[7]

Galileo's epistemology would be grounded, therefore, on a combination of experiments, tests and rational thinking, which followed the Aristotelian theory of knowledge in the construction of a real epistemology based on regressus.

This hypothesis, however, has never been substantiated, and scholarship has doubted the reliability of Randall's conclusions, which seem more likely to be based on resemblances between the writings of Galileo and those of his contemporaries than any proven influence.[8] Many scholars have denied that Galileo maintained a strict adherence to the regressus theory for the whole of his life, but rather turned his methodology towards mathematics very early in his career.[9] Finally, other historians of science have challenged what seemed to be an anachronistic interpretation of regressus, which transformed a mental examination into an experiment, implausibly tracing the *periculum* back to a process of controllable experimentation and measurement.[10] These scholars, aligned with what we have said in Chapters 2 and 3, pointed out that in none of the regressus theorists is there a 'hint that contrived experiment, or indeed any sort of elaborate or systematic appeal to observation, plays a role in scientific inquiry'.[11] Indeed, for the Aristotelians of the time, the intermediate stage of regressus remains something strictly mental, logical, methodological and rational.

In Galileo, however, the situation is much more complex. In his thought, we can clearly perceive a transition from an Aristotelian position to an original perspective on the intermediate stage of regressus.[12] In his epistemological journey, it is possible to appreciate the appropriation and transformation of the Aristotelian negotiation of the intellect, from his early reflections to the *Two New Sciences*.[13] It is my particular belief that Galileo identified the crucial opaque problem in Aristotelian logical writings and found a solution in transforming the negotiation into a new powerful investigative tool, thanks to the application of mathematics to the understanding of natural phenomena. Taking this stance, it becomes implausible to maintain the idea that in his logic Galileo was 'an Aristotelian all his life', or that his early attempts provide 'a clue to the reasoning processes Galileo employed throughout his life',[14] unless we consider an Aristotelian someone who takes Aristotle as his point of departure in elaborating something novel. Looking for consistency in Galileo's methodological thought is to diminish his ability to adopt new solutions and develop new epistemological ideas. He struggled for many years to find the right logic for scientific discovery, and his various attempts should not be considered failures or a lack of method, but rather as steps towards his innovative solution in a new epistemology, capable of reconfiguring the Aristotelian approach.

It is in the manuscript Galileiano 27 that we can find Galileo's early epistemology ideas.[15] Its contents have been viewed by some as no more than rather basic scholastic exercises, only trite scholastic exercises without any real philosophical or scientific value, more useful for insights into what he studied or what his professors taught than for understanding Galileo himself. Others, however, have seen here the origin of the epistemology that Galileo used in his search for new sciences of the heavens and mechanics, and have tried thus to use them to identify his sources. Who his immediate source was remains an open question: some suggest Ludovico Carbo, others Paulus Vallius, and others again Johannes Lorinus, whose positions on the intermediate stage of regressus were explored in Chapter 3. Whoever it was, the evidence remains weak, and there is no smoking gun because none of these philosophers, at least in the specific time interval in which Galileo is alleged to have known them, offer in their works a detailed and exhaustive treatment of the regressus theory and – more particularly – negotiation. Galileo's knowledge of the intermediate stage of regressus might derive just as credibly from a direct reading of Zabarella's work as from the wider circulation of ideas on the topic at the time. Yet my own sense is that a detailed examination of the doctrine of Ms. Galileiano 27 reveals much more about Galileo's understanding of the Aristotelian tradition and his own epistemological development.

For our purpose Ms. Galileiano 27 is interesting in respect of two main issues. The first concerns the origin and knowability of the first principles, while the second has to do with Galileo's theory of regressus. Galileo discusses the origin of the first principles in the first four *quaestiones* of the treatise titled *On Foreknowledges and Foreknown in General*. His aim is to answer the question whether first principles must be previously known in cases of demonstration. His answer is that since the conclusion depends on first principles, if one does not know these principles one cannot reach the conclusion. Therefore, the first principles in one way or another must be foreknown. There are many ways of knowing first principles because they come in very different kinds. For instance, a principle such as 'the whole is greater than its part' is knowable only by an understanding of the terms that compose the proposition. However, in other cases principles like 'fire is hot' are discovered through sensation. What is important for Galileo is that the proper principles of a demonstrative science must be foreknown actually because they are considered as 'the efficient cause of the science'.[16] Without them, scientific knowledge would be impossible.

Such principles, however, cannot be demonstrated within the same science in which they are employed, or else in using them to prove themselves a circular argument is incurred. For this reason, there must be a science which demonstrates these principles, and this science for Galileo is metaphysics: 'first principles in general pertain to metaphysics but when applied to this or that matter they pertain to special sciences'.[17] Galileo is particularly insistent on this topic, stating that first principles are not proven but supposed in special sciences like natural philosophy. Rather 'it is the task of the metaphysician to prove principles', and he adds 'if Aristotle sometimes proved them in physics, he proved them as a metaphysician and not as a physicist'.[18] The implication is that at this stage of his epistemological development Galileo believed that the reliability of natural philosophy or physics depends on metaphysics. The basis for this conception was likely his reading of *Physics* 184 b 26–185 a 5:

> For just as the geometer has nothing more to say to one who denies the principles of his science – this being a question for a different science or for one common to all – so a man investigating *principles* cannot argue with one who denies their existence. For if what exists is just one, and one in the way mentioned, there is a principle no longer, since a principle must be the principle of some thing or things.

However, here Aristotle does not mention metaphysics, but speaks of a different science or a science common to all. Scholars have contested that in these passages Aristotle is in fact referring to metaphysics. But Aristotle would have taken the contrary position, against the possibility of a science capable of demonstrating universal scientific principles – indeed, this accords much more closely with Platonic dialectic, at least as characterized in *Republic* 511 C-D, in which it was characterized as universal science.[19] Galileo was probably drawing this conception from Zabarella's *Commentary to Posterior Analytics*.[20] Zabarella mentions the Aristotelian passages included by Galileo in his *quaestio* and summarizes Aristotle's position by stating as follows:

> Aristotle does not deny that a metaphysician can prove the principles of the other sciences – for that cannot be denied; rather he only denies that this can happen within those sciences of which they are principles. Indeed, from metaphysical principles geometrical principles can be proved – as can, in fact, the very subject matter of geometry – not however in geometry itself but in metaphysics, as we have explained elsewhere. Thus Aristotle in the first book of *Physics* aims to dispute against the deniers of the principles of natural science … by dressing like a metaphysician.[21]

Zabarella's words 'by dressing like a metaphysician' very much anticipate Galileo. This idea is pivotal to the Renaissance discussion on regressus. Indeed, it was a doctrine defended by Agostino Nifo and attacked by Pietro Pomponazzi. By Pomponazzi's own account, he was 'the first to make an objection' to regressus on this very basis during a class with Neritonensis. His objection was that through regressus the principles and subject of a science would be demonstrable within the terms of its own framework. In support of his objection, Pomponazzi lists a series of passages from Averroes. According to these passages, 'a metaphysician cannot demonstrate the subject of natural philosophy'. Furthermore, he goes on, if a metaphysician could prove the principles of natural philosophy such as 'matter and form', he had no logical inference capable of producing such proof, neither negotiation nor demonstration.[22]

Zabarella, in this instance, agrees with Nifo, maintaining that metaphysics is thus the science which demonstrates in itself the principles of all other sciences. This does not mean, in his eyes, that all sciences are servants to metaphysics – rather they are autonomous – but the foundations of their principles are metaphysical. In so doing, Zabarella abandons Aristotle's conception of metaphysics, conceiving it as more like a universal science, a first philosophy (*prima philosophia*), capable of demonstrating the proper principles of all sciences while remaining external to them.[23] Furthermore, he denies on two grounds that dialectic, as opposed to metaphysics, can occupy the position of that universal science that demonstrates the principles underpinning other sciences: (1) dialectic is not a science, but an instrument; (2) it cannot demonstrate because it is based on probable arguments.[24] Zabarella's devaluation of dialectic is a direct consequence of his conception of Aristotelian metaphysics 'as a rigorous demonstrative science that has as its model mathematics' and has as its method regressus.[25]

But why does Zabarella abandon Aristotle and opt for this theory? Zabarella could have found considerable support for his theory (that metaphysics was a universal science and employed a form of regressus from demonstrating the principles of other sciences) in Francesco Barozzi's edition of Proclus's *Commentary on the First Book of Euclid's Elements* (1560). In this work, Barozzi explains in the marginal titles that the demonstration of the proper principles for each science does not pertain to natural science or mathematics, but rather to a 'divine science (*divina scientia*)' or to a 'first philosophy (*prima philosophia*)' according to what Aristotle and Plato established.[26] From this common and universal science, mathematics proceeds with 'convenient measure and right order',[27] but every kind of principle – and not only mathematical ones – are proven by means of it. Barozzi's edition also makes clear that the adjective μαθηματική cannot be referred in absolute terms only to mathematics, but it is applicable to all demonstrative disciplines, and that μάθησις should be translated as *discipline*.[28] Proclus adds somewhat ambiguously, as we shall see in Chapter 7, that this universal science follows 'the method of proceeding from things better known to things to know, and the reverse path from the latter to the former, that is analysis and synthesis'.[29] The method of metaphysics as a universal science is that of regressus, and it is the same as is adopted by mathematics. In these reflections, Zabarella finds confirmation of the facts that (1) metaphysics is the universal science that demonstrates the proper principles of other sciences; (2) its method is regressus; and (3) the method is common to all the contemplative sciences, as we will further elaborate on in Chapter 6.[30]

One particular item in Ms. Galileiano 27, the treatise *On Demonstration*, is pivotal for understanding Galileo's early idea of a logic for scientific discovery. For Galileo, demonstration is one of two instruments that serve the mind in the acquisition of scientific knowledge; the other is definition. Indeed, perfect – that is, clear and distinct – knowledge of its nature or one or other of its properties. It is clear, therefore, that Galileo does not endorse explicitly and directly Aristotle's definition of knowledge as knowledge of causes: he deals with the nature and properties of a thing. Indirectly, knowledge of causes is ineluctably involved, for definition provides cognition of essence – that is, of the formal cause – and demonstration is an illative process based on the four causes. According to Galileo, however, precisely because demonstration deals with four causes and not just one, as is the case with definition, it is a superior instrument for acquiring scientific knowledge.[31]

Galileo points out that scientific knowledge does not so much concern real truth as rational truth, which pertains to cognitive powers rather than to actual things. Indeed, rational truth can itself be either simple or complex. It is simple where the knowing occurs through the first operation of the intellect – that is, as apprehension – and 'is nothing more than the conformity of what is apprehended with the thing that is outside the knower'.[32] It is complex where it concerns judgement or discourse – that is, the second and third operations of the intellect – and 'it is nothing more than the conformity of the proposition made by the intellect with the unity that is outside the knower'.[33] Demonstration has to do with complex rational truth, as in the case 'Man is a rational, risible animal' for the reason that the object outside the mind that is a man is really genuinely conjoined with the attribute of being rational and risible. In logic, Galileo deals only with this rational type of truth, which provides an explanation for how things are in the world, without claiming to say anything about their real being.

Galileo's theory of truth is based on three aspects: (1) necessity, (2) universality and (3) essentiality. These three aspects are indispensable for understanding his theory of regressus. Galileo characterizes necessity as 'a kind of condition bringing exchangeability to things',[34] and there are two kinds (of necessity): unqualified and natural. Unqualified necessity 'cannot be impeded by any power, not even by divine absolute power'.[35] This kind of necessity is of the type 'God necessarily exists'.[36] The second kind of necessity – natural necessity – can be overturned by God's will, but follows the ordinary laws that God gave to nature. Galileo remarks that natural necessity can be broken down in two further ways: there is an absolute natural necessity where things have an intrinsic connection (*intrinsecum ordinem inter se*), as in the case of 'man' and 'rational'; and there is a relative natural necessity when the relation between two things is only extrinsic, as in the case of 'swan' and 'whiteness'.[37] These two kinds of natural necessity are fundamental to understanding Galileo's idea of universality or his better predication of every instance (*de omni*).

Galileo's explanation of what can be said of every instance begins with the distinction between prioristic and posterioristic statements. The former designates a '*universal*' proposition in which the predicate invariably goes with the subject and with everything contained under it, *but abstracting from its inherence at all times*', while the latter characterizes 'a proposition in which the predicate goes with the subject and with all things contained under it, if there are any, always and at any time whatever' (emphasis added).[38] Prioristic statements concern the mere form of reasoning, while posterioristic statements consider matter and truth. Galileo gives 'God is unchangeable' as an example of a posterioristic proposition. This distinction functions as a way of introducing other kinds of statement said of every instance, the posterioristic statement and the universal statement. These represent three different ways of understanding how a predicate inheres to the subject in a universal way. According to Galileo, every essential proposition is said of every instance, as in the case 'every man has sensible knowledge', but this is not valid vice versa; indeed, not every statement said of every instance is essential, as is clear in the example 'every man is two-legged', because the property of having two legs does not characterize the essence of what a man is. Universal statements, in contrast, are narrower in scope in comparison to a statement said of every instance and in relation to an essential proposition. Indeed, that 'every man is risible' is universal and essential, but 'every man has sensible knowledge' is not universal even if it is essential.

This argument serves Galileo in explaining that scientific demonstration – that is, demonstration *propter quid* – can be made only of propositions that are said of every instance and are necessary, because they provide knowledge of the cause of a thing by showing the properties that pertain to it necessarily. For scientific knowledge – for Galileo just as with the Aristotelians – is to have cognition of a thing that cannot be otherwise. Galileo is aware of the fact that natural philosophy deals with contingent effects and phenomena, and therefore it seems impossible to have scientific knowledge of them. His response is that demonstration is concerned with things 'that have a true connection for all eternity in the divine mind'.[39] This eternal connection represents an essential connection between a subject and a predicate, whose relation resides in God's mind. Galileo suggests several counterarguments to his thesis, and he responds in each case using examples employed in the regressus theory. A possible objection might be that an eclipse is demonstrated of the moon, even though an eclipse does not take place on the moon always, necessarily and at all times, or that there are some predicates that are demonstrated of a subject although they are not present in them always. In these phenomena, therefore, there is no eternal and essential connection. Galileo explains that a statement said of every instance can be understood in three ways: (1) when the predicate is attributed to the subject invariably and always; (2) when the predicate inheres to the subject necessarily, but not always; (3) when the predicate pertains to the subject most of the time, but not invariably and not always.

Once he has established what 'necessary' and 'said of every instance' means in a demonstration, Galileo focuses on the modes of speaking essentially of something. He states that

> predicates are either essential to the thing of which they are predicated, and so they constitute the first mode; or they are accidental, and if so, they are either common and rejected as being of little value, or they are proper, and then they make up the second mode of speaking.[40]

There are, therefore, only two ways of speaking essentially of something. The first characterizes propositions in which 'the predicate is the definition of the subject, or it is the ultimate genus or the ultimate differentia, or it is a remote genus or a remote differentia'.[41] In the first mode, Galileo also includes propositions in which the predicate is not exactly a definition but rather an extrinsic entity, through which the subject is nonetheless defined. This is the case with the soul, which is the form of the body and defines it. The second mode characterizes propositions 'in which a property is predicated convertibly with the subject', as in the case 'man is risible'. In both cases, the mode of speaking essentially is fundamental to the character of convertibility. In the second mode, there are ways in which not all predicates are convertible with the subject, but nonetheless they are still essential. For instance, 'man is rational' is convertible and therefore the relation is defined as primary, while 'an animal is a living body' is essential, but not convertible because not every living body is an animal, and vegetables are living bodies. The relation is thus secondary.

In order to better specify the quality of essentiality in the second mode, Galileo discusses four ways of understanding the concept of 'proper'. First, proper is what belongs to every instance, as being two-legged does in the case of the human being. A second instance is where proper relates solely to a subject, as where being a philosopher inheres only in man. In the third case, it pertains to every instance and solely, though not always, as with the fact that man can become grey-haired. The final and most perfect way is the fourth, in which proper is something that belongs to every instance, solely and always, as in the case of 'man is risible'.[42] The perfection of the fourth also stems from its containing because it contains all four causes. Indeed, according to Galileo, 'causes involve an intrinsic relationship to what they cause'.[43] This is a bold statement because it means that between cause and effect there is a necessary and essential relation – that is, the two terms are convertible. Galileo excludes the possibility of a cause being contingent in relation to its effect. It can be considered contingent only if one focuses on the predication, but in no way it is 'and this our concern, if one focuses on the connection of the predicate with the subject and the causal relationship between the two'.[44] Galileo explains this with a compelling example. In general, he states, a proposition like 'Plato is rational, risible, etc.' does not belong to the two modes of predicating for the reason that the subject 'Plato' is individual and contingent. However, he adds, if the proposition is 'considered with respect to the connection of the predicate with the subject, such that the word "is" indicates not existence but the connection of the subject with the predicate', then it is 'in the first or in the second mode depending on whether the predicate pertains to the essence of the subject or vice versa'.[45] This is an important remark because Galileo is emphasizing that his epistemology works in the realm of rational truth, where the word 'is' designates a logical copula, rather than a mode of existence.

The discussion of necessity, universality and essentiality is crucial for Galileo in establishing the characteristics of the propositions for the most perfect demonstration. Indeed, demonstration *potissima* must be made from essential, necessary and universal propositions, and should proceed from a true and proper cause. Following Zabarella, then, Galileo believes that the most powerful demonstration and demonstration *propter quid* are the same, being based on the same kind of propositions.

These preliminary observations on the properties of propositions and on the characteristics of the most powerful demonstration lead Galileo to the last *quaestio* of the *Treatise on the Demonstration*, titled 'Is There Demonstrative Regressus?' Here Galileo revisits all the classic topics on the effectiveness of regressus in natural philosophy and the question of its non-circularity. Though dealing with traditional issues of logic, Galileo surpasses his possible sources in originality. Therefore, even if he did not endorse this kind of epistemology, as has been suggested, we can nonetheless say that this treatment is at least his own.

Galileo defines demonstrative regressus as 'a progression of reasoning in demonstration, which is made from effect to cause and vice versa for the more perfect development of the sciences'.[46] It, therefore, represents the chief instrument for acquiring scientific knowledge. For regressus to be efficient, 'the proving part and the part proved must be connected with each other' in such a way that a necessary argument can be inferred from the one to the other. Furthermore, he specifies, what is more known – that, is the proving part should come first in the demonstration. Moreover, in regressus the cause is necessarily connected with the effect, and according to Galileo this is useful in two situations. The first case, the commonest one, is when one knows the effect and not the cause, 'and consequently one might prove the existence of the cause from the existence of the effect'.[47] The other interesting instance is where someone discovers a cause but is unaware of the proper way to produce its effects, and the 'proper way might then be manifested through the demonstrative regressus'.[48]

With these applications in mind, for Galileo regressus 'is most frequently used in physics because for the most part physical causes are unknown to us'.[49] Conversely, in mathematics, regressus 'has almost no use, because in such disciplines causes are more known both with respect to nature and with respect to us'.[50] If mathematical principles have already been established and are therefore known, and these represent the causes of the demonstration, then – as Zabarella asserted – the first stage of regressus becomes futile. In purely mathematical investigation, therefore, regressus is not applied.

The most innovative aspect of Galileo's reflections is the identification of six essential conditions for regressus. The first of these is that there are two progressions – that is, two demonstrations, 'one from the effect to cause, the other from cause to effect'.[51] On this specific point, Galileo follows the orthodox Aristotelian doctrine of the time: two forms of concatenated reasoning constitute regressus. The second condition is that the starting point is demonstration from the effect and not demonstration from the cause. Otherwise, the former would be pointless, because in knowing 'the proper reason for the effect', knowledge of 'its existence' is implicit.[52] The third condition is that the effect must be more known than the cause, which is unknown at the outset, or else the Aristotelian principle stating that scientific reasoning should proceed from what is known to what is unknown would be inapplicable. The fourth condition deserves careful examination:

> Having made the first progression we do not begin the second progression immediately, but wait until we come to have formal knowledge of the cause we first know only materially. The reason: because we cannot formulate a demonstration *propter quid* if we do not have prior formal knowledge of the cause. You object: therefore it would follow that demonstration *propter quid* would be made to no purpose, because it is made only that we may have formal knowledge of the cause. I deny the inference: because, granted that one who has formal knowledge of the cause also has virtual knowledge of the reason why the property inheres in the subject, one does not understand it actually unless one effects a true demonstration. And from this it follows that the regress is circular in an improper sense, since in it one progresses from an effect to material knowledge of the cause, and then from formal knowledge of the cause to the proper reason for the effect.[53]

Between the two forms of demonstration, according to Galileo, there is not continuous flux; rather, there is interposed a necessary intermediate stage, which leads from a material knowledge of the cause to a formal knowledge.[54] Material knowledge is a rather confused type of cognition arising from sensation, providing only a generic idea of the cause, in particular of its existence. Formal knowledge, on the other hand, is a clear and distinct cognition of the cause by means of the form.[55] Galileo, like all the other Aristotelians of the time, is aware that having a merely formal knowledge of the cause is not sufficient for scientific knowledge, which is based rather on the apprehension of the necessary connection between cause and effect, or, in other words, on the law that governs their relationship. Formal knowledge of the cause can only provide virtual knowledge because the real connection with the effect has not yet been proved; for this reason, the second kind of reasoning – that is, the demonstration *propter quid* – is not superfluous. In this way, circular reasoning is avoided and the accidental knowledge grounded in experience becomes universal and necessary science.

What makes possible the transition of demonstration from effect to that from cause – that is, of confused to distinct knowledge – is without doubt the mental examination or negotiation of the intellect proposed by the Paduan Aristotelians, even if scholarship has tended to neglect this point. Ernan McMullin, for instance, points out that in the demonstrative regress, the inference from effect to cause is logically invalid, 'except where the major premise can be shown to be convertible – that is, where the cause can be shown to be the only possible cause of this effect'. He adds that 'convertibility requires a separate argument', and that Zabarella, 'the authority on whom Valla principally relies, was never really able to circumvent this difficulty, and Galileo himself makes no attempt to address it'.[56] However, looking carefully at Galileo's appropriation of demonstrative regress, we can appreciate his understanding of the negotiation of the intellect as being just as Zabarella understood it – a means of solving the problem of convertibility:

> Fifth condition: that it be made in convertible terms, because if the effect were broader in scope than the cause, this would impede the first [progression]. Hence the following is not valid: there is light, therefore the sun. If, on the other hand, the cause was broader in scope than the effect this would impede the second progression, as is evident. For, although it is valid to argue: something breathes, therefore it has a soul, one cannot do the reverse, because respiration requires organs that may be lacking in things that nonetheless have a soul. Finally the reasoning must be in the first figure.[57]

The intermediate stage – that is, the negotiation of the intellect – which is the fourth condition of regressus, determines that the cause is formally, universally, necessarily and essentially connected to the effect – that is, they are in a bi-univocal relation, which is co-implication in Aristotelian terms. This bi-univocity is the basis of the convertibility of the terms in the two forms of demonstration. And it is this convertibility, as we have seen in Balduino and Zabarella, that guarantees to physical investigation the same degree of certainty as mathematical knowledge.

Galileo is aware that natural philosophy could lead in the first stage of regressus to an inconvertible proposition – that is, to a situation in which cause and effect are not in reciprocal implication. Balduino and Zabarella had aimed through the intermediate stage of regressus to reduce these accidental connections to essential relations – that is, to formal definitions that would describe the underlying formal cause, or in a specific case an efficient cause – in such a way that there could be total convertibility and a degree of certainty equal to that of mathematics. How is it possible to have perfect convertibility between cause and effect? What transforms an accident into an essential and constitutive part of the essence of a thing? What is the correct use of the intermediate stage of regressus? And how does Galileo solve the problem? In the *Treatise on Demonstration*, there is no specific answer to all these questions. Indeed, Galileo's position seems to be twofold: (1) scientific knowledge is possible only through regressus or through demonstration *potissima* when cause and effect are convertible; (2) when cause and effect are not convertible, knowledge is only probable.

From the reading of this manuscript, it is clear, however, that Galileo firmly believed that 'one can use logical laws for inferring causes from effects with very nearly the same degree of rigor' as mathematics, which guarantees the convertibility of the two terms.[58] In studying regressus, Galileo becomes aware of the intrinsic limits of certain kind of the Aristotelian methodology when investigating natural effects, and he is drawn towards mathematical method more and more for its capacity to produce certain demonstrations.

In the following epistemological elaborations, Galileo attempts to introduce into the intermediate stage of regressus mathematical devices that would lend a degree of certainty to natural philosophy commensurate with that of mathematics – exactly what the Aristotelians were seeking. Indeed, only mathematics can make clearer and more distinct that knowledge coming from sensation and experiments.[59] Concomitantly, once Galileo begins his explorations into mathematical method for the investigation of natural phenomena, he tends to abandon the Aristotelian regressus, or at least a strict application of it.

*

A first sign of Galileo's conception of the transformation of intermediate stage of regressus into a mathematical form of reasoning appears in *The Older Works on Motion* (*De motu antiquiora*),[60] which was probably composed after Ms. Galileiano 27 and the publication of Pappus's *Collection* in 1588, in response to and in dialogue with Pisan Aristotelianism.[61] In this work, Galileo states that his aim is to identify the causes of the swiftness and slowness of natural motion, and that 'what we seek are causes of effects, which are not taught us by experience (*quaerimus enim effectuum causas, quae ab experientia non traduntur*)'.[62] Here we are clearly in the situation outlined in his *quaestio* on regressus, namely, the seeking of the cause from effects. Does Galileo apply regressus in this case?

In addressing the deficiency of experience, reason provides arguments, which – according to Galileo – undermine Aristotle's opinion. Galileo aims to investigate the reason why mobiles constituted of the same material but of different size (*mole*) move through the same medium at the same speed – that is, a larger piece of wood does not fall more swiftly than another. This kind of investigation is only carried out mentally, like the intermediate stage of regressus for the Aristotelians. Galileo explicitly states twice in the space of a couple of lines that we should 'conceive in our mind', and this act of conceiving means to elaborate thought examples which are able to make the cause evident. For instance, in considering two pieces of wood of different sizes floating in water, the fact that they move with the same speed becomes less surprising if we conceive of them in relation to the quantity of water in which they move. This is not the only thought example that Galileo produces, but he proceeds from a variety of different suppositions. These pages are full of references to the conceptuality of suppositions and hypothesis:

> Let the following be *presupposed*: namely, if there are two mobiles, one of which is moved faster than the other, the combination of the two is moved more slowly

than that part which was moved faster than the other, but more swiftly than the remaining part, which, alone, was carried more slowly than the other: as, for example, if we understand two mobiles, such as a piece of wax and an inflated bladder, both of which are carried upward from deep water, but the wax more slowly than the bladder, we ask that it be conceded, that if they are combined, the combination will go up more slowly than the bladder alone, but more swiftly than the wax alone. Indeed this is very clear: for who doubts that the slowness of the wax will be diminished by the speed of the bladder, and, on the other hand, that the speed of the bladder will be retarded by the slowness of the wax, and that a certain motion intermediate between the slowness of the wax and the speed of the bladder will result? Similarly, if on the other hand two mobiles go down, one of which is carried more slowly than the other, as, for example, if one is wood, the other a bladder, which go down in air, the wood more swiftly than the bladder, we *presuppose* this: if they are combined, the combination will go down more slowly than the wood alone, but more swiftly than the bladder alone. For it is manifest that the swiftness of the wood will be retarded by the slowness of the bladder, while the slowness of the bladder will be accelerated by the speed of the wood; and similarly a certain motion intermediate between the slowness of the bladder and the swiftness of the wood will result. This having been *presupposed*, I argue as follows: by proving that mobiles of the same species, of unequal sizes, are carried with the same swiftness. Let there be two mobiles of the same species, the larger a, and the smaller b; and, if it can be done, as our adversaries hold, let a be moved more swiftly than b. There are then two mobiles one of which is moved more swiftly than the other; hence, according to what has been *presupposed*, the combination of the two will be moved more slowly than the part, which alone, was moved more swiftly than the other. If then a and b are combined, the combination will be moved more slowly than a alone: but the combination of a and b is larger than a alone: hence, contrary to our adversaries' view, the larger mobile will be moved more slowly than the smaller; which would certainly be unsuitable. What clearer indication do we require of the falsehood of Aristotle's opinion? (emphasis added).[63]

The passage is long but it is testimony to Galileo's way of thinking based on mathematical reasoning and to the reduction to the impossible of the Aristotelian opinion. His conclusion is as follows:

There exists no cause per se as to why mobiles of the same species should be moved with unequal speeds, but there certainly is one relating to why they should move with equal speeds. But if there were some accidental cause, such as, for example, the shape of the mobile, it must not be classified amongst the causes per se.[64]

The process by which he reaches this conclusion is exclusively mental. This conclusion is valid if we imagine eliminating the accidental causes, namely, the possible impediments to the motion, which is in this way idealized. He admits that 'the conclusion must be understood as concerning those mobiles where the heaviness and size of the smaller one is large enough that it is not hindered by that small tenacity of the medium'.⁶⁵ What is interesting in this demonstrative process is not that it is compliant with the Aristotelian regressus, as Wallace believes. Indeed, there is no real proof, but only an idealized possibility that, setting aside accidental causes, two bodies of the same material fall at the same speed in the same medium.⁶⁶ It is interesting because Galileo asserts that he has discovered the cause and that he did this using mathematical reasoning. The mathematical reasoning in this case is no doubt the theory of ratios coming from Euclidian geometry, which shapes his understanding of the mathematization of the world.⁶⁷

Similar passages in which Galileo applies mathematics and the theory of ratios to find the true cause of an effect are scattered throughout the entire work. But in many instances, it is not possible to turn back from principles to effects:

> These, then, are the universal rules (*universals regulae*) of the ratios of the motions of mobiles, whether of the same species or not, whether in the same medium or in different media, whether moved upward or downward. But it must be noted that a very great difficulty arises here: it will be found that these ratios are not observed by one who has made a test (*periculum*). For if one takes two different mobiles, which have such properties that one is carried twice as swiftly as the other, and then releases them from the top of a tower, it will certainly not hit the ground faster, twice as swiftly: what is more, if one makes the observation, the one which is lighter at the beginning of the motion will precede the heavier and will be faster.⁶⁸

This passage seems – against Wallace – to advocate the impossibility of a regress on the grounds that there is no way of proving the universal rules in the experience – in other words, showing or testing it with an effect. In *The Older Works on Motion*, there are many passages in which mathematics is applied in order 'to resolve' the demonstration of a fact to 'its proper principles'. However, this resolution works only supposing that there are no complicating accidents or impediments, and in each case no regressus is possible.⁶⁹

One of the most important examples of the application of mathematical method for discovering causes is in the discussion as to why the falling motion accelerates toward the end of the fall. Galileo explicitly employs what he calls a resolutive method (*resolutiva methodo*):

> But, *leaving aside the ways of thinking of the others, in order that we may track down what we believe* (credimus) *to be the true cause of this effect, we will make use of the following resolutive method.* Since, then, a heavy mobile (let us however speak about natural downward motion, coming from heaviness: for, that being known, we will judge the case of upward motion by proceeding in reverse) in going down is moved more slowly at the beginning, it is therefore necessary that it be less heavy at the beginning of its motion than in the middle or at the end; for we know with certainty, from the things demonstrated in the first book, that speed and slowness follow heaviness and lightness. If, then, it is found out how and why a mobile is less heavy at the beginning, the cause for which it goes down more slowly will certainly

have been found. But the natural and intrinsic heaviness of the mobile is certainly not diminished, since neither its size nor its density is diminished: it remains, therefore, that that diminution of heaviness is against nature and accidental. Hence if we have found in what way the heaviness of the mobile is diminished against nature and extrinsically, what we need will surely have been found. But that heaviness is not diminished by the heaviness of the medium, for the medium is the same at the beginning of motion as at the middle: it remains, therefore, that the heaviness of the mobile is diminished by some violence that is extrinsic and comes from outside (for it is only in these two ways that a mobile gets to be light by accident). If then, again, we find out how a mobile could be lifted by an extrinsic force, the cause of slowness, again, will have been found. Now the force impressed by a thrower not only at times diminishes the heaviness of a heavy thing, but it often even renders it so light that it flies upward with great speed: hence let us see and *search attentively* (*diligenter perscrutemur*), whether perhaps this force is the cause of the diminishing of the heaviness of the mobile at the beginning of its motion. And, I say, it certainly is that force impressed by the thrower which renders natural motion weaker at the beginning: so let us hasten to make clear by what method it is able to accomplish this. For a heavy mobile to be able to be moved upward violently, an impelling force greater than the resisting heaviness is necessary; otherwise the resisting heaviness would not be able to be overcome, and, consequently, the heavy thing would not be able to be carried upward. Hence, the mobile is carried upward, provided the motive impressed force is greater than the resisting heaviness. Now since this force, as has been demonstrated, is continuously diminished, it will finally become so diminished that it will no longer overcome the heaviness of the mobile, and then it will not impel the mobile any further: but that impressed force will not therefore have been annihilated at the end of the violent motion, but it will only be diminished to the point that it no longer surpasses the heaviness of the mobile, but it will be equal to it; and, to put it in a word, the impelling force, which is lightness, will no longer dominate in the mobile, but will have been reduced to parity with the heaviness of the mobile: and then, at the ultimate point of the violent motion, the mobile will be neither heavy nor light. But, moreover, as the impressed force diminishes in its own way, the heaviness of the mobile begins to predominate; and hence the mobile starts to go down. But since at the beginning of such a descent a great deal of force that impels the body upward, which is lightness, still remains (even though it is no longer greater than the heaviness of the mobile), it comes about that the proper heaviness of the mobile is diminished by this lightness, and, consequently, that the motion at the beginning is slower. And, moreover, since that extrinsic force is further weakened, the heaviness of the mobile is increased by having less resistance, and the mobile is moved still faster (emphasis added).[70]

In this passage, Galileo proves through the theory of ratios that the true cause (*vera causa*), which means cause *per se*, of natural downward motion of body is the *impetus*. The demonstrative process he adopts in this case is to find a possible condition – that is, a cause according to which a specific effect happens. He conjectures or supposes a cause. Admitting the cause, Galileo deduces a series of consequences that truly correspond to the effect, and this confirms the hypothesis that the cause is not a mere conjecture, but can form the principle of the following demonstration. In other words, from the comparison between cause and effect, as in the intermediate stage of regressus, this analysis identifies the true cause of the effect. The intermediate stage characterizing this 'resolutive method' is, as Enrico Berti has suggested, a form of 'mathematical analysis which proceeds from the unknown (hypothesis) to deduce some known consequence', and is located 'just in the intermediate stage between the two demonstrations of which regressus is constituted'.[71] The mathematical analysis that Galileo suggests for the intermediate stage of regressus seems to match that described by Euclid in the Scholium after the fifth proposition in the thirteenth book of his *Elements*:

> Analysis is an assumption of that which is sought as if it were admitted [and the passage] through its consequences to something admitted [to be] true. Synthesis is the assumption of that which is admitted [and the passage] through its consequences to the finishing or attainment of what is sought.[72]

These definitions of analysis and synthesis were attributed in Henri Estienne's Renaissance edition of Euclid's *Elements* (1517) to Theon of Smyrna, as is testified in François Viète's *Introduction to the Analytic Art* (1591).[73] This resolutive method as mathematical analysis is introduced in the second stage of regressus. Galileo clearly considers this kind of reasoning to be mathematical, but the mathematical at its most demonstrative and certain. Indeed, he points out that there might be an easier method (*faciliori methodo*) of obtaining the same conclusion, proceeding from natural considerations; but such a conclusion would be based on the rupture of the distinction between natural and violent motion, making of these only one continuous motion – a way of thinking that runs against common opinion.[74]

One might posit that Galileo later abandoned the argument of *The Older Works on Motion* in his mature works, and that thus he had learned that this manner of thinking was invalid. Here he is exploring the cause of the acceleration of falling bodies, at a time when he lacked quantitative description for such acceleration; and he maintains that the cause of acceleration is their weight, and considers the way that this weight is affected by various conditions, such as the original upward force that brought the body to the point of falling. This is undoubtedly the case according to the famous passage in the *Two New Sciences*,[75] but only from the standpoint of its content – that is, of the physical doctrine, not the methodology, as we shall see.

*

Galileo seems to use mathematical analysis in other writings such as the *Treatise on the Sphere* (*ca.* 1602). This quintessentially sixteenth-century work on spheres is deeply influenced by Alessandro Piccolomini's bestsellers on topics including *Of the Sphere of the World* (1540) and *The First Part of the Theories or Speculations of Planets* (1558).[76] In the latter work, Piccolomini writes a short digression titled *Of Whether the Representations Invented by the Astrologers to Save the Appearances of the Planets*

Are Based on Anything Real in Nature, and his answer cannot be other than negative. Indeed, he states that Ptolemaic planetary theory is useful to the practical astronomer to describe the celestial motions 'but represents nothing real'. He is very critical of astronomers who 'represent eccentrics and epicycles as existing in the celestial spheres because they really believe those spheres to be arranged thus'.[77] He maintains that the role of astronomers – that is, mathematicians – should be that of preserving appearances, nothing more. Their task is merely to calculate, compute and predict the various positions of the heavenly bodies; not to provide a causal explanation for the bodies themselves. The question as to whether such mathematical representations are true or not pertains neither to mathematicians or astronomers, but to natural philosophers. Mathematicians 'had little concern whether the things they were imagining were more necessary than probable or false'.[78] Nonetheless, he appreciates their way of thinking because just as logicians can infer correct conclusions from false premises, 'so an effect can be inferred and deduced from a pretended cause'.[79] Thus, Piccolomini explains that

> granted that eccentrics and epicycles are not in the nature of things and that the appearances of the planets derive from other proper and true causes which we do not know, nevertheless if they actually existed such same appearances would necessarily be inferable from them.[80]

Piccolomini is sceptical as to the possibility of knowing the real causes of celestial motions – in other words, of providing a causal explanation in the Aristotelian sense. He contents himself with describing 'how' a natural phenomenon happens, rather than explaining 'why' it happens. Supposing, he says

> that we should see a stone strike a wall and with great force, and not knowing the origin of such fury we should imagine that the stone had come from a bow or a crossbow. And suppose that our representations were false and that, as chance would have it, the stone had come from a sling shot. Nevertheless, it would have struck the wall with the same fury if it had come from the imagined bow. For the aforesaid fury of that stone could have derived from more than one cause. Thus again, though the real causes of the many appearances which we see in the planets in the sky are hidden to us, still it is enough for us that, supposing these representations to be true, these appearances which we see would just the same derive from them. This for us is more than sufficient for the calculations and for the predictions and for the notices which we must have for the positions, places, magnitudes, and motions of the planets.[81]

Mathematicians provide only a numerical representation to fit with the motions, and this description is sufficient to know how things happen in the world. For this reason, no scientific knowledge is possible from mathematics, to the extent that for Piccolomini – celestial bodies should be assigned only 'probable causes, or such that if they had been true those effects would necessarily have followed'.[82] As he puts it, 'although an effect cannot have more than one proper, real, and necessary cause', in mathematical reasoning, 'an effect can derive not only probably but also necessarily from more than one cause, not on account of the nature of the causes, but by force of supposition and inference', as is demonstrated in the case of the stone that strikes a wall.[83] Mathematical knowledge, therefore, is the most certain in describing how astronomical events happen, but provides only probable and not scientific – that is, casual – knowledge of these events. Piccolomini deals with two different kinds of certainty, involving two distinct ideas of necessity – a real or causal necessity and a logical necessity.[84] This distinction allows him to safeguard God's omnipotence in producing effects and to avoid admitting a strict determinism in the world. In the physical world, things can happen in a variety of ways, and what mathematicians provide is no more than a reliable description of how things happen, but not how they are caused.

Like Piccolomini in his *Treatise on the Sphere*, Galileo endorses the distinction between natural philosopher and cosmographer. A cosmographer or an astronomer is essentially a mathematician, who provides a 'description of the world', considering only certain aspects of it describable through numbers, calculations and geometrical figures. Indeed, the proper task for the natural philosopher is 'the consideration of substance and of its qualities'.[85] For this kind of investigation, Galileo renounces the exploration of the cause at the ontological level. The mathematician's epistemology in investigating nature, and in particular celestial motion, follows four different stages:

> The first ... contains appearances, otherwise called phenomena: and these are none other than sensed observations, which everyday we see, as for instance the rising and setting of the stars ... In the second stage there are hypotheses: and these are none other than suppositions concerning the structure of the celestial orbs, and such that they correspond to appearances ... Then, in the third place, geometrical demonstrations follow, with which through the properties of the circle and of straight lines, are demonstrated the particular accidents that follow from hypotheses ... And finally, what has been demonstrated through lines, calculating by arithmetical operations, is reduced and arranged in tables.[86]

In these methodological precepts, there is no hint of regressus, but there is an application of mathematical analysis to observations. This is explicitly attested in Galileo's sentence – most of the time overlooked by scholars – in which he states that 'we will deal only with hypotheses, striving to confirm and prove them with appearances'.[87] The first stage of Galileo's methodology could be seen as the first stage of the Aristotelian, at the end of which the cause of the knowable phenomenon is conjectured in a form of hypothesis.[88] However, in this sense any kind of collection of observations from which a general conclusion may be derived can be understood as such. In Galileo's case, this formulation of a hypothesis has an overlap between what is assumed as true or admitted as such by mathematical analysis, and the generic conclusion yielded at the end of the first stage of regressus. This overlap allows Galileo to introduce mathematical analysis in order to test the truth and certainty of the hypothesis through consequences, and this encompasses the third stage of the methodology. The fourth part recalls Alessandro Piccolomini's practice of making tables in his treatise *On the Sphere of the World*. The third stage of regressus is missing. Is this the epistemology that Galileo employs in this treatise?

Although Wallace qualifies it as a strict form of regressus, the example of the moon's illumination and of its various shades in the *Treatise on the Sphere* seems to reflect the methodological procedure described by Galileo at beginning of his work, and it employs mathematical analysis. Galileo's objective is to assign a cause to the effect, but this assignment is rather a hypothesis to be confirmed with a view to recording appearances rather than discovering the cause. Galileo states that since 'we want to assign the cause to an effect, it is necessary that we suppose some things, and we have to reduce to memory others'.[89] After several suppositions, such as the moon is an obscure sphere, or only one half of it is illuminated, Galileo suggests taking note – that is, committing to memory – the various movements of the moon, and calculating the various accidents or effects that follow from the hypothesis. The third stage of regressus is missing but, as Berti suggests, this could be understood as being implicit because Galileo is focusing on the discovery phase of the method. However, if mathematical analysis is applied at the intermediate stage, what was necessary for the Aristotelian at the third stage – that is, the inference from the causes to the effects – has already been accomplished with the progression from the hypothesis through the consequences. If there is to be a third stage after the mathematical analysis, this should explain something more than the relation between causes and effects, which has already been established during the correct passage from hypothesis to consequences. Indeed, as we shall see, other philosophers who employ mathematical analysis, – notably Robert Hooke and Isaac Newton – will conceive the third stage as the application of a principle, originally a hypothesis to other effects that had not been considered at the beginning. In this sense, mathematical analysis furnishes proof of something, leading subsequently to the explanation of new facts and phenomena. What is evident in the epistemology advanced in the *Treatise on the Sphere* is that Galileo is working more as a mathematician than a natural philosopher, and the discussion about causes remains in the background, lingering as a potential source of embarrassment for the overlapping differences of methodology.

The immediate application of the epistemology developed in the *Treatise on the Sphere* is perceivable in the *Starry Messenger* (1610), where we can find at least three different methodological methods which correspond to regressus but do not apply Aristotelian forms of reasoning.[90] In the demonstration of the earthshine – which is the proof that the earth reflects its light onto the moon when sunlight cannot illuminate it – Galileo explicitly states that his aim is to 'assign a cause to an appearance worthy of wonder', with the same terminology employed in *Treatise on the Sphere*.[91] Galileo recounts all the possible hypotheses for the explanation of the phenomenon, from the reflection of the light of Venus, of the stars, of the moon itself, and of course the earthshine.[92] This set of plausible hypotheses are tested through their consequences and effects. Observation by observation – that is, comparing hypotheses through a consideration of the effects – Galileo eliminates all possible explanations, left only with one idea of the cause – that the earth reflects light onto moon.[93] In this way, Galileo arrives at the true explanation of the effects, and promises to prove elsewhere 'with many arguments and experiments' the validity of this principle.[94]

Also in the *Starry Messenger*, Galileo behaves like an astronomer and a mathematician, providing description of what he was seeing through his telescope.

A first attempt to breach the solid boundaries between mathematics and natural philosophy is made in 1612 with the *Discourse on Bodies in Water*. In *Various Fragments Concerning the Treatise on Bodies in Water*, Galileo openly faces his Aristotelian opponents by emphasizing the importance of mathematics for the understanding of natural phenomena:

> Here I expect a terrible rebuke from one of my adversaries, and I can almost hear him shouting in my ears that it is one thing to deal with matters physically, and quite another to do so mathematically, and that geometers should stick to their fantasies and not get entangled in philosophical matters – as if truth could ever be more than one; as if geometry up to our time had prejudiced the acquisition of true philosophy; as if it were impossible to be a geometer as well as a philosopher – and we must infer as a necessary consequence that anyone who knows geometry cannot know physics, and cannot reason about and deal with physical matters physically! ... Let my adversaries see whether I treat the material in the same terms as Aristotle ... and then let them have the kindness to desist from their bitter enmity toward geometry.[95]

Galileo is confident in the mutual assistance of mathematics and natural philosophy, and this explains his way of proceeding in *Discourse on Bodies in Water*. His main aim is to explain 'the true, intrinsic, and total cause of the ascending of some solid bodies in the water and therein floating'.[96] To do this, he follows 'a different method' from those applied by Aristotle, Archimedes and Francesco Buonamici. His new method consists of observing effects and then

> reducing the causes of such effects to more intrinsic and immediate principles, in which also are discovered the causes of some accidents ... and because the demonstrative progression so requires, I shall define certain terms and afterwards explain some propositions, of which, as of true and known things, I make use of to my present purpose.[97]

The terms 'progression' and 'cause' directly recall what Galileo wrote in relation to regressus in his *Treatise on the Demonstration* but – as Berti correctly points out – Galileo's reference here is to Pappus's definition of analysis, which seems to be applied to what should constitute the intermediate stage of regressus:

> Now analysis is the way from what is sought – as if it were admitted – through it the things that come next in order to something admitted by synthesis. For in analysis we suppose that which is sought to be already done, and we inquire *from what it results*, and again what is the antecedent of the latter, until on our backward way we arrive at something already known and being first in order.[98]

Pappus's mathematical analysis offers a clear way back to the first principles. However, in Commandino's version of Pappus that Galileo read, the passage sounds quite different in the second part of the definition:

> For in analysis we suppose that which is sought to be already done, and we inquire *what results from it* (*quid ex hoc contingat*), and again what is the antecedent of the latter, until advancing in this we find something already known or that falls within the list of principles.[99]

It is evident that there are two different subjects of investigation in the two versions. In Commandino's version what is sought are the consequences, while in the original version it is the antecedent. But Commandino's ambiguous text also suggests that through this analysis one of three things occurs: we establish the hypothesis as a principle, or we find something manifest and already known, or, again, we discover a principle. In all three cases, there is established as true a cognitive antecedent to the consequences. These would be the premises of the synthetic or compositive method. Mathematical analysis thus conceived seems to encompass two movements: one forwards and the other backwards, like testing a conjecture. This was at least what Galileo could have read in his time.

In the *Discourse on Bodies in Water* Galileo's main idea is to start with a hypothesis admitted as a cause and, working through its effects, to reduce it to an intrinsic principle, which is either something more primary than the hypothesis, or it is the hypothesis itself but established now as principle. This process of reduction was mathematical analysis, and it discovers that 'the excess or defect of gravity of the solid, unto the gravity of the water, is the true and proper cause of floating or submersion'.[100] It is important to emphasize that Galileo asserts that gravity is the cause, but he does not explain what it is. In other words, he assigns a causality to gravity without saying what this cause is. This is sufficient, however, in the context of regressus theory to proceed from the intermediate to the third stage: indeed, one should know the existence of a cause and that it is the real cause of the effect. After the discovery of such true causes, Galileo asserts that it would be 'good to proceed in a way of discovering demonstratively those particular accidents that do attend these effects'.[101] And this would represent the third stage. The *Discourse on Bodies in Water* makes very clear the role of mathematical analysis in what was the regressus theory, and it shows how Galileo is clearly departing from the Aristotelian tradition.

Scepticism about knowing the cause is evident in the *History and Demonstrations concerning Sunspots and their Phenomena* (written in 1612, but published in 1613). In the third letter to Mark Wesler, Galileo writes that

> in our speculating we either seek to penetrate the true and internal essence of natural substances or content ourselves with a knowledge of some of their properties. The former I hold to be as impossible an undertaking with regard to the closest elemental substances as with more remote celestial things … Hence I should infer that although it may be vain to seek to determine the true substance of the sunspots, still it does not follow that we cannot know some properties of them, such as their location, motion, shape, size, opacity, mutability, generation and dissolution. These in turn may become the means by which we shall be able to philosophize better about other more controversial qualities of natural substances.[102]

Knowledge of substances and essences seem to be unachievable or nearly impossible to Galileo. However, it is feasible to acquire scientific knowledge of qualities that can be mathematized like shape, motion, location and so on without discovering the essence of the cause. Through these properties, which can be known through mathematics, and which can be tested, one can then derive further knowledge. Galileo is opening the path to the epistemologies of Isaac Barrow, Robert Hooke and Isaac Newton.

The *Reply to Oppositions of L. delle Colombe* (1615) represents a further step in Galileo's understanding of the Aristotelian notion of the negotiation of the intellect. Galileo states that in the demonstration *propter quid* many philosophers make serious errors, whether because they presuppose 'principles which are less certain than conclusions', or because they are the same as 'what one tries to demonstrate' and differ only 'in terms and names', or again because they deduce 'conclusions from things which have nothing to do with them'.[103] It is quite clear that Galileo identifies the error in the wrong application of the logical process entailed in the intermediate stage of regressus. Indeed, anyone who failed to apply the negotiation adequately, or who took as true the conclusions of the demonstration *quia* – which are of themselves uncertain – or who, by applying division, explained the same conclusion with different terms, but without verification, in the end arrived at a wrong conclusion unwittingly. In the *Reply to Oppositions of L. delle Colombe*, therefore, Galileo expresses his conviction that it was not the theory of regressus in its entirety that was inadequate,[104] but rather it was only the intermediate stage that should be revised in order to provide a robust scientific foundation for knowledge. The nature of the Aristotelian error is clear from the following statement:

> Using mainly, but not well, the *resolutive method* (which, if well used, is the best method of discovery), they take the conclusion as true and instead of going on deducing from it this and then that and then that other consequence, until they come across one that is manifest either by itself or because it has been demonstrated, from which then the intended conclusion is reached by the compositive method; instead, I say, of making good use of such a graduation (*gradazione*), they form with their imagination a proposition that squares immediately with the conclusion they intend to prove, and without falling back even a single step, they take it as true, though as false or equally doubtful as conclusion, and immediately they construct on it a syllogism, which leaves us without any gain in our original uncertainty.[105]

Galileo in describing this resolutive method is proposing almost *verbatim* mathematical analysis as Pappus outlined it: admitting something as true, one proceeds through consequences until a more intrinsic and immediate principle is found. Galileo harshly criticizes those who conceive 'with their imagination a proposition that squares immediately with the conclusion', a position he shares with Bacon, as we have seen. By contrast, he suggests introducing an intermediate stage, which he calls 'graduation', through which a gradual – that is, continuing and regular – comparison between the cause and the effects is able to reveal their necessity and how the cause is the genuine cause of that effect. Galileo's analysis of cause and effect works according to 'laws of logic (*leggi logicali*)', which were – as has been correctly pointed out – 'the scholastic rules of inference: presence and absence, and concomitant variations',[106] which played an important role for Bacon too. The grounds for this reasoning appear to lie in Aristotle's rule for the predication of properties, according to which to argue from greater and lesser degrees means

> seeing whether a greater degree of the predicate follows a greater degree of the subject ... for if an increase of the accident follows an increase of the subject, as we have said, clearly the accident belongs; while if it does not follow, the accident does not belong. You should establish this by induction.[107]

What scholars have failed to notice is that this process suggested by Aristotle in the *Topics* was employed[108] – as we have seen – by all Renaissance Aristotelians in the intermediate stage of regressus as a key procedure for determining the essential relation between cause and effect, namely, to establish whether a given property belonged to a given subject.[109]

For Galileo, there is no contradiction between mathematical analysis and Aristotelian regressus because he includes, when needed, the former in the intermediate stage of the latter. Galileo provides a piece of evidence for this in the *Two Chief Systems*. First of all, he manifests dissatisfaction with the idea that the demonstration *potissima* is capable of providing scientific knowledge.[110] Salviati, who personifies Galileo's thought, criticizes Simplicius's claim that Aristotle would use a method proceeding from causes and principles to effects – in other words an a priori argument. This would be the demonstration *potissima*, but it does not deal with discovery and invention, but rather with the order of teaching. On the contrary, Salviati states,

> I think it certain that he [Aristotle] first obtained it [knowledge] by means of the senses, experience, and observations, to assure himself as much as possible of his conclusions. Afterwards he sought means to make them demonstrable. That is what is done for the most part in the demonstrative sciences; this comes about because when the conclusion is true, one may by making use of *resolutive method* hit upon some proposition, which is already demonstrated, or arrive at some already known principle.[111]

In this passage, Galileo shows how he implements mathematics within the Aristotelian regressus. This resolutive method, once again, is none other than the mathematical analysis applied to the results of the demonstration *quia*. Mathematical analysis works on this conclusion through consequences finding true principles. However, Galileo points out that 'if the conclusion is false, one can go on forever without ever finding any known truth', and this conclusion should be considered as a hypothesis inadequate for explaining phenomena. Only after the application of mathematical analysis and the establishment of principles is it possible to proceed with the demonstration *propter quid* or the a priori argument. This is the only passage in Galileo's writing in which he clearly outlines his contribution to reforming Aristotelian epistemology. Galileo's method is outlined as follows:

Galilean method	Aristotelian regress
Resolutive method: senses, experience and observations	Induction and demonstration *quia*
Mathematical analysis and then further tests and observations	Negotiation of the intellect
Compositive method	Demonstration *propter quid*

That mathematics for Galileo plays a crucial role in the intermediate stage of regressus is clear from his criticism of William Gilbert's experimental method of investigation on magnets. He praises and admires Gilbert in comparison to many other illustrious men who worked on the same topics because of

> the many new and sound observations which he made, to the shame of the many foolish and mendacious authors who write not just what they know, but also all the vulgar foolishness they hear, without trying to verify it by experiment.[112]

Galileo sets great store by experience and experiments in collecting observations, yet he is critical of basing scientific discoveries only on this method. Indeed, what he might wished for

> in Gilbert would be a little more of the mathematician, and especially a thorough grounding in geometry, a discipline which would have rendered him less rash about accepting as rigorous proofs those reasons which he puts forward as true causes for the correct conclusions he himself had observed. His reasons, candidly speaking, are not rigorous, and lack that force which must unquestionably be present in those adduced as necessary and eternal scientific conclusions.[113]

What makes conclusions universal, necessary and true after observations and experiments – that is, what generates scientific knowledge from experience – is no doubt mathematics. Mathematics becomes in the *Two Chief Systems* the means that Aristotelians and also experimental philosophers were not able to conceive of as a way of providing scientific explanations. The inadequacy and inability of Aristotelians is on display in Simplicius, who not only supports the distinction between mathematics and natural philosophy asserted also by Galileo in his early writings but also declares the useless character of geometrical tools for understanding nature:

> Simp. ... I should agree with Aristotle that he plunged into geometry too deeply and became too fascinated by it. After all ... these mathematical subtleties do very well in the abstract, but they do not work out when applied to sensible and physical matters. For instance, mathematicians may prove well enough in the theory that *sphaera tangit planum in puncto*, a proposition similar to the one at hand; but when it comes to matter, things happen otherwise.[114]

Simplicius maintains the idea that the imperfection of matter 'prevents things taken concretely from corresponding to those considered in the abstract'.[115] Therefore, mathematics is of little use for understanding nature. But what was Galileo's opinion on this? In the same passage, Sagredo states that 'it must be admitted that trying to deal with physical problems without geometry is attempting the impossible'.[116] The question is how mathematics deals with natural phenomena. As in his early writings, Galileo is forced to admit that mathematics works only in idealized cases. Indeed, 'the philosopher geometer, when he wants to recognize in the concrete the effects which he has proved in the abstract, must deduct the hindrances of matter'.[117] Only in this way, according to Galileo, are material things in agreement with mathematical computations. In order to apply mathematics, Galileo has to renounce the perfect overlap between mathematics and natural philosophy, despite his famous profession that the book of nature is written in mathematical characters – that is, a combination of numbers and figures.[118]

This sense of renunciation is not to be considered a failure, because at least for Galileo it is the pathway to scientific knowledge. This is the leitmotiv of the *Two New Sciences* and of Galileo's late reflections. The new science of mechanics 'has its foundation in geometry', in which there are only 'properties of circles, triangles, cylinders, cones and other solids'.[119] However, this kind of investigation works only 'by abstracting all imperfections from matter and supposing it perfect, unalterable and free from all accidental variations'.[120] The powerful tool of mathematics is particularly important for Galileo in the phase of the invention of new knowledge, so much so that Simplicius begins 'to understand that while logic is an excellent guide in discourse, it does not, as regards stimulation to discovery, compare with the power of sharp distinction which belongs to geometry'.[121] Geometry replaces the common analytic method of the first stage of regressus, and therefore logic in this sense becomes useless. Logic can only teach 'how to test the conclusiveness of any argument or demonstration already discovered and completed'.[122] Logic seems to be involved only in what was the third stage of regressus, and also in the order of teaching knowledge, but in nothing more than this.

Galileo makes clear that the powerful tool of geometry is mathematical analysis. The starting point is always a hypothesis:

> First of all it seems desirable to find and explain a definition best fitting natural phenomena. For anyone may invent an arbitrary type of motion and discuss its properties; thus, for instance, some have imagined helices and conchoids as described by certain motions which are not met with in nature, and have very commendably established the properties arguing by supposition (*ex suppositione*), even if these motions do not occur in nature.[123]

The method proceeds from something admitted as true through its consequences. This is mathematical analysis, and it works even if there are no similar effects or consequences in nature. The reference is to Archimedes, as is clear in the letters to Pierre de Carcavy and Giovan Battista Biancani. Galileo's situation is different and since nature does employ a certain kind of acceleration for descending heavy things, we decided to look into their properties (*passiones*) so that we might be sure that the definition of accelerated motion which we are about to adduce agrees with the essence (*essentia*) of naturally accelerated motions. And at length, after continual agitation of mind (*mentis agitationes*), we are confident that this has been found, chiefly for the very powerful reason that the properties successively demonstrated by us correspond to, and are seen to be in agreement with, that which natural experiments (*naturalia experimenta*) show forth to the senses.[124]

For Galileo, it is evident that before determining whether the supposition of the definition of the accelerated motion agrees with the essence of naturally accelerated motion, it is necessary to interpose a continuous working of the mind, which employs a mathematical description of the effect. The cause describes the effect because there is agreement and correspondence – which he supposes to be natural – between the observations and the mathematical model. In this passage, Galileo does not state that he has found the essence of accelerated motion, but just that he has established the congruence between his mathematical description and the effects. Indeed, Galileo openly states that for the purpose of his discussion it is not proper 'to investigate the cause of the acceleration of natural motion'.[125] He prefers to investigate the properties of accelerated motion. How to do this? Again through a mathematical analysis, by supposing a cause and checking whether effects work according to that cause:

> At present it is the purpose ... to demonstrate some of the properties of accelerated motion (whatever the cause of this acceleration may be) – meaning thereby a motion, such that the moments of its velocity go on increasing after departure from rest, in simple proportionality to the time, which is the same as saying that in equal time-intervals the body receives equal increments of velocity; and if we find the properties which will be demonstrated later are realized in freely falling and accelerated bodies, we may conclude that the assumed definition includes such a motion of falling bodies and that their acceleration goes on increasing the time and the duration of the motion.[126]

This is a clear application of the mathematical analysis as it is expressed in Pappus. In examining these properties, Galileo explicitly renounces knowledge of the causes. Mathematical analysis, however, is not sufficient for Simplicius, who asks for some experiment that can agree with the established conclusion. When Salviati is ready to prove his theory of uniformly accelerated motion, Galileo provides the experiment of the inclined plane. It consists, on the one hand, of the elimination of all conditions that interfere with the hypothesis and, on the other, the replication of those conditions that make possible the quantitative determination of the effect. It leads to the discovery of proportional causality, which combines speed and time, and which confirms the principle of uniformly accelerated motion by providing convertibility between one factor and another, thus identifying an essential and reciprocal connection between cause and effect.[127] Using 'sensed experiences', this process, according to Galileo, confirms the previously supposed principles, which now become 'the grounds for all the subsequent structure'.[128] By structure he means the following scientific argument, which for the Aristotelians encompasses the third stage of regressus. Mathematical analysis does not replace the demonstration *propter quid*, but the former intervenes before the latter. In the same period, in a letter to Pierre de Carcavy, Galileo emphasizes as follows:

> I argue *ex suppositione*, picturing to myself motion with respect to a point from which [a thing] leaving from rest goes accelerating, increasing its speed in the same proportion with which time increases, and in this way I conclusively demonstrate many properties; *then* I add that if experience shows that such properties are found verified in the motion of naturally descending heavy things, we can without error affirm this [natural motion] to be the same motion that I defined and assumed; if not, my demonstrations founded on my assumption lose nothing of their force and conclusiveness, just as it in no way prejudices the conclusions proved by Archimedes about the spiral that no naturally moving body moves spirally in that manner. But in the motion I defined it is found that all the properties that I demonstrate are verified in the motion of naturally descending heavy bodies – these are verified, I say, in that when we make experiments upon the earth and at heights and distances practicable to use, no sensible difference is discovered (emphasis added).[129]

The argument *ex suppositione* represents the introduction of mathematics into the intermediate stage of regressus, and it verifies through a series of tests and observations the assumption or the conjecture, but with a different tool from that employed by Aristotelians. Both Aristotelians and Galileo were looking for an essential connection between causes and effects, but while, for Aristotelians, this concerned the substance or the essence of the cause and was mainly qualitative, for Galileo, it was a mathematical relationship, and the essential conditions were determined not only qualitatively but also quantitatively. From the letter to de Carcavy, we can deduce that for Galileo, like his Aristotelian fellows, the intermediate stage of regressus overcomes the unclear and still uncertain knowledge of sensation and experience. This intermediate stage, as with the Aristotelians, is first of all mental and rational in bent, rather than empirical and experimental. Unlike the Aristotelians, for the Pisan scientist mathematics plays a crucial role. Galileo clearly states that at the beginning he pictures in his mind the conjectures, and that only thereafter does he verify in experience whether what he conceived was either right or wrong. In this second step, we can sense a hint of the subsequent movement towards experimentalism in the conception of the intermediate stage of regressus, becoming much more completely evident only in the mature works.

In conclusion, Galileo takes from the Aristotelians of his time the logical structure of his scientific method, and clearly identifies the *vulnus* and the opacity of their theory concerning the negotiation of the intellect. In order to resolve this weakness of scientific reasoning, Galileo – unlike the Aristotelians – applies mathematics as a new method for discovering and achieving scientific knowledge. This kind of method, operating during the intermediate stage of regressus and coming after an effort that is specifically grounded in experience, is in the first place mental and rational, and only subsequently experimental in proving the hypothesis initially mentally conceived at the end of the analysis. With this process, what is hypothetical becomes a necessary and universal law of nature, written in mathematical characters. Following on from it, a geometrical-deductive process based on solid grounds and principles provides scientific conclusions.

Within a century, there is a radical shift in the conception of the intermediate stage of regressus, a shift that parallels the emergence of mathematics as the most powerful tool for interpreting reality – and here, of course, we can see the turning point and radical break with the Aristotelian tradition. This shift, however, has the effect of renouncing cognition of causes and substance for scientific knowledge in order to focus on properties that were mathematically describable. But this is compatible with Galileo's commitment to working on rational truth rather than on material truth. Galileo is developing an epistemology and not an ontology, and in this sense he is paving the way for a new understanding of nature.

5

Thomas Hobbes

Oxford. 1601. Thomas Hobbes entered Magdalen Hall at the very moment that Paduan Aristotelianism achieved the highest consensus among university professors, especially vis-à-vis Ramism. Fifty years later, in his masterpiece *Leviathan* (1651) Hobbes was very critical of his philosophical education: what he had studied was not philosophy but 'Aristotelity'. According to Seth Ward, the only thing that pained Hobbes at that time was 'the desire that Aristotelity may be changed into Hobbeity'.[1] But in order to realize his wishes, the task was to overturn Aristotelian philosophy, and in particular logic, from within. Indeed, logic and method were the key to accessing philosophy.

Hobbes's epistemology, and in particular his conception of method, has been the subject of a number of studies, most of all in relation to the geometrical method and its sources. Some scholars argue that Hobbes applied the geometrical method in his natural considerations; others charge him with inconsistency, employing alternative ways of demonstrating physical phenomena. Some scholars believe that his idea comes from Paduan Aristotelianism and from Jacopo Zabarella;[2] others maintain the independence and divergence of his method from the Aristotelians.[3] There have been some staunch attempts to show how the methods of physical investigations and mathematics for Hobbes may be viewed as having the same degree of necessity and certainty. They identified the crucial problem of the convertibility of cause with effect in the natural investigations as the point for comparison with mathematics.[4] However, they did not reach a satisfactory solution to the question, admitting varying degrees of certainty, which is against what Hobbes himself actually held, that is 'the certainty of all sciences is equal or else they would not be sciences, since to know does not admit of greater or less'.[5] Furthermore, they failed to recognize how this question was related to the Renaissance epistemological debates.

Scholarship has placed great emphasis on the fact that Hobbes and the School of Padua maintained two entirely incompatible approaches to science. This incompatibility was based on different conceptions of the nature and function of logic; on different ways of acquiring and transmitting knowledge; on different ways of conceiving order and method, because Hobbes's aim was to integrate the order of knowledge with the natural order.[6]

I want to expand and revise this narrative, showing how Hobbes owed his method to a substantial re-elaboration of Renaissance Aristotelian epistemology, without diminishing his original contribution to the history of theory of knowledge in relation to the logic of scientific discovery. Indeed, the aim is to provide a new reading of Hobbes's epistemology by focusing on mental examination. Previous studies were unable to identify this process in Hobbes because they lacked a clear idea of what mental examination exactly entailed. Stewart Duncan, who has revived the old thesis of the influence of the Paduan school's Aristotelianism, argues for the existence of something analogous to the mental examination in Hobbes, without offering any substantial proof of what it is or how it works, to the extent that in the end the two conceptions seem more to diverge than to overlap.[7] However, I shall demonstrate the centrality of this notion in Hobbes for understanding the causation of physical phenomena, and likewise for the use of a rigorous method that is the equal of mathematical method, following the specific trend of Renaissance Aristotelians.

Hobbes discusses the problem of method in the *Elements of Philosophy the First Section, Concerning Body* in ch. 6. In order to understand what method is for Hobbes, we should recall the definition of philosophy given in ch. 1, sect. 2:

> Philosophy is the knowledge we acquire, by rigorous reasoning, of phenomena, or of apparent effects, from the conception we have of some possible production or generation of the same; and of such production, as has been or may be, from the cognition.[8]

This definition makes clear that philosophy is a kind of knowledge of the various ways in which a phenomenon is generated, in other words of the cause of an effect. Indeed, method is 'the shortest way of finding out effects by their known causes, or of causes by their known effects'.[9] That Hobbes's argument is within the Aristotelian framework is clear because for him science is science τοῦ διότι, that is of causes, while any other kind of cognition by sensation and imagination is knowledge τοῦ ὅτι. We are dealing with knowledge and not with demonstration, because for Renaissance Aristotelians – Zabarella in particular – and Hobbes, the way in which we acquire knowledge τοῦ ὅτι is not necessarily demonstrative.

Scholars see a strong difference between Aristotelians and Hobbes because for the former knowledge τοῦ ὅτι would be necessarily demonstrative, based on the demonstration *quia*, and it would be 'an integral part of the entire scientific demonstration known as regressus',[10] so much so that they would have used the term resolution 'strictly in the sense of a demonstrative induction or an a posteriori proof'.[11] For Hobbes, rather, knowledge τοῦ ὅτι would be something that is not at all demonstrative.

In the *Examination and Emendation of Modern Mathematics* (1660), Hobbes writes that all demonstrations are scientific – that is, they produce a necessary knowledge of the cause. So while the argument τοῦ διότι is a true demonstration because it leads to this kind of knowledge, the argument τοῦ ὅτι can never be characterized in this way. Arguments τοῦ ὅτι are a posteriori reasonings based on hypotheses, suppositions and possible things, but they are not scientific, and only seem so because after going without seeing a contradictory instance for a long period the mind assents to them as if they were science. In contrast, demonstrations are a priori reasoning τοῦ διότι and are based on definition.[12]

Hobbes clearly enlarges the first part of regressus in comparison to Aristotelians like Zabarella, not narrowing it to a form of demonstrative reasoning. Contrary to Hobbes, for Zabarella there is a demonstration called τοῦ ὅτι, and in speaking of the first part of regressus he always deals with a form of demonstrative reasoning. However, we may add that in Zabarella, as with Hobbes, knowledge τοῦ ὅτι can be acquired in multiple ways, but it concerns both cognition coming from sensation, the so-called *experientia imperfecta* – which deals with the question of the 'an sit' and makes use of observations rather than scientific explanations – and also that of *experientia singular*, which is used to confute past theories or general assertions, two peculiar forms of knowledge particularly dear to the Paduan logician in his logic and natural philosophy.[13] Knowledge τοῦ ὅτι has its own independent status both in Zabarella and in Hobbes, even if it is not yet scientific knowledge. Yet in Zabarella this knowledge τοῦ ὅτι in the regressus theory must have the form of demonstrative reasoning in order to show the reciprocity of cause and effects in the first and third stages. Nonetheless, the cause of knowledge τοῦ ὅτι can be known by other means, for instance, by experience, according to Zabarella, and this can then be applied to a demonstration τοῦ διότι. In such a case, we have scientific knowledge without having a perfect regress.

There is not necessarily an incompatibility between the two approaches – indeed in Hobbes we find no clue that knowledge τοῦ ὅτι cannot also be acquired through syllogism from effects. The fact that Hobbes deals with knowledge τοῦ ὅτι in terms of sensation and imagination does not exclude the possibility that it is acquirable by other means, like syllogism. However, for Hobbes, as we shall see, syllogism from effects is not a demonstration. Furthermore, Hobbes himself states that only one kind of knowledge leads to science, the other (τοῦ ὅτι) not. This is because knowledge τοῦ διότι goes from causes to effects, that is, from what is first by nature to what is first for the mind, while knowledge τοῦ ὅτι goes from what is first for the mind to what is first by nature, from effects to causes. Finally, for both Zabarella and Hobbes, knowledge τοῦ ὅτι is only provisional, conjectural, confused, obscure and not scientific, based as it is on the particular, which is subject to confutation. The argument from effects to causes is not sufficient for scientific knowledge, and for this reason in the eyes of both philosophers there is necessarily another step: the transition from causes to effects.

From the definition of philosophy, we can understand that these causes or principles of generation are conceived – that is, they are a product of the mind, not a product of experience or deriving immediately from it. Indeed, out of experience – that is, from sensation and imagination – comes only knowledge of the existence of these causes and principles, not why and what they are. In philosophy, however, we are at a higher level of knowledge τοῦ διότι in comparison to knowledge τοῦ ὅτι. It is important, therefore, to understand how the mind conceives this cause. Within this process it is possible to glimpse the logical operation of the mental examination working on the experiential content, but in a completely rational and mental way. Cause, as we have said, is not known by experience, but it 'is necessary the work of reasoning' to find it. It is legitimate to ask what kind of reasoning Hobbes is talking about in discussing the discovery of the cause.

He immediately provides the answer by stating that he dealt with this manner of reasoning in ch. 1, sect. 2, where we find the famous passage in which reasoning is considered as a variety of calculus or computation:

> By reasoning, I mean computation. Now to compute is either to collect the sum of many things that are added together, or to know what remains when one thing is taken out of another. Reasoning, therefore, is the same with addition and subtraction ... all reasoning is comprehended in these two operations of the mind.[14]

Any kind of reasoning is therefore addition and subtraction, but what exactly is added and subtracted? In ch. 1, sect. 3, Hobbes explains that a silent cogitation is involved, by means of which it is possible to add or subtract mentally without using words or verbal discourse.

Hobbes gives an example inspired by *Posterior Analytics* II.19, which concerns the case of seeing something from afar. The first thing that we can say about this object is that it is a body. As it comes closer, then the idea of the thing changes, adding the realization that it is an animated body. Once this animated body comes closer still – so much so that we can make out the voice and speech – then we can say that this thing is a rational, animated body. To the idea of something seen from afar, therefore, the properties of being a body, animated and rational are added. In this way, these three properties do not characterize three distinct things, but the idea is conceived distinctly and entirely as being one entity – and only one entity – and to this idea we can attribute the name of human being. The addition in this case helps to determine those essential elements which characterize human beings.

Conversely, when we see a man at close quarters, we have a distinct idea, but once he goes away, the properties of his being a rational, animated body are no longer distinguishable, and only a confused idea remains. In this way, properties are not added, but subtracted. In subtracting properties, the idea of what there was vanishes. If in subtracting something the idea does not vanish, then what has been subtracted is not a property of the thing, but a mere accident. Indeed, according to Hobbes, the property is the means through which we can distinguish a thing from another. Adding a property makes the idea more clear and distinct – in other words, we have a transition from confused and obscure to clear and distinct knowledge, which is exactly the process guaranteed by the mental examination.

It is noteworthy that in the *Elements of Philosophy*, ch. 6, sect. 1, Hobbes deals with this addition and subtraction as a kind of reasoning by composition and division (*compositione et divisione*). Composition and division, it is important to emphasize, are the same processes, referred to with the same terms by Nifo and Zabarella in dealing with mental examination for finding the cause of an effect. We can argue, therefore, that the reasoning for finding the cause is similar to that of mental examination employed by the Aristotelians for the same purpose. If this is so, then we should read in a different way the end of the ch. 6, sect. 1, where Hobbes states that 'the method, by which we find out the causes of things, is either compositive or resolutive, or partly compositive and partly resolutive'.[15] Up to now scholarship has understood in this definition a perfect parallelism between Hobbes's method and the Aristotelian conception of analysis and synthesis. However, very little attention has been paid to Hobbes's entire sentence in which he clearly says that the method is not only compositive (or synthetic) and resolutive (or analytic), but it is also compositive and resolutive at the same time.

This 'partly compositive and partly resolutive' is another kind of reasoning, which is different from the first stage of analysis and the last stage of synthesis. Indeed, it is the mental examination which combines in itself this twofold process which is partly compositive and partly divisive. In Hobbes, therefore, we can find all three stages of regress as are in Paduan Aristotelianism, mental examination included, even if they are listed in a different order. I suggest reading 'partly compositive and partly resolutive' as one single process, that of mental examination, and as different from analysis and synthesis.

Previously I have said that composition and division, as addition and subtraction, add and remove properties. It is necessary, however, to understand how mental examination does this within Hobbes's method. Hobbes states that the true method proceeds from known to unknown or, better, from what is first for the mind – that is, the effects – to what is first in nature – that is, the causes. This would argue in favour of the idea that for Hobbes the logic of scientific discovery is opposed to the order or the method of teaching, which proceeds from what is prior according to nature (the causes) to what is prior for the mind (the effects). Nevertheless, the method of teaching proceeds in a compositive and synthetic way, like the final step of regress, simply because the causes are already known.

In keeping with the idea of proceeding from what is most known by the mind to what is most known by nature, what is known first of all for Hobbes is the generic, confused and obscure idea of a phenomenon – that is, that it exists – and then, following on from this, all of its properties, as is the case with the Aristotelians. The sum or addition of these properties constitutes the cause of the phenomenon – that is, its nature – as is clear from the *Elements of Philosophy*, ch. 9, sect. 3:

> A cause absolutely (*simpliciter*), or an entire cause (*causa integra*), is the aggregate of all the accidents both of the agents how many soever they be, and of the patient, put together; which when they are all supposed to be present, it cannot be understood but that the effect is produced at the same instant; and if any one of them be wanting, it cannot be understood but that the effect is not produced.[16]

In order to know the cause of an effect, it is necessary 'to conceive and to understand in the mind the notion or perfect idea'[17] of this cause, that is, the entire cause, and to do this, a mental consideration – which proceeds by division and composition of the accidents that constitute the cause – is required.

To recall: Hobbes explains the work of this process as being partly compositive and partly resolutive in ch. 6, sect. 10. In the conclusion of the same section, Hobbes states that 'in the searching out of causes, there is need partly of the analytical, and partly of the synthetical method'.[18] Here Hobbes is not providing a generic summary of his concept of method proceeding from known to unknown and for which both analysis and synthesis are necessary. He is dealing specifically with the process that clarifies what the cause is, before determining whether this cause is really the cause of the effect by means of the compositive process of the demonstration τοῦ διότι. This process, 'partly compositive and partly resolutive', which is reminiscent of mental examination in the Aristotelian tradition, serves 'to conceive how circumstances conduce severally to the production of effects' and is a system 'for the adding together and the compounding of what they can effect singly by themselves'.[19] From Hobbes's definition, it is clear that composition and division so conceived could not be equivalent to demonstration τοῦ ὅτι or demonstration τοῦ διότι, and it is also evident that this twofold process fulfils the same function as the Aristotelian negotiation of the intellect. In order to discover the cause through this twofold process,

> we must examine singly every accident that accompanies or precedes the effect, as far forth as it seems to conduce in any manner to the production of the same, and see whether the propounded effect may be conceived to exist, without the existence of any of those accidents; and by this means separate such accidents, as do not concur, from such as concur to produce the said effect; which being done, we are to put together the concurring accidents, and consider whether we can possibly conceive, that when these are all present, the effect propounded will not follow; and if it be evident that the effect will follow, then that aggregate of accidents is the entire cause, otherwise not; but we must still search out and put together other accidents.[20]

Such a method was proposed by Capivacci and Zabarella and is at this stage merely a mental process, not an experimental technique. However, the example Hobbes provides makes clear that an appeal to experience is possible

> if the cause of light be propounded to be sought out; first, we examine things without us, and find that whensoever light appears, there is some principall object, as it were the fountaine of light, without which we cannot have any perception of light; and therefore the concurrence of that object is necessary to the generation of light. Next we consider the medium, and find that unless it be disposed in a certaine manner, namely, that it be transparent, though the object remain the same, yet the effect will not follow; and therefore the concurrence of transparency is also necessary to the generation of light. Thirdly, we observe our own body, and find that by the indisposition of the eyes, the brain, the nerves, and the heart, that is, by obstructions, stupidity and debility we are deprived of light, so that a fitting disposition of the organs to receive impressions from without is likewise a necessary part of the cause of light. Again, of all the accidents inhaerent in the object, there is none that can conduce to the effecting of light, but onely action, (or a certain motion,) which cannot be conceived to be wanting, whensoever the effect is present; for, that any thing may shine, it is not requisite that it be of such or such magnitude or figure, or that the whole body of it be moved out of the place it is in, (unlesse it may perhaps be said, that in the Sun or other body, that which causeth light is the light it hath in it selfe; which yet is but a trifling exception, seeing nothing is meant thereby but the cause of light; as if any man should say that the cause of light is that in the Sunne which produceth it;) it remaines therefore that the action by which light is generated, is motion only in the parts of the object. Which being understood, we may easily conceive what it is the medium contributes, namely, the continuation of that motion to the eye; and lastly what the eye and the rest of

the organs of the sentient contribute, namely, the continuation of the same motion to the last organ of sense, the heart. And in this manner the cause of light may be made up of motion continued from the original of the same motion, to the original of vitall motion, Light being nothing but the alteration of vitall motion, made by the impression upon it of motion continued from the object.[21]

By virtue of this twofold process, Hobbes guaranteed that the set of accidents constituted the entire cause of the effect, and 'an entire cause is always sufficient for the production of its effect, if the effect be at all possible'.[22] We may ask whether the contrary is valid: that is, that at any time we observe that particular effect, this effect has been generated by that entire cause. Hobbes's answer is affirmative:

> In whatsoever instant the cause is entire, in the same instant the effect is produced. For if it be not produced, something is still wanting, which is requisite for the production of it; and therefore the cause is not entire, as was supposed. And seeing a necessary cause is defined to be that, which being supposed, the effect cannot but follow; this also may be collected, that whatsoever effect is produced at any time, the same is produced by a necessary cause.[23]

Hobbes states that whenever that effect occurs, there has been a prior set of accidents that constitute a specific cause – otherwise one should admit a plurality of causes for the same effect. This would mean not acquiring real knowledge of the true cause of the effect – that is, the cause would not have been sufficiently determined in an unequivocal manner. The entire cause of the effect is conceived as the necessary and sufficient cause of the effect, only in circumstances where if there is a cause, then there is an effect, and – equally – if there is the effect, then there is the same cause. To sum up, there is the total convertibility between cause and effect required also by Aristotelians in the mental examination. This convertibility is what makes it possible for knowledge in natural philosophy to have the same level of certainty as mathematics.

The nature of this convertibility becomes clearer in the *Examination and Emendation*. In commenting on the *Posterior Analytics* 71 b 16–22, Hobbes compares Wallis's mathesis with the Aristotelian ἐπιστήμη and states that 'mathesis is knowledge of the truth by demonstration'.[24] Hobbes grounds his interpretation on *Posterior Analytics* 71 a 1–4:

> All teaching and all intellectual learning (μάθησις) come about from already existing knowledge. This is evident if we consider it in every case; for the mathematical sciences are acquired in this fashion, and so is each of the other disciplines.

As is evident from this passage, Aristotle simply says that intellectual knowledge is acquired by means of something pre-existing: this happens in mathematics as in all other disciplines. He does not equate επιστήμη with μάθησις. Indeed, in *Posterior Analytics* 71 b 9–11, Aristotle famously states that to know scientifically is to know the reason why something is, but not with any other form of intellectual knowledge. In the Aristotelian sense, μάθησις is not always scientific knowledge. Hobbes pushes his interpretation even further, concluding that according to this definition of mathesis 'all sciences are mathematics'.[25] But why? How is it possible that they are mathematical?

Hobbes needs this equation in order to award the same certainty of mathematics to natural philosophy. Looking at the problem of convertibility, he would have come across *Posterior Analytics* 78 a 9–12, where Aristotle states that 'in mathematics things convert more because they assume nothing accidental – and in this too they differ from argumentations – but only definitions'. As we have seen, with the negotiation of the intellect Renaissance Aristotelians attempted exactly this in reducing the discovery of the first stage of regressus to the form of a definition. However, it is evident that for Aristotle himself only the mathematical sciences necessarily imply a convertibility based on definition. Not all sciences, however, have this facility. In Chapter 3, we saw how Zabarella departs clearly from the Aristotelian standpoint and interprets that passage quite differently, stating that for mathematical sciences Aristotle meant all 'demonstrative sciences' – that is, 'all contemplative sciences had the same evidence as those of mathematics'.[26] In Zabarella, there is an equation between mathematics and natural philosophy – that is, all other sciences – exactly as for Hobbes.

For Hobbes, if the cause uniquely characterizes the effect, then it will give the reason for its being – that is it will be in one way or another its 'essence', even if, as we shall see, this essence is not described by means of a formal cause. We can talk of 'essence' because without the cause, the effect cannot be what it is. This is reinforced by the fact that for Hobbes the cause of the effect should constitute the definition of the effect itself:

> Definitions of things, which may be understood to have some cause, must consist of such names as express the cause or manner of their generation, as when we define a circle to be a figure made by the circumduction of a straight line in a plane.[27]

If the cause is not expressed in the form of a definition, then for Hobbes science is not possible, because 'besides definitions there is no other proposition that ought to be called primary' or 'be received into the number of principles'. In other words, as he writes in the *Examination and Emendation*, if the cause is not a definition then it is impossible to proceed with a demonstration τοῦ διότι:

> The reason why I say that the cause and generation of such things, as have any cause or generation, ought to enter into their definitions, is this: the end of science is demonstration of the causes and generations of things, which if they be not in the definitions, they cannot be found in the conclusion of the first syllogism, that is made from those definitions, and if they be not in the first conclusion, they will not be found in any further conclusion deduced from that; and, therefore, by proceeding in this manner, we shall never come to science; which is against the scope and intention of demonstration.[28]

Hobbes excludes from the categories of the principles of demonstrations the axioms and postulates of Euclidean geometry because they are not definitions. For this reason, the geometrical method is not, and cannot be deemed to be, in any manner equivalent to discovery,[29] despite their resemblance. Hobbes himself states in ch. 6, sect. 19, that the art of geometricians – that is, logistics – is not properly speaking a method of discovery, because it proceeds from unknown suppositions to known things, whereas method, as we have previously emphasized, proceeds from the known to the unknown. Indeed, if the synthesis of the geometric method proceeds like that of discovery, the analysis, instead, 'is ratiocination from the supposed construction or generation of a thing to the efficient cause or coefficient causes of that which is constructed or generated'.[30] Mathematical analysis and that involved in the discovery are different, because the starting points are different. In the analysis of the discovery, the starting point is the phenomenon and a given effect, while in the mathematical analysis it is a supposition conceived by the mind.[31]

*

The process described here by Hobbes works in much the same way as the mental examination of the Paduan Aristotelians. The real difference between the two processes is as follows: for Zabarella, the convertibility between cause and effect was granted mainly by virtue of the formal cause expressed by the definition, or from the efficient and final causes if they implied coexistence and inseparability of cause and effect, in such a way that the method of natural philosophy had the same certainty as mathematics; for Hobbes, this convertibility is expressed always by a definition based on the sole efficient cause. In *Some Principles and Problems in Geometry Thought Formerly Desperate* (1674), Hobbes makes clear that

> definitions are of two kinds, the first of which merely indicates the nature of the thing, while the second explains the cause or means of generation. But those definitions are the most useful for advancing knowledge which contain the causes and means of generation of the things defined. This saying of Aristotle's is true: 'to know is to know through causes.' The other definitions, which only declare the essence of the thing defined are usually less fruitful, for nothing follows from them that was not previously contained in them.[32]

There is a radical shift in the understanding of the notion of the cause in Hobbes's epistemology. Hobbes shares with the Aristotelian tradition the idea that science is knowledge through causes yet – while for the Aristotelians the formal cause was the eminent source of knowledge – for the English philosopher only the efficient cause can explain natural phenomena. The kind of cause of generation that Hobbes has in mind is efficient because the cause of every effect is mainly motion, since

> the variety of all figures arises out of the variety of those motions by which they are made; and motion cannot be understood to have any other cause besides motion; nor has the variety of those things we perceive by sense, as of colours, sounds, savours, etc. any other cause than motion, residing partly in the objects that work upon our

senses, and partly in ourselves, in such manner, as that it is manifestly some kind of motion, though we cannot, without reasoning, come to know what kind.[33]

In other words, the cause–effect relationship is always dynamic and concerns something's change of state.[34] According to Hobbes, the cause must be sought in a kind of motion and the task of the mental examination, which he calls reasoning, is that of determining which kind of motion it is. Indeed, every phenomenon known by sensation 'is determined and made to be of such and such quality and quantity by compounded motions, every one of which has a certain degree of velocity, and a certain and determined way'.[35] This was clear from the fact that what had to be sought was a cause understood as the principle of generation. Before any investigation of the specific motion – that is, the specific cause of an effect – it is necessary, according to Hobbes, to study the manner of the motion *simpliciter*, and this is the task of geometry. Only once this investigation has taken place is it possible to search for 'the ways of such generated motions as are manifest and the ways of internal and invisible motions',[36] those investigated by natural philosophers. Anyone operating in the field of natural philosophy would study in vain if they did not start from the standpoint of geometry. This is not to say that the method of discovery in natural philosophy is that of geometry, or that the two coincide. In geometry, the principles of generation are easily identifiable because the elements are constructed by convention; explaining the principles of the generation of the human being – understood as a 'rational, animated body' – is much more complex, for the effect 'human being' is given by experience and its properties are sought by means of the twofold compositive and divisive method. Nonetheless, a process of reasoning – or as the Aristotelians would say – a mental examination, is necessary. Both Zabarella's and Hobbes's methods aim to mathematize the mental examination – that is, to invest the process – and natural philosophy – with a rigour and certainty equal to that of mathematics. This rigour is sought by Zabarella in finding an essential connection between the cause and the effect, and by Hobbes, instead, in trying to reduce all causes to a kind of motion, a motion that is quantitatively explicable. From this perspective, we should look again at statements like that 'in Zabarella's view the natural philosopher can acquire absolutely certain knowledge', while 'Hobbes considers physics a purely hypothetical science'.[37] For Zabarella, the certainty to which the natural philosopher can aspire is equal to that of the mathematician, yet it is ontologically different because the starting points are different: for the natural philosopher it is a given, for the mathematician it is a product of the mind. As it is for Hobbes, so it is for Zabarella: the certainty of the knowledge of natural philosophy depends, as we shall see in the conclusion, on the power of the mind.

If my reconstruction is correct, Hobbes incorporates and re-elaborates Aristotelian epistemology – that is, regressus according to the needs of his new natural philosophy – in an original manner which was without doubt different from that of the Aristotelians of his time. The major departure is the transition from the pre-eminence of the formal cause in the Aristotelian worldview to the primacy of the efficient cause in Hobbes. Methodologically, however, the two approaches are not fully incompatible. According to Jan Prins, the main difference lies in the nature and function of logic:

Whereas Zabarella considers logic an instrument for the arrangement of notions that, thanks to its metaphysical foundation, enables the scientist to reveal objective reality, to Hobbes logic is a technical science or scientific technology of language by means of which the scientist can construct and consolidate truth itself.[38]

The core of the difference is the claim that Zabarella was a realist, while Hobbes was a conceptualist and a nominalist. However, recent studies have shown that Zabarella was far from being a realist; rather, he was more of a conceptualist than any other logician before him.

Zabarella's logical system is based on the distinction between first notions (*primae notiones*) and second notions (*secundae notiones*). Logic deals exclusively with second notions, which are also called mental concepts.[39] The first notions are subject to investigation *in primis* by the natural philosopher. These are concepts which immediately designate real things independently of the power and activity of the mind.[40] First notions, nonetheless, are always concepts and never things, and this is a decisive consideration because the natural philosopher never deals directly with things, even if he describes how things are. The natural philosopher deals with concepts which directly characterize things, and aims for strategies that explain natural things – that is, the reasons why things are as they appear. It is eminently a conceptualist position, whose realist aspect can be salvaged by Zabarella only through a defence of direct correspondence between things and concepts.

Second notions are particularly relevant in Zabarella's definition of logic as being a habit concerning second notions and an instrument of science.[41] If logic is an instrument of science, and has as its subject the second notions derived from the first notions, which are the subject of science, we can say that both logic and science have the same subject, but consider it in different ways.[42] For the natural philosopher, second notions are the known subject of demonstration, whereas for the logician they are the subject of logical operations and must be produced. In other words, for the natural philosopher, concepts or concepts of concepts are what is known, while for the logician they are the subject of mental operations. Nonetheless, what is known as the subject of a demonstration cannot disregard what is subject to its operations, because the subject of the demonstration is only known by virtue of those concepts that are ordered by the methods of knowledge generation. This means that if the first notions embody the ontological foundations for the second notions, then the epistemological validity of the first notions depends on the second notions. The first notions make us aware that the thing exists, and give us ontological certainty; however, the knowledge of the thing is guaranteed not by the first notions, but by the second notions. Thus, according to Zabarella, it is the business of the mind to guarantee the epistemological validity of the world, not the world in itself, and he emphasizes the importance of the constructional nature of knowledge, by which he means the primacy of mind over the world. The consequence of this for Zabarella is that 'in science, truth rests on discourse rather on the objective truth external to the mind'.[43] But such truth is grounded not on things in themselves, but on the concepts that the mind furnishes of them: to put it another way, existence is a given, while truth is discovered in the mind. In Zabarella we can find a clear 'centrality of the active role of the subject in the construction of knowledge'.[44]

Read in this way, Zabarella's logic is much closer to Hobbes's philosophical standpoint than might have been expected.[45]

Dismantling this idea of Zabarella as a realist means that any apparent incompatibility with Hobbes falls away. Prins suggests that in Hobbes synthesis follows immediately on from analysis, while in Zabarella the connecting of these two steps requires the mental examination. Mental examination for Hobbes would be superfluous because

> the knowledge of causes of artefacts does not require a special consideration just because we construct them ourselves, while the causes of natural phenomena can never be known for sure as we do not produce them. Therefore the procedures of the natural philosopher will never lead to knowledge of true causes in the ontological sense, but at best to an adequate description of the phenomena concerned. So, according to Hobbes, we can not expect the natural philosopher to find the truth but only to construct a truth compatible with our experience.[46]

Yet this is exactly what Zabarella says. Mental examination is essential in order to determine that the discovered cause is able effectively to give a truthful, causal explanation of reality, so much so that there is a correspondence between what is conceived in the mind and what exists in reality; otherwise no knowledge would be possible. Also for Zabarella, as for Hobbes, mental examination, like resolution generally speaking, is not required in the case of completely invented or constructed objects of the mind. It would be useless in such instances. This, however, is not pertinent regarding the knowledge of phenomena where the correspondence between what is in the mind and what exists should be admitted, unless we strongly follow conventionalist epistemologies, misconstruing their idea of science, and judging them to be purely idealist.[47]

Hobbes is without doubt different from Zabarella in that his philosophy of mind is qualitatively distinct and palpably influenced by recent advancements coming from the application of mathematics to the study of the motion of bodies. However, the direction and logical manner of the reasoning process is the same as in the field of scientific discovery.[48]

It would be wrong, however, to reduce Hobbes to his sources and recently it has been argued that 'Hobbes's method of knowledge does not need to get its content from Paduan Aristotelianism; the definition of ratiocination as computation is enough for its constitution'.[49] Looking at Hobbes's epistemology from within, this statement is correct. However, Hobbes himself left some clear, substantial and irrefutable traces concerning the origin of his doctrine in the *Elements of Philosophy*, ch. 6, and his epistemology is better understood within the framework of discussions on regressus, notwithstanding the need to avoid flattening his thought into a series of regurgitated Aristotelian doctrines. *Elements of Philosophy*, ch. 6 represents a precious indication of how Hobbes incorporates Aristotelian epistemology, and at the same time transforms it, opening up new ways for philosophical research. We cannot deny that there are some differences between Hobbes and Zabarella, or more generally the Aristotelian tradition; however, these do not make the two systems entirely incompatible. There are doctrinal divergences even among Aristotelians living at the same time and within the same intellectual context, as we have seen in our analysis of the multiple interpretations of the negotiation of the intellect.[50] What is more interesting is how these divergences away from this Aristotelian background and framework lead to a new philosophy which will come to shape early modern thought.

6

René Descartes

Leiden. 1637. In writing his masterpiece *Discourse on the Method*, René Descartes reflected on his student experience at the Jesuit Collège at La Flèche, 'one of the most famous schools of Europe'. His curriculum was mainly Aristotelian. The first year was devoted to the study of the *Organon* and of the *Nicomachean Ethics*; the second was focused on natural philosophy, the final year mainly on mathematics.[1] He studied Coimbrans, Franciscus Toletus and Antonious Rubius in philosophy, Christoph Clavius in mathematics.[2] But he declared himself dissatisfied with his education:

> As soon as I had completed the course of study ... I found myself beset by so many doubts and errors that I came to think I had gained nothing from my attempts to become educated but increasing recognition of my ignorance.[3]

His idea was to provide an innovative method for discovering knowledge and finding truth. His new epistemology, rising from the ashes of the Middle Ages and the Renaissance, has always been conceived as a breakpoint with the old, prevalently Aristotelian, Scholastic philosophy,[4] which had dominated the philosophical scene in the previous centuries.[5] Descartes himself sanctioned the break with his harsh criticism of Aristotelian logic from his early works to his mature writings.[6] On the basis of his own admissions, scholarship understood Descartes's method to be anti-Aristotelian.

His anti-Aristotelianism aimed to solve the crucial problem of Aristotelian epistemology – that is, the discovery of the certainty of the foundations or principles of scientific knowledge – a problem which I outlined in Chapter 1. In Cartesian terms, the Aristotelian issue can be stated as follows: if *nihil est in intellectu quod prius non fuerit in sensu*, knowledge of the principles comes from experience, induction and demonstration from effects to causes – that is, it relies on something accidental and provisory, which cannot constitute the premise of a scientific demonstration. How, therefore, is it possible to acquire scientific knowledge if the demonstration is not based on valid, certain and truthful premises?

As we saw in Chapters 2 and 3, the Aristotelian answer was not very convincing, and many philosophers attempted to confront the problem from different perspectives. Descartes shifts the interest to the cognitive and psychological dimension of the problem, moving away from the Renaissance debate on method, and this shift will lead subsequent philosophers like Spinoza, Locke and Leibniz to find the solution within their philosophy of mind.

In this chapter, I will examine (1) how, through his methodology, Descartes solved this crucial problem in an original way, making an exceptional contribution to the philosophical debate of his time, overcoming the Aristotelian tradition from within;[7] (2) how in addressing this difficulty, Descartes gives precious clues in relation to the method for establishing the only criterion for clearness and distinctness;[8] and (3) how his answer led to the transformation of Aristotelian epistemology into a form of facultative logic, which foreshadows more elaborate attempts in Locke and Leibniz.

Descartes's early epistemology and understanding of scientific knowledge in the *Rules for the Direction of the Mind* has been proven to be influenced by the Renaissance debate on method.[9] In Rule 4, Descartes states that 'we need a method if we are to investigate the truth of things',[10] while in Rule 6 he establishes that the secret of his method consists entirely in 'noting in all things that which is absolute in the highest degree', and for absolute he means 'whatever has within it the pure and simple nature ... that is whatever is viewed as being independent, a cause, simple, universal ... and qualities of that sort'.[11] In other words, the method lies in the recognition of those characteristics through which 'we can intuit straight off and *per se* (independently of others)'[12] the subject matter under investigation. Why are simple natures so important for Descartes's epistemology? Because, 'it is possible to have experience which is certain only of things which are entirely simple and absolute',[13] he answers. By definition, these simple natures are ideas that do not need other ideas to be understood: they are conceived directly and immediately for themselves and in themselves.

Scholarship usually holds this to be a new and original idea, yet this conception in Renaissance discussions on method finds a forerunner especially in Jacopo Zabarella and also in other authors of the Aristotelian tradition. In the third book *On Methods*, as we briefly saw in Chapter 3, the Paduan logician writes that what is 'known *per se*, or known naturally is of two types ... for [known *per se*] not only is that which is most evident to everyone ... but also that which, even if it is not known, nevertheless, if it is set out, is known without need of any middle, either at once or after some consideration of it'.[14] Known *per se* is not only what is evident, that is what is known by its 'own light and its own evidentness', but also what is self-evident and explainable for itself and by itself. If what is known *per se* is not actually known *per se* by everyone, it must be set out through division, which does not provide knowledge of the known *per se*, but it prepares it to be known. Division is a kind of clarification, and Descartes elaborates this Aristotelian conception in Rule 12, where he associates simple things with division: they are 'those things which we know so clearly and distinctly that they cannot be *divided* by the mind into others which are more distinctly known' (emphasis added).[15] Thus, for Descartes, simple natures are not analysable further; they are the objects of cognition that are ultimately unanalysable and they seem to be the result of a division. Finally, simple natures are qualities.

Simple natures, therefore, are all those things which are conceived by themselves, and which independently of their origin are understood as self-evident by the mind. Not all things in nature are simple, however, and in order to know them clearly and distinctly, a precise method is prescribed in Rule 5:

To reduce complicated and obscure propositions step by step to simpler ones, and then starting with the intuition of the simplest ones of all, try to ascend through the same steps to a knowledge of all the rest.[16]

Usually scholarship schematizes Descartes's method in two steps: (1) reduction – that is analysis – and (2) ascension – that is synthesis. At first glance, one would be tempted to establish an immediate parallel between Descartes's method and the Aristotelian theory of regressus, which has usually been understood, as we have seen, as being composed of two stages: (1) analysis or resolution (demonstration *quia*, induction) and (2) synthesis or composition (demonstration *propter quid*).[17]

My point is that scholarship has neglected the important intermediate stage between analysis and synthesis, and is therefore guilty of interpreting the Aristotelian and the Cartesian methods as comprising only two stages. As I have shown, in Chapter 2, scholars of the Aristotelian tradition usually fail to emphasize the importance of the intermediate stage between the demonstrations *quia* and *propter quid*, and typically intuition – outlined in the Rule 5 – is not understood as forming a part of the Cartesian method. I suggest that the Rule 5 be read as being composed of three stages – (1) reduction, (2) intuition and (3) ascension – in the same way as the Aristotelian method is composed of three stages – (1) analysis or resolution (demonstration *quia*, induction), (2) intermediate stage and (3) synthesis or composition (demonstration *propter quid*). However, as I will demonstrate in due course, the three stages of the Cartesian method do not correspond to the three stages of the Aristotelian regressus, and for this reason, the association of the Aristotelian analysis and synthesis with Descartes's conception, generally invoked by scholarship, is entirely misleading. Indeed, it is my contention that Descartes worked with inferences from effects to causes (analysis) and from causes to effects (synthesis) in a way that is similar to the Aristotelians but very different from what is sketched out in Rule 5.

The first stage of the Cartesian method is reduction. In Rule 12, Descartes identified this reduction to simple things with a divisive procedure.[18] In Rule 13, Descartes states that there exists a logical process that divides everything into the simplest thing possible – that is, it reduces all things to their simplest terms and abstracts from every superfluous conception – and this process is called enumeration, which he describes in Rule 7.[19] In Rule 8, Descartes exemplifies the process of enumeration in discussing the problem of the anaclastic lines.[20] He writes that in the case where someone

> is unable to discern at once what the nature of light's action is [that is to know immediately] ... in the hope that knowledge of some other natural power will help him understand this one, if *only by way of analogy* ... having done that, he will investigate the way in which a ray passes through the whole transparent body. Thus he will follow up the remaining points in due order, until he arrives at the anaclastic itself (emphasis added).[21]

Not only can enumeration infer with 'manifest certainty',[22] but it can also render 'the truth of conclusions more certain than any other kind of proof allows',[23] by reducing the thing to its essential conditions, having removed all those elements superfluous to the explanation of the thing itself and of its existence. Descartes establishes that all of these essential conditions, which are not clearly known at the beginning, 'must be delineated only by way of something else which is already known'.[24] In other words, the process is possible only by means of comparison, and indeed 'the chief part of human endeavour is simply to *reduce* these proportions to the point where an equality between what we are seeking and what we already know is clearly evident'.[25] In Rule 7, Descartes characterizes enumeration as induction and as a process of comparison[26] which identifies the essential conditions and proceeds by means of division and distinction:

> These conditions should, in our view, be gone into right from the very outset. We shall do this if we concentrate our mind's eye on intuiting each individual condition distinctly, looking carefully to see to what extent each condition delimits the unknown object of our inquiry.[27]

In this process of distinction, division and determination of the essential conditions, the mind can be deceived either by leaving a condition out of the definition of the simple thing, or in assuming as a condition something beyond the essential characteristics required to define it.[28] If enumeration proceeds correctly, its role is so fundamental that – save for knowledge obtained through simple and pure intuition of a simple thing – all human knowledge results from a comparison between two or more things, with the effect that almost the 'entire business' of the human mind consists in this operation (*tota fere rationis humanae industria in hac operatione praeparanda consistit*). Yet 'when the operation is straightforward and simple', according to Descartes, we have no need of an enumeration 'to help us intuit the truth which the comparison yields'.[29] This is the case with mathematical knowledge.

How this process of comparison works in natural philosophy is clear in the case of the magnet.

> If the question concerns the nature of the magnet ... someone who thinks that nothing in the magnet can be known which does not consist of certain self-evident, simple natures: he is in no doubt about how we should proceed. First he carefully gathers together all the available experiences concerning the stone in question; then he tries to infer from this what sort of mixture of simple natures is necessary for producing all the effects which the magnet is found to have. Once he has discovered this mixture, he is in a position to make the bold claim that he has grasped the true nature of the magnet, so far as it is humanly possible to discover it on the basis of given observations.[30]

In order to discover the nature of a thing, it is necessary to recognize the simple natures that constitute it. To do this in the case of the magnet, one should undertake a number of observations and experiments to identify which of these simple natures are necessary for a mere stone to be classed as a magnet. In Aristotelian terms, one should find the essential conditions for the appearance of the effect and this is not possible without the aid of experience. Descartes outlines the process of recognition and identification in Rule 14 as a type of comparison:

If the magnet contains some kind of entity the like of which our intellect has never before perceived, it is pointless to hope that we shall ever get to know it simply by reasoning; in order to do that we should need to be endowed with some new sense, or with a divine mind. But if we perceive very distinctly that combination of familiar entities or natures which produces the same effects which appear in the magnet, then we shall credit ourselves with having achieved whatever it is possible for the human mind to attain in this matter ... Indeed, it is by means of one and the same idea that we recognize in different subjects each of these familiar entities, such as extension, shape, motion and the like ... This common idea is carried over from one subject to another solely by means of a simple comparison, which enables us to state that the thing we are seeking is in this or that respect similar to, or identical with, or equal to, some given thing. Accordingly, in all reasoning it is only by means of comparison that we attain an exact knowledge of the truth.[31]

As in the case of the anaclastic lines, the process of comparison proceeds by analogy, recognizing what are the essential conditions of the effect – that is, the simple natures – and additionally, and above all, what relates to previous knowledge. This process of comparison and division, leading to the discovery of the essentials of a thing, is properly speaking what distinguishes Cartesian enumeration, and also what establishes the criterion for clarity and distinctness in relation to simple natures. If all the essentials are identified, then a clear and distinct intuition is possible.[32]

Reduction, however, constitutes only the first stage. It does not provide us with perfect knowledge. Indeed, being based on experience and sensation, the object of our inquiry can still 'elude us', but at least through enumeration 'we shall be wiser', which is more or less what Aristotle wrote about division being a non-demonstrative tool, though nonetheless capable of clarifying our knowledge.[33] When things are not simple, enumeration is necessary. This means that enumeration in such cases is always ancillary to any kind of intellectual intuition. Not only this, we can push our thesis even further and say that it is impossible to have an intuition of what comes from sensation without a prior intervention of enumeration. As a matter of fact, only what is produced by the intellect – as in the case of mathematical objects – can be intuited immediately, while everything else is known or intuited only after enumeration. In this sense, therefore, the scope of intuition is drastically reduced and Descartes's alleged rationalism resized: a strong component of realism and empiricism is necessary to achieve the simple natures of material bodies.[34] Indeed, in this way Descartes does not go beyond the Aristotelians.

*

In the second step of Rule 5, the mental intuition of simple natures clearly follows reduction. Descartes introduces it in Rule 3:

By [mental] 'intuition' I do not mean the fluctuating testimony of the senses or the deceptive judgment of the imagination as it botches things together, but the conception of a clear and attentive mind, which is so easy and distinct that there can be no room for doubt about what we are understanding. Alternatively, and this comes to the same thing, intuition is the indubitable conception of a clear and attentive mind which proceeds solely from the light of reason.[35]

Descartes distinguishes mental intuition from both sensation and imagination, traditionally the two powers of the mind related to experience and sensible knowledge. Mental intuition is different from sensation because the latter does not provide firm, fixed and complete knowledge. It differs from imagination, which can provide false images and deceptive knowledge. Mental intuition, by contrast, provides a clear and distinct conception of the thing – that is, a concept, so very clear and distinct in fact that any doubt about the thing vanishes. This conception is a product of the intellect which is a pure and logical operation where the mind gives special attention to its subject matter. By definition, it is evident that Descartes opposes sensible to intellectual knowledge: the former offers an inadequate, confused and obscure apprehension of the thing, while the latter provides an evident, certain, clear and distinct cognition of it.

Mental intuition, however, does not replace sensation and imagination, and it is not a second source of knowledge. The content of knowledge of a natural object always comes from experience, but – as for the Aristotelians – the understanding of it differs. Indeed, mental intuition corrects sensible knowledge coming from experience. It provides a conception – that is, it conceptualizes or intellectualizes what comes from sensible knowledge or, in other words, it reduces the manifoldness of experience to one single apprehension, whose evidence, necessity and certainty are beyond doubt. Both these operations of supplying what is lacking in sensible knowledge and of providing a form of clear and distinct knowledge are those traditionally attributed to the negotiation of the intellect in the Aristotelian tradition.

In Rule 9, mental intuition is understood as a kind of vision and attentive mind:

> If one tries to look at many objects at one glance, one sees none distinctly. Likewise, if one is inclined to attend to many things at the same time in a single act of thought, one does so with a confused mind. Yet craftsmen who engage in delicate operations, and are used to fixing their eyes on a single point, acquire through practice the ability to make perfect distinctions between things, however minute and delicate. The same is true of those who never let their thinking be distracted by many different objects at the same time, but always devote their whole attention to the simplest and easiest of matters: they become perspicacious.[36]

Mental intuition is thus characterized as a kind of distinct intellectual perception, a kind of seeing with the mind, a visual or steadfast mental gaze,[37] which has as subject matter one single, simple thing. Often mental intuition is characterized as the natural light of the mind, or as an innate light, with an explicit allusion to the semantic constellation of vision with reference to the natural power of the intellect, an intellectual vision. What is interesting in this passage is that Descartes relates clear and distinct knowledge to the expertise and work of a craftsman, suggesting that intuition, while acting immediately, does not come out of the blue, but that a previous working of the mind is required.

In Rule 2, Descartes establishes that 'all scientific knowledge (*scientia*) is certain and evident cognition',[38] and for this reason he rejects all merely probable and conjectural cognition – that is, that coming from arguments based on opinions. This probable knowledge is for the majority of the time provided by experience, which is often 'deceptive'.[39] Besides experience, Descartes places deduction, which 'can never be performed wrongly by an intellect which is in the least degree rational'.[40] Deduction's weak point lies in the fact that human minds 'take for granted certain poorly understood experiences, or lay down rash and groundless judgments'.[41] In other words, experience can diminish the strength of deductions if the principles – which are derived from experience and serve as premises – lack a strong foundation. The latter is necessary according to Rule 3, where deductive argument is characterized as 'the inference of something as following necessarily from some other propositions which are known with certainty'.[42] At this stage, according to Descartes, all knowledge starts from experience, but not all knowledge derives from experience. There are two different ways of processing the same content, one following the other: deduction can be based on empirical content and, to be valid, experience must be reliable and certain.

The problem is how to determine that knowledge coming from experience can serve as a principle of deduction – that is, rephrasing in Aristotelian words, how the conclusions of the analysis can constitute the premises of the synthesis. Descartes is reiterating the basic problem of the Aristotelian theory of regressus: specifically, how to determine the certainty and evidence of the principles of scientific demonstration. Descartes is quite clear on this, stating in Rule 3 that in natural philosophy 'we ought to investigate what we can clearly and evidently intuit ... and not what other people have thought or what we ourselves conjecture ... for scientific knowledge (*scientia*) can be attained in no other way'.[43] The role of mental intuition is to supersede the various opinions and conjectures coming from experience in order to attain clear and distinct knowledge, as in the stage of the negotiation.

Furthermore, Descartes remarks that mental intuition is not necessary in the case of geometry and arithmetic, where the empirical and experiential moment is missing; indeed

> they alone are concerned with an object so pure and simple that they make no assumptions that experience might render uncertain; they consist entirely in deducing conclusions by means of rational arguments.[44]

In the case of mathematics, deduction relies on a priori principles or hypotheses generated by the mind itself and does not lead to hypothetical knowledge. Descartes is aware that not all human knowledge can achieve the same degree of certainty as mathematics, but nonetheless maintains that natural philosophy should aspire to this level, and it is the main aim of the *Rules* to articulate the right and truthful path for this inquiry.[45]

In the same vein, Descartes is anxious to maintain the importance of deduction,[46] which might seem superfluous once one has acquired clear and distinct knowledge by mental intuition. Nonetheless, deduction remains necessary because 'very many things which are not self-evident are known with certainty, provided they are inferred from true and known principles through a continuous and uninterrupted movement of thought in which each thing is clearly intuited'.[47] Deduction, therefore, is a discursive process, while intuition is a more immediate operation, being a single act of the understanding which concerns the self-evidence of what is intuited. Mental intuition differs from deduction because 'we are aware of a movement or a sort of sequence in the latter but not in the former, and also because immediate self-evidence is not required for reduction, as it is for intuition'.[48] The subject matter of mental intuition must be self-evident and deduction can be considered as a kind of 'intuition-on-the-move'.[49] Furthermore, deduction can be identified neither with the final stage of regressus – that is, with synthesis – nor with analysis. Otherwise, deduction would no longer be considered the inference of something following on necessarily from other propositions known with certainty, by virtue of the fact that the grounds of the analytic arguments are empirical and accidental, with a low degree of certainty.[50]

The importance of mental intuition is such that in Rule 8, Descartes establishes that 'if in the series of things to be examined we come across something which our intellect is unable to intuit sufficiently well, we must stop at that point'.[51] In the same rule, discussing the problem of the anaclastic lines in optics, Descartes clearly outlines the role of mental intuition within natural philosophy:

> Who eagerly seeks the truth on any question that arises ... he will discover when he goes into it that the ratio between the angles of incidence and the angles of refraction depends upon the changes in these angles brought about by differences in the media. He will see that these changes depend on the manner in which a ray passes through the entire transparent body, and that knowledge of this process presupposes also a knowledge of the nature of the action of light. Lastly, he will see that to understand the latter process he must know what a natural power in general is – this last being the most absolute term in this whole series. Once he has clearly ascertained this through mental intuition, he will ... *retrace* his course through the same steps.[52]

Descartes is very consistent, and the final sentence is a rephrasing of the general method he advanced in Rule 5, which is that one can proceed backwards after intuition. What is interesting, however, is that mental intuition occurs after a laborious working of the mind that we have previously identified with enumeration. The real apprehension of the thing, however, does not happen through enumeration or from experience: indeed, 'if someone proposes to know ... it from experience, he will achieve nothing'.[53] Knowledge of what is first and absolute is possible only by means of mental intuition – that is, through the pure activity of the intellect.

According to Descartes, mental intuition can be sharpened and its perspicacity improved by continuous apprehension. Such momentum is gained that, from the knowledge of simple natures and evident principles, the mind can proceed to intuit the most complex propositions.[54] In this activity, the mind acquires a *habitus* that is also capable of intuiting the relationships between things or propositions, usually the subject matter of deduction:

Everyone ought therefore to acquire the habit of encompassing in his thought at one time facts which are very simple and very few in number – so much so that he never thinks he knows something unless he intuits it just as distinctly as any of the things he knows most distinctly of all. Some people of course are born with a much greater aptitude for this sort of insight than others, but our minds can become much better equipped for it through method and practice.[55]

The almost explicit reference is to the Aristotelian notion of νοῦς, understood as a habit or intellectual acquired virtue in the *Nicomachean Ethics* VI.6, 1141 a 8–9, which is able to understand first principles, ultimate (i.e. absolute, simple) things and primary definitions.[56]

Descartes attributes the same function of the Aristotelian νοῦς to mental intuition. Rule 11 establishes that in logic mental intuition is capable of grasping and conceiving distinctly the essential connection between two simple and absolute elements. It does not differ from deduction because the proposition – that is, the composition of two simple elements – is intuited in a clear and distinct way, but rather because the whole proposition is 'understood all at once, and not bit by bit'. Whereas deduction 'does not seem to take place all at once, inferring one thing from another involves a kind of movement of our mind'.[57] According to Peter A. Schouls, another kind of intuition is operating here, so much so that he suggests that we should distinguish an intuition$_1$ from an intuition$_2$. Both have in common the fact that the act of intuition occurs immediately, excluding any successive mediation typical of deduction. However, they differ in the following consideration: whereas the object of intuition$_1$ is not subject to a further division or analysis, the object of intuition$_2$ can be complex and thus further divided or analysed.[58] This distinction is functional in asserting the autonomy of intuition$_1$ from Descartes's later precept by which all true and certain knowledge follows from the mind's knowledge of God.[59] Descartes, at least in the *Rules*, has not yet conceived this precept and, furthermore, does not seem to envisage two kinds of intuition – indeed, the act of intuiting seems to be the same in both cases. What differs is the subject matter: there is a shift from simple and absolute things in themselves to their connections and relations.[60]

The intuition of the essential connection between two simple or absolute elements – shared by the Cartesian mental intuition with the demonstrative induction of the Aristotelians – is important for explaining how mental intuition is supported by enumeration, with the effect that 'they seem to coalesce into a single operation, through a movement of thought, as it were, which involves carefully intuiting one thing and passing on at once to the others'.[61] The complementarity of mental intuition and enumeration lies in the fact that the various enumerated parts can be conceived as a unique *quid* – that is, their relation is such as to be considered as essential and necessary – so much so that they are considered as a whole, as one single thing. According to Schouls, ' "the whole" thus held in a single intuitive grasp is, again, the kind of compound which, qua object of intuition, is divisible, and intuition is intuition$_2$'. Moreover, he adds that 'intuition$_2$ comes about only after deduction is an accomplished fact'.[62] Yet we should emphasize that deduction always happens only after intuition$_1$, and thus intuition always follows intuition$_1$. In other words, intuition would seem to be that *habitus* of the mind that allows us immediately to know essential connections.

Intuition$_2$ works not only for propositions with a necessary connection between subject and predicate or cause and effect – that is, where the two terms are reciprocally implied – but also for all those parts that can be intuited at once as a whole. What happens to a simple proposition happens also to a chain of propositions to the point that

> by reflecting on the mutual dependence of simple propositions, we acquire the *habit* of distinguishing at a glance what is more, and what is less, relative, and by what steps the relative may be reduced to the absolute (emphasis added).[63]

Wherever an essential and necessary connection is recognizable, it is possible to acquire an immediate, clear, distinct and purely intellectual cognition of the thing, as with the Aristotelian demonstrative induction.[64] The connection is necessary when one thing 'is somehow implied in the concept of the other', as in a definition or an analytic proposition.[65] For Descartes, if the concept of one thing implies the concept of another, either a clear and distinct knowledge is impossible – because a distinct idea is separated from the other and defined by itself excluding the other[66] – or these two things are indeed conceived clearly and distinctly but as simply a single thing and not as two separate entities. The latter is the case where a proposition or a series of propositions can be intuited through a necessary connection, because they are not known as a combination of two terms or more, but only as a single clear and distinct idea, as one term.

Once these simple natures and their eventual necessary connections have been established, Descartes is able to conclude that 'the whole of human, scientific knowledge consists uniquely in understanding distinctly how all these simple natures contribute to the composition of other things',[67] and that knowledge consists simply 'in the putting together of self-evident things'.[68]

If my reconstruction is correct, and the core of the Cartesian method lies in Rule 5, the epistemology of the young Descartes is characterized by (1) the discovery of simple natures by means of enumeration and (2) the apprehension of necessary connections through mental intuition. These processes, I have suggested, aim to solve a typical problem for Aristotelian epistemology, that of the reliability of the first and basic principles of knowledge.[69]

Looking backwards, the first process is mainly described in *Posterior Analytics* II.19, where Aristotle discusses the formation of the universals and first principles. This process is called επαγωγή – that is, induction: what Descartes called enumeration. The similarity of the two processes is evident in *Topics* 108 b 7–12, where induction is understood as a kind of comparison just as Descartes undertook in his example of the magnet:

> The examination of likeness is useful with a view to inductive arguments … It is useful for inductive arguments, because it is by means of an induction of particulars in cases that are alike that we claim to induce the universal; for it is not easy to do this if we do not know the points of likeness.

This is a common procedure employed with division in the intermediate stage of regressus to make clear the confused knowledge of the cause.

The second process is once again set out in *Posterior Analytics* II.19 in the use of the term νοῦς, that is, intellection, which corresponds to Cartesian mental intuition.[70] Aristotle deals with this intellection also in *On the Soul* III.6 in considering the intellection of indivisibles, which is directly reminiscent of the Cartesian discussion of absolute, simple natures.[71] The intellection of the indivisible is anti-predicative – that is, precedent to every judgement of truth and falsehood which pertains to another activity of the according to Aristotle and concerns the synthesis or division of notions. Descartes himself establishes this distinction, stating that 'simple natures are all self-evident and never contain any falsity', and that 'this can easily be shown if we distinguish between the faculty by which our intellect intuits and knows things and the faculty by which it makes affirmative or negative judgments'.[72] In this context, discussing mental intuition, Descartes employs the Latin verb *attingere*, a verb that occurs, this time in Greek, in the form of θιγεῖν,[73] in *Metaphysics* IX.10, where Aristotle provides a more detailed explanation of the intellection of indivisibles.[74] These indivisibles or incomposites (ἀσύνθετα) – that is, Cartesian simple natures – have a special relationship with truth: intuiting (θιγεῖν) gives truth, while failure to intuit (μὴ θιγγάνειν) is ignorance. Descartes with his simple natures seems to follow Aristotle, who establishes that there is neither 'truth' nor 'falsehood' in the intuition of what is simple, but merely the knowledge of it if cognition occurs, or otherwise ignorance.

Aristotle furthermore confines the fields of indivisible and non-composite objects to essences – that is, 'what is' by itself.[75] But the characteristic to be such pertains also to Cartesian simple natures, and the similarity is thus quite close.[76] Moreover, before Descartes, Aristotle distinguishes the stage of discovery of simple things, of essences and definitions, from their apprehension. In *Metaphysics* VII.17, Aristotle writes that 'in the case of simple things no enquiry nor teaching is possible; but we must enquire into them in a different way'.[77] One could argue as to what this different kind of enquiry regarding simple things is. The answer is in *Metaphysics* VI.1 where Aristotle states that simple things and essences cannot be acquired by sensation, nor can they be hypothesized, but we need 'another kind of clarification'.[78] In the same passage, however, Aristotle adds that 'the thought that leads to the knowledge of the essence of a thing is the same that leads to the knowledge of the existence of a thing'.[79] The process that leads to the knowledge of simple and necessary things, which – speaking in Cartesian terms – is mental intuition, also leads to the knowledge of their existence. Even in this respect, the parallel between Aristotle and Descartes is strong, on the simple grounds that through mental intuition we know not only the essence of the thing but also of its existence. This aspect will be crucial for Descartes in inferring the existence of the 'I think' from its essence, and it lays the basis for the ontological realism suggested by Marion.[80]

Finally, from Rule 9 a third shadow of the Aristotelian tradition on Descartes emerges in the concept of the intellect as a habit capable of sharpening its capacity to intuit.

To sum up, Rule 5 outlines an epistemology according to which the young Descartes introduces a new departure within Aristotelian discussion of the intermediate stage of regressus, proposing a new and, to his eyes, more functional solution. What appeared to be a threefold process reflecting the three stages of the Aristotelian regressus, is instead a Cartesian reappraisal of the negotiation of the intellect. Outside of this theoretical framework and historical context, both his early epistemology and the *Rules* are destined to be understood as a story of a failing project, rather than an important step in the evolution of a tradition, which will leave substantial traces even in foundational texts of early modern philosophy such as the *Meditations*.

*

In the *Meditations*, Descartes reflects on the function of negotiation by focusing on the notion of *inspectio mentis*, usually translated as mental scrutiny or mental inspection. This is an understudied and marginalized subject in the scholarship, but it represents the core of Descartes's epistemology. Indeed, there exists no specific study on the topic despite its importance for the reason that this process makes possible scientific knowledge.[81]

An attempt to contextualize and give historical understanding to this epistemological process and its function is still conspicuously lacking. My intention is to show that Descartes's *inspectio mentis* (1) is not traceable back to the notion of intuition, (2) is not anti-Aristotelian, (3) is not related to the theory of innate ideas; and (4) does not advance the idea of rationalism. My intention, rather, is to demonstrate the centrality of this mental inspection for Cartesian epistemology, and also how Descartes has recourse to the Aristotelian problem, and recasts it in order to explain the acquisition of scientific knowledge from experience.

Descartes introduces the process of mental inspection in the *Second Meditation* on the nature of the human mind. Yet one of the central topics of the *Second Meditation* is how to acquire certain and scientific knowledge regarding natural objects. In particular, in discussing the nature of a piece of wax Descartes says that the knowledge, or the act of getting to know the same, is not

> vision or touch or imagination, nor has it ever been, despite previous appearances, but only a mental inspection (*solius mentis inspectio*), which can be imperfect and confused, as it was before, or clear and distinct as it is now, depending on how carefully I fix my mind on what the wax consists of.[82]

Here we are clearly operating on an epistemological level relating to our knowledge of the nature of a thing: Descartes uses the Latin word *perceptio* to characterize both the act and the content of knowledge.[83] In this epistemological enquiry, he first of all denies that the act of understanding the nature of a thing – this perception – is a kind of sensation or product of the imagination, and thus, he concludes, it must be an act of the intellect. *Quartum non datur*, Descartes is still working strictly within the Scholastic framework of Aristotelian psychology using the three faculties of the mind outlined in *On the Soul*. The exclusion of sensation and imagination as operations capable of understanding the nature of things in favour of an act of understanding seems to recall Rule 3 on mental intuition. However, as I will show, this is not the case.

Indeed, in the *Second Meditation*, Descartes explains that the nature of bodies is not known 'by the senses or the faculty of imagination but by the intellect alone, and that this knowledge derives not from their being touched or seen but from their being conceived' by the intellect.[84] Scholarship agrees that by this means Descartes argues for an intellectual understanding of the nature of natural objects.[85] Secondly, he emphasizes that such knowledge, or perception, is at first imperfect and confused and then clear and distinct. Thus, the understanding of the nature or the 'what is' of the natural object should involve a process of clarification. Finally, this clarification depends on the quality of attention paid by the mind to the object – that is, how much the mind focuses on the object. This implies that the act of the intellect in knowing and making clear the nature of the natural object is not immediate, but rather a discursive process intentionally directed by the mind.

So, why and how is the nature of the piece of wax understood only by mental inspection? Descartes achieves this conclusion by exclusion and elimination. He compares the knowledge of the same piece of wax before and after contact with something hot like fire. Even if all sensible qualities vanish, the same wax remains. This leads him to ask what was it in the wax that we understood with such distinctness? He answers: 'Evidently none of the features which I arrived at by means of the senses; for whatever came under taste, smell, sight, touch or hearing has now altered'. Nonetheless, he says, 'the wax remains'.[86]

Reading this passage from an Aristotelian standpoint, what seems to remain is the substance (or essence), and what vanishes are the accidents, as Robert Pasnau has correctly pointed out.[87] Descartes does not state this openly, but it is clear that, in the case of the nature of a natural object like wax,[88] where some features of the wax vanish but the wax remains, this means at the very least that those features were not essential to the nature of the wax, and that the nature of the wax must be something other than what one may sense or imagine.

Descartes proposes examining the case of wax a little further, and he suggests taking away 'everything which does not belong to the wax, and seeing what is left'.[89] This means taking away what is accidental and maintaining what is essential. Descartes writes that distinguishing the wax from its outward form is the equivalent of taking its clothes off and considering it naked – that is, to engage in an abstraction.[90] This is a procedure for reduction to the essence or definition of a thing, which he shares with the Aristotelians of his time.

How is it possible to take away what is accidental and identify what is essential? After the reduction, Descartes explains, what is left of the piece of wax is that it is extended, flexible and changeable. These seem to be the essential elements or qualities of wax, and they are general. They are not known by means of sensation, as we have seen; they cannot even be conceived of by the imagination. Otherwise, we would only be able ever to imagine that this piece of wax is capable of changing from a round to a square shape or to some other such form. In any case, it would be impossible to grasp the countless and immeasurable number of changes that characterize the essence of wax. Sensations and images are limited in number and cannot provide support for understanding the nature of the wax, so a more general and universal concept is required. This is a very Aristotelian consideration, and it is one of the reasons why – according to Aristotelian epistemology – the transition from images to concepts is necessary for scientific knowledge; by no other means can the manifoldness of experience be understood in all its possibilities. If sensations and images are limited, the procedure of a division starting from them is not in itself helpful, because its conclusions – as we have previously seen – are not necessary, but merely accidental and contingent. The logical conclusion for Descartes, therefore, is that the nature of the piece of wax is revealed purely by the working of the intellect alone.

Descartes emphasizes that he is 'speaking of this particular piece of wax', and indeed 'the point is even clearer with regard to wax in general'.[91] This is instrumental, he claims, in explaining that he is not dealing with a universal in general, which is traditionally conceived by the intellect, but with the nature of a singular object, which is being made universal. Descartes also emphasizes that the wax 'conceived only by the intellect or the mind', is the same wax which has been previously seen, touched and imagined – that is, it is not a different object, but rather something that we have known from the start. But while at the start the mind knew that particular object in that particular state, now the particular object is known through its own nature. This means that the mind had already had knowledge (or perception) of the particular object, but not of its nature; or rather the knowledge that the mind had was so confused and imperfect that the object's nature was concealed. The role of the mental inspection is to make clear and distinct this nature that the mind already knew, even if indistinctly: that is, to distinguish what is essential and what is not. In this sense, the nature of the wax was always present to the mind – Descartes would say it was innate to the mind – but not in the form that it should have had, for there was no distinction between essential and unessential properties.

Descartes gives another illuminating example. It is the same as when we look out of the window and see human beings. Indeed, looking out of the window we see no more than hats and coats, which could very well be hiding ghosts or automatons. Nonetheless, Descartes argues that through the intellect, which is the faculty of judgement, we can say that they are not ghosts or automatons, but real human beings.[92] It is quite clear that for the faculty of judgement here Descartes means the faculty which distinguishes the essential characteristics of each thing, and which is capable of referring to one single object what pertains to its nature, and also of taking away what is unessential. But in order to do this – that is, to identify the nature of what we perceive, in this case human beings – the mind should rely on previous knowledge and experience, which teach that under hats and coats there are persons and not ghosts or automatons. The mind, therefore, should be accustomed to understanding the nature of a natural object coming from sensation, and to identifying the essence or substance beneath the characteristics. The mind should have the habit of grasping this nature.

At this stage, one should question whether the whole of this process of clarification is merely intellectual, or whether, in one way or another, it also involves experience. We have already advanced a possible answer by saying that mental inspection – being a process that proceeds from knowledge that is confused and indistinct to that which is clear and distinct – cannot be merely intellectual, even if the understanding of the nature of a thing is purely so. Indeed, even the discovery by reduction of the three essential characteristics of the wax was possible only after testing under different conditions, helping thus to identify the real nature of the wax.

If in the *Meditations* we do not find a specific passage which explains how this reduction works, we can find a clue in the *Rules for the Direction of the Mind*, without which Cartesian mental inspection would remain obscure and somewhat abstract. Garber has correctly shown that Descartes's methodology underwent very significant changes after the *Rules*, and that the many ideas contained in that manuscript work were later abandoned.[93] However, the *Rules* are helpful in understanding the evolution of Descartes's thinking and the general historical framework within which he understood this enactment of logic. In both works, indeed, we can find a process of clarification from a confused and obscure starting-point, and this makes a comparison which is at least fruitful for an appreciation of Descartes's development.[94]

In Rule 13, as we have seen, Descartes establishes that there is a logical process that divides and reduces all things to their simplest elements and that goes to the essence of the thing. This process is enumeration or induction,[95] briefly introduced in Rule 7 as a kind of discursive process which 'renders the truth of our conclusions more certain than any other kind of proof'.[96] Indeed, if mind infers one fact from another immediately, this 'comes under the heading of true intuition'.[97] In Rule 8, as I have shown, Descartes explains how enumeration works also in relation to experience. He offers a further glimpse of how this enumeration works in the experimental method featured in the explanation of the rainbow in the *Meteors*,[98] the only place where the French philosopher applied his method to a specific case, as he himself reveals to Antonie Vatier.[99] In general, experience and experiment for Descartes are mandatory for finding the right path to channel our reasoning in the reduction.

*

In the letter of 23 December 1630, Descartes confessed to Marin Marsenne that he had found a system for collecting useful experiences in Francis Bacon's *experientia literata*, since these experiences are 'needed for the search for truth'.[100] The study of experience, however, should not be reduced to a mere collection of curiosities; it requires reason and method to provide certain knowledge: the experimental and empirical part requires systematization by the intellect.

In the *Meteors*, in connection with the question as to what causes the rainbow, Descartes uses an experiment to make clear the nature of light – that is, the essence of a natural object – which is then grasped by intuition. The process, working from the phenomena to reveal the nature of the natural object – that is, the reduction – finds precious support in the experiment, for it suggests 'how things depend on one another', 'what question we might turn to next',[101] and points 'the way to correct deduction, and eliminates incorrect deductions'.[102] In general, experiment helps us find the correct 'chain of causes, by making the phenomena more precise, and suggesting how it is that the phenomena are actually produced in this world'.[103] Of course, this movement towards the discovery of the cause is not sufficient without a rational explanation of effects through causes, which provides a general theory of phenomena.

The procedure from effects to causes and from causes to effects recalls that of regressus. Descartes is perfectly aware of this, and in discussing the *Meteors* in the *Discourse on the Method* writes that his argument is not circular, which was a common criticism of regressus's theorists. For Descartes, the twofold process of proving causes from effects, and explaining effects from causes, is based on experience, which 'makes most of these effects certain', since 'the causes from which I deduce them serve not so much to prove them as to explain them'.[104] The discovery of the causes rests on experience, and the procedures that have been outlined in this chapter make the cognition of the causes certain. The following stage from causes to effects serves to explain the effects from these causes scientifically. In the twofold process, Ernan McMullin detects traces of the influence of Paduan Aristotelianism and of Jacopo Zabarella in particular.[105] Indeed, this association is reinforced by Descartes's use of the term supposition. Since this kind of knowledge is based on experience, supposition borrows contingency, and for this reason the result of this knowledge is merely hypothetical and conjectural. Descartes explicitly claims to use the term 'supposition' in this context in the same way as Aristotelians such as Nifo did in the regressus theory, as a general acquisition from experience to be tested or explained by a rational and intellectual procedure.

This conception of hypothesis or supposition is connected with that developed in the *Principles of Philosophy* and from which Newton will distance himself. In sect. 204 Descartes points out that what cannot be perceived by the senses can be established through hypothesis, bearing in mind that the actual nature may be very different from that supposed. Descartes gives the famous example of the two clock mechanisms:

> Although this method may enable us to understand how all the things in nature could have arisen, it should not therefore be inferred that they were in fact made in this way. Just as the same craftsman could make two clocks which tell the time equally well and look completely alike from the outside but have completely different assemblies of wheels inside, so the supreme craftsman of the real world could have produced all that we see in several different ways. I am happy to admit this; and I shall think I have achieved enough provided only that what I have written is such as to correspond accurately with all the phenomena of nature. This will indeed be sufficient for application in ordinary life, since medicine and mechanics, and all the other arts which can be fully developed with the help of physics, are directed only towards items that can be perceived by the senses and are therefore to be counted among the phenomena of nature. And in case anyone happens to be convinced that Aristotle achieved – or wanted to achieve – any more than this, he himself expressly asserts in the first book of the *Meteorologica*, at the beginning of Chapter Seven, that when dealing with things not manifest to the senses, he reckons he has provided adequate reasons and demonstrations if he can simply show that such things are capable of occurring in accordance with his explanations.[106]

This passage is particularly important principally because Descartes refers to the same Aristotelian passage that Nifo mentions in explaining the kind of knowledge obtained after the first stage of regressus, clearly indicating causes as things not manifest to the senses.[107] Descartes's emphasis on explaining effects through qualities like motion, magnitude and figure in such a way that all the phenomena may correspond to the hypothesis without referring to the real nature or the cause of the effects themselves reminds to what Galileo argues in the *History and Demonstrations Concerning Sunspots and Their Phenomena*.

In the letter of 13 July 1638 to Jean-Baptist Morin, Descartes explains the nature of this supposition and twofold argument, in a way that recalls regressus. Against Morin, who accused him of arguing in circle, he disagrees that 'it is circular to explain effects by a cause, and then prove the cause by the effects'.[108] Descartes introduces a distinction between providing proof and explanation. While proving is associated with discovery and with an argument from effects to cause, explaining is related to an argument that moves from causes to effects. Tackling the central question of regressus, Descartes explicitly argues that only the first part of regressus leads to scientific discovery, while the second part merely has an explanatory role. A central role in the transition from one argument to another – from proving to explaining – is played by experience. Indeed 'experience renders most of these effects quite certain and so the causes from which I deduce them serve not so much to prove them as to explain them'.[109] This means that experience is essential in providing certain effects and finding causes, and that the causes play an explanatory role.

Descartes objects to Morin that it is not so easy to fit a cause to effects, 'unless it is the true cause which produces them'.[110] Indeed many effects can be generated by many causes. But through experience one can discover that that particular cause refers to that particular effect, and vice versa. There is only one case in which it is possible to infer immediately a direct relation between an effect and its cause, and Descartes claims that he deals with causes that belong to this class. This class of causes for the Aristotelians would imply a kind of univocity between effect and cause, and Descartes agrees with them since 'the effects which I explain have no other causes than the ones from which I have deduced them'.[111]

Experiment, through enumeration, thus becomes fundamental in the process of clarifying knowledge in the mental inspection of natural objects,[112] establishing the connection between cause and effect; otherwise, it would be impossible in many cases to establish the necessary characteristics of a natural object. In this context, Descartes does not make explicit the role of enumeration and experiment, but, as Daniel Garber correctly points out, it is difficult to imagine how the nature of the rainbow can be discovered without the appeal to experience: we cannot do it 'on the basis of the seeds of truth alone'.[113] As Garber suggests, 'experiment can lead us to facts', but it is only the mental inspection of the phenomena that provides us with a clear and distinct knowledge of natural objects. Therefore, despite Descartes's emphasis on intellectual knowledge, mental inspection could not work without the assistance of experience and experiment.[114]

If mental inspection were merely intellectual, Descartes' concept would not differ from Platonic division, which he attacks for its sterility. In the *Seventh Replies* to Pierre Bourdin, Descartes exemplifies his opposition to division with the story of a peasant:

> The fellow had just seen a wolf for the first time, though from a long way off, and stopped to ask his master, a well-born young man whom he was following, about what he had seen. 'What is it?' he asked. 'It is undoubtedly an animal – it moves and walks. But what animal can it be? It must be one of those I know already. What

are they? Ox, horse, goat and donkey. Is it an ox? No; it does not have horns. Is it a horse? No; its tail is not big enough. Is it a goat? A goat has a beard, but this animal does not, so it is not a goat. So it must be a donkey, since it is not an ox or a horse or a goat!' ... 'Come now,' said the young master, 'you may just as well conclude that it is a horse as a donkey. Look: is it an ox? No; it does not have horns. Is it a donkey? Certainly not: I see no ass's ears. Is it a goat? No; it has no beard. So it is a horse.' The peasant was a little upset by this new analysis, and exclaimed 'Wait: it is not an animal. For the animals I know are the ox, the horse, the goat and the donkey. It is not an ox, or a horse, or a goat, or a donkey. And so ... it is not an animal; and hence it is something that is non-animal!'[115]

Descartes shows that from experience we cannot infer necessary conclusions, especially if the definition of the nature of a natural object is to be discovered by division, which is unable to provide new cognition. Indeed, division is not a method of discovery, and it is useless in general if one does not yet know the object under investigation; and if one knows the object perhaps erroneously or not completely, the inference of the division will be false.

In the *Search after Truth*, Descartes makes another compelling criticism against this kind of divisive process. The natural thing in this case is the human being:

> Were I for example to ask Epistemon himself what a human being is, and were he to reply, as is done in the Schools, that a human being is a rational animal, and if, in addition, in order to explain these two terms which are no less obscure than the first, he conducted us by all the steps which are termed metaphysical, we should be dragged into a maze from which we would never be able to emerge. For from this question two others arise: the first is what is an animal? The second, what is rational? And further, if in order to explain what an animal is he were to reply that it is a living thing which has sensations, that a living thing is an animate body, and that a body is a corporeal substance, you see that the questions would go on increasing and multiplying like the branches of a genealogical tree; and finally all these wonderful questions would finish in pure tautology, which would clear up nothing, and would leave us in our original ignorance ... Indeed, whatever Epistemon may say, it strikes me that there is a great deal of obscurity in these metaphysical levels. If, for example, we say that a body is a 'corporeal substance,' without making clear what a corporeal substance is, these two words do not tell us any more than does the word 'body.' In the same way, if we assert that to be living is to be an 'animate body,' without first explaining what a body is and what it is to be animate – and likewise with all the other metaphysical levels – we are uttering words and putting them as it were in a certain order, but we are not really saying anything. The words do not convey anything that can be conceived, or form any distinct idea in our mind.[116]

Descartes is explicit in emphasizing the futility of this process for discovery. It is a rational procedure, which does not add anything to the concept that we had at the beginning, and leaves the idea of the nature of a thing completely obscure and indistinct. The conclusion would be the same as the premise, and the inference would lead to a kind of erroneous tautology. This Cartesian argument also seems to suggest that a rational process in itself is not enough to acquire new knowledge and make clear and distinct the idea of the object of knowledge. Experience plays a fundamental role in the cognitive process, beyond all possible rationalist interpretation of Descartes's epistemology.[117]

Inspectio mentis is, therefore, first of all a process for clarifying something that is already known, but that at the outset is obscure and confused and then, by means of the same inspection, becomes clear and distinct. It leads to clear and distinct knowledge of the 'what it is', that of the nature of the object under investigation. Just to be clear, in dealing with the nature of things Descartes is not discussing how things are in reality, but only how they are relative to the mind. A thing's nature is its perceived essence in the mind. We are at an epistemological level, as Williams correctly points out,[118] and this makes the project of the *Second Meditation*, at least at this stage, comparable with that of the *Rules*.[119]

To sum up, according to Descartes mental inspection (1) is mental, even when the content of knowledge is previously acquired by sensation; (2) focuses on what is already known – that is, the thing is already present and known in the mind, but unclearly and indistinctly; (3) is a process of clarification, making knowledge that was previously confused and obscure, clear and distinct; (4) reveals the nature of the thing – that is, its essence and essential parts; (5) is mainly applied to natural objects, since products of the mind alone cannot be first confused and obscure and then clear and distinct; and (6) is a process through which we exert the mind's acuity in considering a thing.

In conclusion, Descartes's *inspectio mentis* (1) requires hard and time-consuming work of the mind, (2) concerns all kinds of ideas, especially adventitious ones and (3) demands a consistent contribution of experience. If my historical reconstruction and contextualization of the *inspectio mentis* are correct, we can dismantle some of the established opinions of the scholarship.

First of all, we said at the beginning that some interpreters – such as Desmond M. Clarke – have identified mental inspection with a later version of the mental intuition outlined by Descartes in his early *Rules*.[120] However, this identification is unjustified, for *mentis inspectio* is not equivalent to *intuitus*, since the *mentis inspectio* is at the beginning imperfect and confused, and subsequently clear and distinct, while the *intuitus* should always be clear and distinct.[121] Not only this, for, as we have seen, *inspectio mentis* is a working of the mind, which takes time and effort, while intuition is a single, immediate act of understanding. Finally, intuition intuits an object of the understanding, while mental inspection makes this object *available* to the understanding, or prepares the understanding for the intuition of the object.

My investigation has also shown that Descartes was not reacting against Aristotelian or Thomistic opinions. *First*, the idea that every act of the intellect is related to a sensation or to an image – that is, to something corporeal – was again introduced and made widespread by Pietro Pomponazzi – the so-called Alexandristic interpretation of Aristotelian psychology. This was one of the main issues used to demonstrate the mortality of the soul against the Thomists, who maintained that there was at least one act of the intellect that was unrelated to, and free of, corporeal dependence, the intellection. Furthermore, in Aquinas there is at least one further act of understanding borrowed from the Platonic tradition which is free of corporeal dependence, and this is the act of self-reflection, through which the mind knows itself as a thinking being.[122] Therefore, in the Aristotelian tradition there is cognitive access to the nature of things independent of sensation. *Second*, understanding, according to Aquinas, is not built on sensation and imagination, but rather the higher faculties incorporate the lower faculties – that is, understanding incorporates sensation and imagination – for only in this way can the intellect be the final form of the human being. Thus, sensation and imagination are dependent on understanding, as in Descartes. *Third*, proper sensibles, common sensibles and universal natures refer to the same object, though known differently: sometimes more distinctly and clearly, at other times more confusedly and obscurely. It is the role of the negotiation of the intellect to clarify the object of knowledge. As in Descartes, the object of knowledge is the same, but known differently. *Fourth*, negotiation of the intellect does not proceed by mere abstraction; rather, it investigates and clarifies the inner nature. Finally, from this last perspective Descartes's model of the mind does not differ from the Aristotelian – at least in the *Second Meditation* – on these key issues.

A third misunderstanding in the scholarship is the notion that, if the intellect alone through inspection can reveal the nature of a physical object, sensation and imagination are indeed useless. But if so, the object cannot be otherwise than something innate to the mind – an innate idea for Descartes, which at the beginning is confused and obscure and at the end is clear and distinct. Williams believes that 'our ideas of sensory qualities are also said to be innate', so much so that 'the conception of wax as pure extension is not something we arrive at by inference from reflection on the changes in sensory qualities, but is rather something we find we have',[123] in opposition to the line which the Aristotelian tradition has usually maintained. Williams grounds his thesis in the *Comments on a Certain Broadsheet*, which, however, seems to suggest something quite other. Here Descartes writes that senses in no case present to us the ideas of things 'just as we form them in our thinking'.[124] This means only that things are known by sensation and by the faculty of thinking in two different ways, something that every Aristotelian would have maintained, being τὰ αἰσθητά and τὰ νοητά – two different mental objects but related to the same physical object. Furthermore, Descartes argues as follows:

> There is nothing in our ideas which is not innate to the mind or the faculty of thinking, *with the sole exception* of those circumstances which relate to experience, such as the fact that we judge that this or that idea which we now have immediately before our mind refers to a certain thing situated outside us. We make such a judgement not because these things transmit the ideas to our mind through the sense organs, but because they transmit something which, at exactly that moment, gives the mind occasion to form these ideas by means of the faculty innate to it. Nothing reaches our mind from an external object through the sense organs *except* certain corporeal motions ... but neither the motions themselves nor the figures arising from them are conceived by us *exactly* as they occur in the sense organs (emphasis added).[125]

Thus, according to Descartes, there is something coming from outside which stimulates the mind to form ideas; these ideas, however, are not understood in the same way that sensations are known. Nonetheless, this is perfectly compatible with the Aristotelian framework, according to which experience occasions the formation of concepts in the mind and accords with the possibility of a moment of clarification as the object of knowledge moves from sensation to intellect, which the Aristotelians called negotiation of the intellect. Accordingly, the nature of a natural object is revealed by the intellect alone, but the object comes from sensation, and from the sensation of that particular object we acquire a quality of knowledge, different from an understanding of its nature, and the main difference lies in the exact understanding of the essence of thing, which experienced by sensation is confused and obscure, and then becomes more exact, clear and distinct.

Finally, everything we have said seems to detach the mental inspection from Descartes's theory of innate ideas, at least according to Williams's interpretation.[126] The essential feature of Descartes's theory of innateness in the *Comments* is, as Clarke correctly pointed out, that 'ideas are irreducible to sensory stimuli'. In other words, what Descartes is saying is that 'mental entities, such as ideas, are different in their nature from the physical patterns in the brain which accompany them'. Not only this, Clarke adds that 'even those ideas which are provoked by and correlated with extramental stimuli are innate in the sense of being irreducible to the type of reality which triggers them in the mind'.[127] This kind of innateness, according to Clarke, concerns 'the immateriality of the mind, the universality of concepts as opposed to images, and the perfection of concepts relative to the imperfect instantiations of them which are encountered in physical reality'[128] – in other words, all the items at the core of Aristotelian philosophy. In this sense, also, every Renaissance Aristotelian, except Pomponazzi perhaps, could be conceived as an innatist or nativist. But this kind of innateness does not preclude the involvement of experience and experiments, rather it implies these, and for these reasons the idea that Cartesian science is fundamentally rationalist deserves revisiting.

Descartes represents a turning point in the history of epistemology, marking the transition from a classic scientific methodology to a philosophy of mind or facultative logic.[129] Descartes's epistemological center-of-gravity shift towards the subject led subsequent philosophers, who challenged his philosophy, to focus more on the various operations of the mind and on their relations with the external world, undermining some hinges of Aristotelian epistemology such as the possibility of knowing essences and causes.

All these epistemological discussions must be understood within the context of Descartes's elaboration of a *mathesis universalis*.[130] On 26 March 1619, Descartes writes a letter to Isaac Beeckman in which he announces a totally new science (*scientia penitus nova*)

> which would provide a general solution of all possible equations involving any sort of quantity, whether continuous or discrete, each according to its nature. In arithmetic, for example, certain problems can be solved by means of rational numbers, while others require irrational numbers, and others again we can only

imagine how to solve, but not actually solve ... I am hoping to demonstrate what sorts of problems can be solved exclusively in this or that way, so that almost nothing in geometry will remain to be discovered. This is of course a gigantic task ... it is an incredibly ambitious project.[131]

He has clearly in mind a universal science different in scope and method but not dissimilar to Ramon Lull's *The Short Art*, as he himself makes explicit. At the end of the same year, he speaks of this universal discipline as *mirabilis scientiae fundamenta*.[132] In 1628, Isaac Beckman records in his *Journal* that Descartes had told him 'that insofar as arithmetic and geometry were concerned, he had nothing more to discover', adding that he had 'arrived at perfect knowledge of geometry' but also claimed to have embraced 'all human knowledge'.[133] The idea of establishing a single, universal science starting from arithmetic and geometry, a science which could encompass all kinds of knowledge, as we have seen, was clearly pointed out by Proclus in Pietro Barozzi's 1560 edition. Descartes gives a glimpse of this science to Beeckman, who records it in his *Journal* and calls it universal algebra (*algebra generalis*).[134] That this universal algebra is much more than a form of mathematics is clear from Rule 4. The reflection is long but worth quoting in full:

> We can have no scientific knowledge without mental intuition or deduction. The method cannot go so far as to teach us how to perform the actual operations of intuition and deduction ... So useful is this method that without it the pursuit of learning would, I think, be more harmful than profitable. Hence I can readily believe that the great minds of the past were to some extent aware of it ... For the human mind has within it a sort of spark of the divine, in which the first seeds of useful ways of thinking are sown ... This is our experience in the simplest of sciences, arithmetic and geometry: we are well aware that the geometers of antiquity employed a sort of analysis which they went on to apply to the solution of every problem, though they begrudged revealing it to posteriority. At the present time a sort of arithmetic called algebra is flourishing, and this is achieving for numbers what the ancients did for figures ... I would not value these *Rules* so highly if they were good only for solving those pointless problems with which arithmeticians and geometers are inclined to while away their time ... But if one attends closely to my meaning, one will readily see that ordinary mathematics is far from my mind here, that it is quite another discipline I am expounding, and that these illustrations are more its outer garments than its inner parts. This discipline should contain the primary rudiments of human reason and extend to the discovery of truths in any field whatever. Frankly speaking, I am convinced that it is a more powerful instrument of knowledge than any other ... I wondered why the founders of philosophy would admit no one to the pursuit of wisdom who was unversed in mathematics – as if they thought that this discipline was the easiest and most indispensable of all for cultivating and preparing the mind to grasp more important science ... the method seems to me to be none other than the art which goes by the outlandish name of algebra ... It was these thoughts which made me turn from the particular studies of arithmetic and geometry to a general investigation of mathematics. I began my

investigation by inquiring what exactly is generally meant by the term *mathesis* and why it is that, in addition to arithmetic and geometry, sciences such as astronomy, music, optics, mechanics, among others, are called branches of mathematics. To answer this it is not enough just to look at the etymology of the word, for, since the word *mathesis* has the same meaning as discipline (*disciplina*) ... I came to see that the exclusive concern of mathematics is with questions of order or measure and that it is irrelevant whether the measure in question involves numbers, shapes, stars, sounds, or any other object whatever. This made me realize that there must be a general science which explains all the points that can be raised concerning order and measure irrespective of the subject-matter, and that this science should be termed *mathesis universalis*.[135]

This passage seems to be a commentary on Proclus. Indeed, from arithmetic and geometry, Descartes aims to erect a more universal science, which would explain all human knowledge. His *mathesis universalis* is not only therefore a form of mathematics, but something far beyond. The reference to the 'first seeds', to *mathesis* as 'discipline', the emphasis on 'order and measure', and finally the priority of geometry and arithmetic not for their content but for their ways of proceeding, suggest that while writing this rule Descartes had on his desk Barozzi's 1560 edition of Proclus. Some words and propositions are taken *verbatim* from this work.[136]

In a letter of March 1636 to Mersenne, Descartes writes that he has ready a work titled 'The Plan of a Universal Science Which Is Capable of Raising Our Nature to Its Highest Degree of Perfection',[137] which will become his *Discourse*. In this work too, Descartes writes that the study of the method of geometers had given him occasion 'to suppose that all the things which can fall under human knowledge are interconnected in the same way'.[138] In the reply to Mersenne in the *Meditations on First Philosophy* (1641), Descartes develops Rule 4 in an original way. In using a geometer's manner of writing, he distinguishes an order (*ordo*) and a method of demonstration (*ratio demonstrandi*). Concerning method, he writes as follows:

> This divides into two varieties: the first proceeds by analysis and the second by synthesis. Analysis shows the true way by means of which the thing in question was discovered methodically and as it were a priori (*& tanquam a priori*) ... Synthesis, by contrast, employs a directly opposite method where the search is, as it were a posteriori (*& tanquam a posteriori*), (though the proof itself is often more a priori than it is in the analytic method), that is it demonstrates the conclusion clearly and employs a long series of definitions, postulates, axioms, theorems and problems, so that if anyone denies one of the conclusions it can be shown at once that it is contained in what has gone before ... It was synthesis alone that the ancient geometers usually employed in their writings, but in my view this was not because they were utterly ignorant of analysis, but because they had such a high regard for it that thy kept it to themselves liked a sacred mystery.[139]

Descartes, contrary to the Aristotelian tradition, inverts the definition of analysis and synthesis. Analysis becomes an a priori argument, while synthesis becomes a posteriori.¹⁴⁰ This inversion becomes even sharper in the French edition revised by Descartes, where he substitutes 'as it were a priori' with 'and shows how the effects depend on the causes' (*fait voir comment les effets dependent des causes*), and 'as it were a posteriori' with 'as in examining the causes through their effects' (*comme en examinant les causes par leurs effets*). This inversion is not surprising if we look at it from the standpoint of the *mathesis universalis* – namely, a universal discipline, capable of providing principles for all the other sciences. Why does Descartes invert the order? The solution is far from simple. The most probable explanation is that he has taken his conception directly from the ambiguous passage of Proclus on the two ways of proceeding.¹⁴¹ Indeed, from Proclus we learn that there is a way of proceeding from things that are better known (*notior*, γνωριμώτερον), but it is not clear what these things are exactly – that is, whether better known by nature or better known by the human mind.

In Aristotle, there is a clear distinction between ἡμῖν γνωριμώτερον, that which is better known to sensation or to the mind, and φύσει γνωριμώτερον, that which is better known to nature, corresponding to what is primary to us, or to the scope of our human sensation, and what primary in absolute terms. But in Proclus this characterization is missing. Since in this passage Proclus is dealing with principles and common axioms, Descartes could with good reason have interpreted the words referring to analysis as the method proceeding from principles to effects, and synthesis the other way around. This seems to be confirmed by another passage where Proclus states that mathematical science, that discipline that have received the principles from the universal science,

> unfolds and traverses the immaterial cosmos of ideas, now moving from first principles to conclusions, now proceeding in the opposite direction; now advancing from what it already known to what it seeks.¹⁴²

The parallelism between the two passages is perfect. Descartes was entirely legitimate in considering analysis as the method working from first principles towards conclusions, at least according to Proclus's definition. Indeed, absolute principles are better known of the things of which they are principles, and only from the latter can the things be known.¹⁴³ If a universal science has already established the principles, and the way of proceeding is from what is better known (i.e. the principles themselves), then, according to Proclus – or at least according to Descartes's reading of Proclus – analysis is the method proceeding from these principles.¹⁴⁴

There is a further problem with the knowledge of these first principles, because in Proclus it was based on the Pythagorean-Platonic remembrance of something always present to the mind:

> According to the tradition, the Pythagoreans recognized that everything we call μάθησις is remembering, not something placed in the mind from without, like the images of sense pictured in the imagination, nor transitory, like the judgments of opinion. Though awakened by sense-perception, μάθησις has its source within us, in our understanding's attending to itself. They realized too that, although evidences of such memories can be cited from many areas, it is especially from

mathematics that they come, as Plato also remarks ... This is why Socrates in the *Meno* uses this kind of argument to prove that learning is nothing but the mind's remembering its own ideas. The explanation is that what remembers is the understanding. This part of the soul has its essence in these mathematical ideas and it has a prior knowledge of them, even when it is not using them; it possesses them all in an essential, though latent, fashion and brings each of them to light when it is set free of the hindrances that arise from sensation ... Consequently when we remove these hindrances we are able to know by understanding itself the ideas that it has, and then we become knowers in actuality, that is, producers of genuine knowledge ... This, then, is what μάθησις is, recollection of the eternal ideas in the soul; and this is why the study that especially brings us the recollection of these ideas is called the science concerned with μαθηματική. Its name thus makes clear what sort of function this science performs. It arouses our innate knowledge, awakens our intellect, purges our understanding, brings to light the concepts that belong essentially to us, takes away the forgetfulness and ignorance that we have from birth, sets us free from the bonds of unreason.[145]

Descartes transforms this idea in his conception of innate ideas. Finally, Descartes's concept of *mathesis universalis* is metaphysical, concerning not only quantities – which are usually considered by mathematics – but also any kind of objects. For this reason, it was a universal algebra and not only a mathematical science. In this sense, Descartes's interpretation of *mathesis universalis* and of Proclus is similar to that of Barozzi and Zabarella. This idea takes its final shape in Descartes's letter to the Abbé Picot in the *Principles of Philosophy*. Here, Descartes suggests starting with the principles of mathematics, which are easer and simpler, but from there it is necessary to ascend, as he did, to the first stage of philosophy, that is metaphysics, 'which contains the principles of knowledge',[146] the roots of the tree of philosophy on which all the other disciplines depend.

Descartes's epistemological ideas on mathematics, universal algebra and on the conception of analysis and synthesis had a profound impact on fellows of the Royal Society such as John Wallis, Isaac Barrow and Robert Hooke, who had not had the opportunity to encounter the *Rules*, which were to be published only at the beginning of the eighteenth century. Some of them incorporated Descartes's doctrines, others revised and refuted Cartesian positions in favour of a new experimental philosophy, setting the tone for subsequent advancements in epistemology, which would come to characterize what has been called the 'inquisitive age'.[147]

7

Isaac Barrow

Rome. 1547. In his *Commentary on the Certainty of Mathematics* (1547),[1] Alessandro Piccolomini questioned not only the validity of the negotiation of the intellect introduced by the Averroists but also the very essence of the Averroistic idea that mathematical demonstration could aspire to be the most perfect kind of demonstration (*potissima*) in terms of certainty. Piccolomini presents two lucid arguments against Averroes's thesis, and both are based on two essential elements of the demonstration *potissima*, namely, that the middle term is (1) the definition of either the subject or the attribute and (2) the immediate cause of the attribute.[2] In mathematics, neither of these conditions is satisfied. Regarding the first condition, in fact Proposition 32 of the first book of Euclid shows that the external angle, used as the means of demonstrating that it is the attribute of a triangle to have the sum of the three interior angles equal two right angles, is 'the definition neither of the triangle itself, nor of the attribute'.[3] Indeed, Piccolomini adds, it is evident that the external angle is neither necessary nor contributes to the definition of a triangle because, without it, the triangle would still remain a triangle. Concerning the second condition, it is clear that mathematicians demonstrate the attributes of a subject assuming different middle terms, and therefore a specific middle term cannot be the immediate cause. Thus, Theon of Alexandria, Proclus and Campanus of Novara would have demonstrated Proposition 17 of the first book of Euclid each in different ways.[4]

Yet even if none of these conditions are present in a mathematical demonstration, this does not mean that mathematical demonstrations are any less certain than the most scientific demonstrations: they simply have a different kind of certainty, which is not scientific in the Aristotelian sense.

According to Piccolomini, mathematical objects are abstractions of the imagination, a 'quantity' of the imagination (*quantum phantasiatum*). This quantity is conjoined with matter, but not in the same way as an accident with the form. Indeed, quantity is not an accident of a form or substance for the reason that it can be separately conceivable in the understanding on its own terms. If something is conceivable on its own terms, it cannot be an accident.[5] Quantity is a true property of a material substance, and it has the power to receive any kind of determination that form gives to matter. It represents the most manifest and evident property of substances, so much so that it is easy to grasp that

> mathematicians can without falsehood and any deception separate with the understanding quantities as their subjects from material substances and on them make demonstration without reference to any substance that is in the world.[6]

The certainty of mathematics comes from these abstractions – that is, from the simplicity of the quantity taken as being the most common and universal attribute of matter. In this sense, for Piccolomini mathematical knowledge is more certain than knowledge of natural phenomena. It is not, however, scientific knowledge because it does not deal with causes and substances, but with abstract things.

This sceptical conclusion about mathematics was shared also by Benet Perera, but for different reasons, in his *On the Principles and Properties Common to All Natural Things* (1576). The quantities are abstracted from material substances and therefore cannot deal with any cause, neither material nor efficient or formal.[7] But if for Aristotle science is knowledge through causes, then, for Perera, 'mathematical disciplines are not properly speaking sciences'.[8] Furthermore, supporting the theory of mathematical objects as abstractions, he regards knowledge of mathematical objects as not proceeding from what is first in nature, but from what is first known by the mind. And for the same reason, mathematical demonstrations cannot be demonstrations *potissimae*.[9]

On the opposite side of the debate, we find Pietro Catena, a firm supporter of the certainty and scientific nature of mathematics.[10] In his *All Places* (1556), Catena denies that mathematical objects are abstracted conceptions from sensation; rather, they are purely intellectual phenomena, whose definitions, however, do not imply their existence. Indeed, many non-existent things are definable. For this reason, this is necessarily a mental act of consideration, through which whenever there is knowledge that a subject has an attribute, then this subject must exist, even if only in mental form. Moreover, scientific knowledge of something non-existent is impossible, and therefore if something is truly known, it follows that it must exist.[11]

This is important for Catena because it enables him to explain how mathematical objects exist even though lacking an empirical reality and a reference to the imagination, where, conversely, Piccolomini situated mathematics. If it were not so, and mathematics depended rather on sensation, then no mathematical demonstration would be able to deliver universal and necessary proof. In order to admit this eminently demonstrative, universal and necessary character of mathematics, Catena introduces a kind of reminiscence as a form of knowledge that supposes the pre-existence of mathematical universals. Reminiscence more usually characterizes Platonic philosophy, but Catena attributes it to Aristotle. If the universal and a priori character of mathematical objects was not granted, then no universal conclusions would be possible. Even when the demonstration starts from a particular, it must refer to a universal if the conclusion is to be scientific.

Catena shows this by revisiting the fifth chapter of the first book of the *Posterior Analytics*, in which Aristotle explains that mistakes are often made in demonstrating. The error lies in mistakenly assuming that an attribute belongs primitively, necessarily and universally to a subject. Aristotle takes as an example the instance whereby if there were no triangles other than the isosceles, having two right angles would seem to belong universally and primitively to it in being isosceles. The same, however, would be the case with equilateral or scalene triangles if they were considered separately. To solve this puzzle – Catena posits – Aristotle introduces a logical procedure in order to test whether an attribute belongs to its subject universally and primitively in an absolute way.

Aristotle's passage is complex and misleading because it takes as its example a physical object with a mathematical property. This object is a bronze isosceles. If the attribute of being made of bronze is removed but the property of having two right angles remains, then the attribute of being made of bronze does not belong universally and primitively to the essential nature of having two right angles. The next step is to remove the property of two sides of equal length. It is evident that, despite this removal, the property of having two right angles remains, and therefore the characteristic of being isosceles does not belong universally and primitively to the nature of the subject. By removing attributes in succession like this one reaches the point where it is possible to understand which attribute belongs universally and primitively to the nature of the subject, for the reason that once this is removed, the subject's distinctive nature is lost. This logical procedure – holds Catena – makes it possible to find out what is universal in the particular, and therefore to infer a universal conclusion.[12] Catena attributes this procedure to mathematical demonstrations, but indeed it is proper to every kind of demonstration if one wants to arrive at universal conclusions. This was particularly clear for the regressus theorists. Zabarella calls this process of removal a 'rule for discerning what is first or not for a subject', and says that it represents the only way of determining the reciprocity of the terms in a demonstration, this being fundamental in the intermediate stage of regressus in order to switch from demonstration *quia* to demonstration *propter quid*.[13]

All these reflections lead Catena to draw some interesting conclusions for the application of mathematics to the study of nature. The first is that the scientific character of a demonstration does not derive from the syllogistic reasoning for which the definitions are demonstrative middle terms that express formal causes. The second is that it is eventually possible to reduce particular physical objects to mathematical objects capable of giving reasons for natural phenomena.

Piccolomini's fiercest opponent is Francesco Barozzi. In his *Question on Certainty* (1559), Barozzi states that it is wrong to support the idea that mathematical demonstrations are not demonstrations *potissimae* simply because they are not proved through causes.[14] Indeed, if it is true that their argument does not proceed through efficient and final causes, it does in any case employ formal and material causes as the middle term of the conclusion.[15] Barozzi contests the validity of Piccolomini's two arguments, stating that using just one example to refute the scientific character of mathematical demonstration amounts to an over-hasty generalization.[16] The certainty and scientific nature of mathematics derive from its ideality. He is against the idea that mathematical objects are abstractions on the grounds that if they were so conceived their demonstration would remain uncertain in being based on sensation. They are purely intellectual objects. However, even if they have no material existence, this does not mean that mathematics cannot be used for providing an understanding of reality. Mathematical demonstrations can explicate reality without explaining causes. They offer ways of concluding conjectures, especially in astronomy.[17] But this application of mathematics to natural effects was merely another way of saving documenting phenomena.

The problem of certainty and the causal nature of mathematical demonstration in England was sparked within the Royal Society by John Wallis's criticism in his *Universal Discipline* (1657) of Martin Smiglecki's opinion that demonstration in mathematics could not reach the pinnacle of perfection usually attribute to it by mathematicians. Smiglecki was the author of *Logic* (1618) – one of the most popular textbooks in Oxford (1634, 1638, 1658) and largely based on Jesuit commentaries on Aristotle. Against Smiglecki – but without referring to the Renaissance background of the question – Wallis's aim was to defend mathematics as a science, and the capacity of mathematical demonstrations to achieve the highest category of perfection. Like Zabarella, Wallis tends to identify the demonstration *propter quid* with Aristotle's most perfect demonstration (κύριος ἀποδείξις, *potissima*), and to maintain that knowledge should aspire to the ideal certainty of mathematics. However, while for Zabarella every demonstration *propter quid* was the most perfect, and every mathematical demonstration was *propter quid* and – therefore – every mathematical demonstration was the most perfect, for Wallis only certain mathematical demonstrations can attain the highest perfection. This conclusion is rejected by Thomas Hobbes in his first dialogue of the *Examination and Emendation of Modern Mathematics* (1660). He maintains that one may properly speak of a demonstration only in the case of a *propter quid* argument, which leads to scientific knowledge. If every mathematical demonstration is a demonstration *propter quid*, then all mathematical demonstrations are able to reach the highest perfection and lead to scientific knowledge.[18]

*

The Lucasian professor of mathematics Isaac Barrow takes the discussion around the certainty and causality of mathematical demonstration to another level, thanks to his mastery of Aristotle and the most important ancient and modern commentators.[19] In comparison to Wallis – at least from what we may gather from his writings – Barrow shows an immense knowledge of the debate from the Renaissance up to his own days.[20] Barrow's genuine interest is not merely mathematical but concerns the very foundations of the possibility of scientific knowledge. He knew the dispute between Wallis and Hobbes intimately, having both works in his library.[21] But his knowledge extended well beyond these two works or others in his collection. Indeed, in his writings, Barrow flaunts his acquaintance with the texts of Jacopo Zabarella, Benet Perera, Giuseppe Biancani and Gerhard Johannes Vossius.

Isaac Barrow's main treatment of the topic is found in the *Lectiones mathematicae* (1683), probably given in 1664.[22] In Lecture 1, he points out that mathematics – deriving its name from μάθησις – is synonymous with 'discipline', following Proclus's commentary on Euclid. According to Aristotle's definition at the beginning of the first book of *Posterior Analytics*, its primary meaning is the attainment of knowledge of something previously unknown from something already known. This kind of knowledge for Barrow is mainly intellectual (μάθησις διανοητική), and it is characterized by its evidence, certitude and constancy.[23] However Barrow rejects any classification of remembrance (ἀνάμνησις) as a form of intellectual knowledge, as outlined by Proclus,[24] being loyal to the Aristotelian perspective that all knowledge starts from sensation. The subject of mathematics is therefore a mental abstraction.

At the conclusion of Lecture 2, having established the subject and division of mathematics, Barrow reflects on the existence of a universal discipline or μάθησις. He calls this universal discipline 'Algebra or Analytics':

> Perhaps some may wonder, that while I am endeavouring to make a perfect Enumeration of all Parts (at least the principal Parts) of Mathematics, I am wholly silent about that which is called Algebra or the Analytic Art. I answer, this was not done unadvisedly. Because indeed Analysis, understood as intimating something distinct from the Rules and Propositions of Geometry and Arithmetic, seems to belong no more to Mathematics than to Physics, Ethics, or any other Science. For this is only a Part or Species of Logic, or certain manner of using Reason in the Solution of Questions, and the Invention or Probation of Conclusions, which is often made use of in all other Sciences. Wherefore it is not a Part or Species of, but rather an Instrument subservient to Mathematics: No more is Synthesis, which is the manner of demonstrating Theorems in Contradistinction to Analysis. But of these we will treat more clearly hereafter, when we come purposely to discourse of the Mathematical Ways of Invention and Demonstration.[25]

This long passage is extremely significant because it shows how Barrow was influenced by Descartes's *Principles* in his mathematical lectures. There is a universal discipline which is prior to mathematics, physics and ethics, and on which all these disciplines are based. This universal discipline for Descartes was metaphysics, and we will see that Barrow concurs with this opinion in Lecture 7. However, in Lecture 2, this discipline is not called metaphysics as in the *Principles*, but algebra – as in Descartes's private discussion with Beeckman and in the *Rules*.[26] We will see that – following Descartes – another member of the Royal Society, Robert Hooke, likewise employs algebra to characterize his own epistemology, but in a very different way. Barrow establishes also an opposition between analysis and synthesis, and analysis seems to coincide with this conception of algebra. Barrow's idea of analysis and synthesis is close to that of the regressus theory. Indeed, analysis is conceived as invention, synthesis as demonstration. Both are understood as instruments of mathematics and other disciplines, rather than being parts of them, and they are conceived as a form of logic. In Lecture 4, Barrow makes it clear, however, that his discussion does not aim to tackle logical issues but to focus largely on 'the mathematical way of reasoning'.[27]

Barrow characterizes the mathematical method as a form of discursive reasoning that investigates the properties, affections and passions of its subject. This reasoning is called demonstration (ἀποδείξις) and has the characteristics of being the most rigorous and necessary form of argument.[28] However, Barrow – like Hobbes – makes the bold statement that demonstration pertains only to mathematical sciences.

What does he mean precisely? That only mathematics is capable of demonstration, or that other disciplines such as natural philosophy can aspire to a demonstrative certainty equal to mathematics? Barrow opts for the second option, but he is perfectly aware that maintaining this perspective means arguing against what Aristotle writes in the second book of the *Metaphysics* – that is, that natural philosophy cannot reach the same high level of rigour as mathematics.[29] Barrow is not alone, and he finds in Jacopo Zabarella a precious ally. In order to address this intrinsic defect that Aristotle attributes to natural philosophy, Barrow follows the path of 'that excellent Logician Jac. Zabarella', who 'diligently pursued Euclid's *Elements* over and over, that he might better understand and explain the Nature of Demonstration'.[30] Barrow ascribes to Zabarella the idea that demonstration is mathematical, and this explicit reference is absent in the cases of Smiglecki, Wallis and Hobbes. He does not opt for any of the other solutions available in his time for proving the scientific character of mathematics. Indeed, neither Catena nor Barozzi were able to transform mathematics into a powerful instrument for understanding reality, and both furthermore adopt a more Platonic approach towards mathematical objects, which Barrow rejects for their innatist character.

For Barrow mathematics uses demonstration more than any other discipline because the things it considers are clearly and distinctly perceived. This means that no analysis is needed in order to provide clear and distinct knowledge. In all the other disciplines, by contrast, knowledge of things coming from sensation is intricate, composite, obscure and confused.[31] Therefore these disciplines employ obscure and confused notions, while mathematics deals with clear and perspicuous concepts. Equally, mathematical reasoning is based on universal and necessary principles and axioms, which do not require proof or explanation because of their intrinsic self-evidentness and absolute truth. An example of these principles is 'the whole is greater than its part'. Natural philosophy, on the other hand, deals with common opinions and principles coming from induction, which are not always true or self-evident. They can be weakened by contrary instances, and most of the time they have limited validity. Examples of these principles are 'nature does nothing in vain', or 'all things aim at good'.[32]

Another important characteristic is that mathematics does not admit postulates or arbitrary hypotheses that cannot be verified, while in natural philosophy the principles derived from experience do not immediately 'explicate the causes of things', but rather they are arbitrary suppositions and hypotheses that fit the argument, and 'hard to be digested, and monstrously immodest'.[33] They do not provide a true knowledge but only an argument which is possible or in various degrees probable. This does not mean – according to Barrow – that they do not achieve the status of true principles, but only that at the beginning they do not have the same kind of certainty as mathematical principles. All of these characteristics of mathematical demonstration should be taken as the marks of a true demonstration. In demonstrations, mathematicians

> only take those Things into Consideration, of which they have clear and distinct Ideas, designing them by proper, adequate and invariable Names, and premising only a few Axioms which are most noted and certain to investigate their Affections and draw Conclusions from them, and agreeably laying down a very few Hypotheses, such as are in the highest Degree consonant to Reason and not to be denied by any one in his right mind. In like manner they assign Generations or Causes easy to be understood and readily admitted by all; they preserve a most accurate Order, every Proposition immediately following from what is supposed

and proved before, and reject all Things howsoever specious and probable which cannot be inferred and deduced after the same manner.[34]

Barrow identifies the mathematical demonstration with what Aristotelian traditions usually call demonstration *propter quid*, and attributes to it a causative power by stating that it is capable of assigning generations or causes (*generationes seu causas*), recalling a terminology very close to Hobbes. This kind of argument is the only one capable of producing scientific knowledge in the human mind.[35]

Barrow is aware that many philosophers do not agree with his idea that mathematical demonstration produces scientific knowledge and deals with causes. In Lecture 5, therefore, he tries to formulate objections against certain opinions. The first position that he tackles is that of those who are against not only the certitude and scientific nature of mathematics, but also the possibility of knowing something certain in general. These philosophers maintain that there cannot be a certain and demonstrative knowledge that proceeds from universal, necessary and true principles for the reason that these are generated by induction or sensation, both of which are fallible because they cannot consider the infinitude of cases that experience provides. In response to this sceptical position, Barrow counters with Aristotle's strategy for demonstrating the principle of contradiction. He points out that there must be something certain and impossible not to assent to, or else the position against the possibility of certainty would be equally untenable and nonsensical. This necessitates that there are at least some certainties on which demonstrations can base their own validity.

What is more interesting is Barrow's seeming indifference, for the moment, about the origin of these principles.[36] In Lecture 5, he states that 'we are not disputing about the manner of acquiring Knowledge, or the Origin of Certitude', which was the usual task of analysis, but rather his focus is on the existence of principles of this kind. He makes a parallel to elucidate his position. Some may be ignorant of how sensation perceives, yet it cannot be denied on this ground that there is something before their eyes. Equally, the certainty of these principles may be known in various ways or may not even be known at all, but this does not diminish their force.

Even if in Lecture 5, the origin of the certainty of mathematical principles is not yet in question, Barrow has to confront the problem of the source of mathematical objects' certainty. He excludes the idea that the certainty of mathematics depends on innate ideas (*connate notionibus*), as Platonists like Barozzi believed, because otherwise the certainty would be immediately evident,[37] but it does not seem so. Its origin, therefore, comes from experience, and while much of the time sensation leads to a certain knowledge of the things, sometimes it is deceitful and produces false judgement. The reason for this, however, is not imputed to the work (*negotio*) of sensation itself (*sensu*), but to the faculty of reason (*rationis*).[38] Barrow consciously avoids lengthy philosophical discussion of the topic and closes the matter by stating that 'the Mind of Man, if rightly disposed (and not out of Tune, as Ideots and Mad People), is by its native Faculty able to intuit universal Propositions'.[39] Principles and ideas are themselves not innate, but only the faculty of grasping them. This kind of intuition is characterized in a very Aristotelian fashion, re-proposing the conceptual couple, νοεῖν-νοῦς:

> Contradictory Propositions cannot be both true, What begins to exist has its Rise from another ... and such like Propositions, which the Mind directly contemplates, and finds to be true by its native Force, without any previous Notion or applied Reasoning: Which Method of attaining Truth is by a peculiar Name stiled *Intellection*, and the Faculty of attaining it the *Intellect*.[40]

There is nothing that can prevent the principles of mathematical demonstration being known by this faculty. This, however, does not deal with the problem whereby, according to Aristotle, first principles come from induction, sensation and experience, from which no strict certainty is possible, only probable conjecture.[41] Barrow answers in Aristotelian manner:

> where any Proposition is found agreeable to constant Experience, especially where it seems not to be conversant about the Accident of Things, but pertains to their principal Properties and intimate Constitution, it will at least be most safe and prudent to yield a ready Assent to it.[42]

This was Zabarella's characterization of demonstrative induction, commonly employed by seventeenth-century Aristotelians and Jesuits. The validity of this argument is based on the fact that there is continuous agreement with 'our Expectations', which are 'accurately answered, after a thousand Researches'.[43] Many experiences, explorations and multiple experiments confirm the principles, and this is sufficient to consider them 'as universally true, and not suspect that Nature is inconstant'.[44] Barrow believes so much in this constancy of nature – as his pupil Newton will do likewise – that 'we may prudently infer a universal proposition even by one Experiment alone'.[45] If so, there is no reason to doubt the principles of mathematical demonstration.

Barrow adds that assent to the universality and certainty of mathematical principles does not rely only on sensation. First, because mathematical objects do not immediately occur in sensation and, second, because it is implausible that anyone would give assent to mathematics from sensorial observation.[46] Yet sensation can be extremely useful because sensation 'on sensible Observation alone may sufficiently attest that every Magnitude treated of by Mathematicians is capable of a real Existence ... one only Discovery of Sense will abundantly evince that all Mathematical Hypotheses are possible'.[47] This reflection on the 'real existence' of mathematical objects serves to reject the opinions of those who believe that mathematicians deal only with ideal and mental being, with no reference to reality.

Barrow's criticism is directed at Biancani and Vossius. He believes that geometrical figures are implicit in every particle of matter, although they are invisible to sensation. The argument supports a form of transduction from visible to invisible, and a peculiar kind of Platonism, which is evident in the example Barrow provides:

> Michelangelo, the most famous Carver, says that Sculpture was nothing else but a Purgation from Things superfluous. For take all that is superfluous from the Wood or Stone, and the rest will be the Figure you intend.[48]

Pushing his Platonism even further, Barrow states that 'whatsoever we perceive with any Sense is really a Mathematical Figure',[49] even if sensation is not sufficient to guarantee its certainty and in Lecture 14 (1665) he provides a justification for this bold statement. First of all, he posits that mathematics – and in particular the 'mother' and 'mistress' of all mathematical sciences, geometry[50] – does not concern only the quantities of magnitudes, but also their qualities.[51] By qualities of magnitudes he means the dispositions of the parts, with what shape they are endowed, and how they are situated. In providing this definition he clearly follows Proclus, who, according to Barrow, would have maintained that mathematics, and in particular geometry, was 'the Knowledge of Magnitudes and Figures, and their Limitations; Also of their Reasons, Affections, Positions, and Motions of every Kind'.[52] Barrow's extension of the reasoning to qualities, and so going beyond quantities is crucial for understanding how he conceives of mathematics as a descriptive tool for reality, and how nature's texture is composed of number, figure, weight and – more generally – measure. Measure is characterized as a 'stated pattern' (*aliquod statum*), 'commonly known and proved, exposed to the Sense, or comprehended by the understanding',[53] through which things are to be examined: everything can be reduced to measure. Indeed, 'since every Thing, every natural Agent, has a determinate Sphere of its Activity, it may as such some Way discharge the Office of a Measure'.[54] In so far as everything has relations with other things, the thing can be considered as a measure, and the knowledge of it is certain. The examples Barrow provides are various:

> Since Fire exerts a Power of Heating to a certain Distance and is perceived to effect nothing beyond that; since a Flower, or any odoriferous Body, disperses its Vapours of Smell to a certain Place; since a visible Object is beheld to a certain Distance, and disappears farther off; since the Sound of the Voice is perceived by the Ears placed within a definite Bound, beyond which it is insensible.[55]

All these sensations can be reduced to measures – in other words, to mathematical objects – with a view to their being perfectly understood and scientifically known. Indeed, according to Barrow a measure properly speaking is something assumed to

> render something more known and determinate to us than it was before, and is expounded with this Design that other Quantities coming into Consideration may, as Quantity, be compared with it, or with one another by means of it; viz. that it may be investigated how often this contains, or is contained in those, or what Proportion those obtain to these; and consequently how they are related one to another, so that their Quantities may be some Way known and determined, which were before unknown and indeterminate.[56]

This kind of knowledge is possible because it does not refer to the substance of the thing but to the relation that the thing occupies with the respect of the others. In order words, this epistemology privileges knowledge of relations over substance: relations have their own consistency.

In Lecture 5, Barrow adds also that if mathematical objects are considered as non-existent because they are exclusively mental, then all ideas in the mind would be nothing other than dreams or fancies.[57] Following these considerations, Barrow concludes that the truth of principles is acquired neither through a Platonic experience of immediate mental intuition, nor by mere sensation, or by innate notions; rather it

> is deduced by a some implicit and quasi virtual reasoning (*discursus*), of kin to an intuitive knowledge, either from the declared signification of the Terms themselves, or from the supposed Generation of the Thing, which is manifestly possible and agreeable to the Suffrage of the Sense.[58]

For Barrow, this kind of reasoning should follow the infallible rules of logic. One of these establishes that 'the Truth of most Axioms may be resolved into Definitions or Explications of the Terms or Words whereby they are expressed'.[59] This is an important remark that aligns Barrow with both Aristotelians and Hobbes – who by no coincidence are mentioned in this passage – in emphasizing the importance of definition for the principles that constitute the premises of the demonstration. According to Barrow, there is a clear difference between mathematicians and natural philosophers: mathematicians do not consider definitions, for the simple reason that they are already self-evident. This means that mathematics does not require resolution or analysis, which remains indispensable in the case of natural philosophy.

The resolution of terms is not the only way of determining axioms, since reasoning directed at the supposed causes of a thing can achieve the same result. Once again, the implicit reference is Hobbes. Indeed, Barrow employs the same example of the generation of a circle:

> As suppose a Circle to arise from the drawing about a Right Line, one of its Extremes remaining fixed which is stiled the Center; hence is immediately inferred that Axiom with scarce any Axiom at all of the mind *That all Right Lines drawn from the Center of a Circle to its circumference are equal.*[60]

The Lucasian professor has also to face a further important issue about mathematical demonstration – concerning neither certainty nor evidence, but rather its dignity or excellence, which had been previously questioned by authors such as Piccolomini. His polemical target is here, however, Benet Perera, and his attempt to prove that mathematical demonstration is not causal. Barrow counters that the conditions that make demonstration capable of scientific knowledge accord with mathematical reasoning. Indeed, the premises are true and necessary, primary and immediate, and more known and more evident than the conclusions. The most important condition, however, is that premises should be the causes of their conclusions, and Barrow sees this as possible in two distinct ways. The first is when the premises 'contain the Reason which necessarily causes the Conclusions to be believed as true, and produces a certain Assent'.[61] This is valid in the case of demonstration τοῦ ὅτι – that is, where the middle term 'obtains a necessary connection with the Terms entering the Conclusion'.[62]

The second is a situation where the middle term is 'a proper Cause of the Attribute or Property, which is predicated of the Subject in the Conclusion'.[63] This generates, properly speaking, a causal demonstration or a demonstration τοῦ διότι. This second way is that of mathematical demonstration, and therefore Barrow conceives it to be demonstration τοῦ διότι or *propter quid* in the Aristotelian sense. The main reason for maintaining that mathematical demonstrations belong to the category of the most perfect form of argument is that

> they only fetch their Conclusions from Axioms which exhibit the principal and most universal Affections of all Quantities, and from Definitions which declare the constitutive Generations and essential Passions of particular Magnitudes. From whence the Propositions that arise from such Principles supposed, must needs flow from the intimate Essences and Causes of the Things.[64]

Lecture 6 is entirely devoted to explaining this causal nature of mathematical demonstration. In order to understand the causality of a demonstration, and in what sense mathematical reasonings are properly speaking demonstrative, Barrow follows Aristotle in considering three aspects: (1) the subject, (2) the axioms and (3) the affections or attributes, which are to be demonstrated of the subject. There are two kinds of attributes: common and proper. Common attributes inhere necessarily but not exclusively in the subject – in other words, they do not come from the nature or essence of the subject. They are not helpful in characterizing the definition of the subject and, more particularly, the subject does not follow from them. In contrast, proper attributes inhere necessarily and exclusively in the subject, so much so in fact that

> the Subject is determined and limited by any of these Affections, and consequently that any of these may be rightly supposed or assumed in defining the Subject, since they are connected together with such an essential, close and reciprocal Tie, that if any one be supposed, the rest necessarily follow.[65]

These proper attributes were those properties that Aristotelians had been looking for with the negotiation of the intellect, with a view to establishing a convertibility in natural philosophy comparable with that of mathematics, where the reciprocity was obvious in being based on definitions. This was the logical procedure that Catena had established as being fundamental for determining whether an attribute belongs primitively and universally to a subject.

According to Barrow – as with the Aristotelians – only by working with proper attributes is it possible to establish real causality in the demonstration: this is a form of mutual dependence that characterizes an intimate connection between subject and attributes. This intimate connection (*intima connexio*) is called 'formal causality (*causalitas formalis*)', representing the reciprocal relation between the attributes and the subject.[66] Barrow seems to inscribe his epistemology within the Aristotelian tradition, yet he denies the possibility of involving in this demonstrative reasoning other forms of external causes, whether efficient or final. His explanation is on the long side but it deserves quoting in full because it questions the epistemological validity of the negotiation of the intellect, and in broad terms the attempt that Zabarella and Hobbes make to reduce efficient cause to formal:

> There can be no such [intimate] Connection of an external, e.g. efficient Cause with its Effect, (at least none such can be understood by us) through which, strictly speaking, the Effect is necessarily supposed by the Supposition of the efficient Cause; or any determinate Cause by the Supposition of the Effect. Nay there can be no efficient Cause in the Nature of Things of a philosophical Consideration which is altogether necessary. For every Action of an efficient Cause, as well as its consequent Effect, depends upon the Free-Will and Power of Almighty God, who can hinder the Influx and Efficacy of any Cause at his Pleasure; neither is there any Effect so confined to one Cause, but it may be produced by perhaps innumerable others. Hence it is possible that there may be such a Cause without a subsequent Effect, or such an Effect and no peculiar Cause to afford any Thing to its existence. There can therefore be no Argumentation from efficient Cause to the Effect, or contrarily from Effect to the Cause, which is lawfully necessary.[67]

Unlike Aristotelians like Zabarella – though following Aristotle's original words – Barrow claims it to be impossible to establish a convertibility or mutual reciprocity between subject and attributes in physics where these are characterized by an effect and an efficient cause. The reason given is that this form of naturalism would prevent God's omnipotence,[68] which has the potential to create the same effect by means of an entirely different cause. In any case, like Zabarella, Barrow reduces the relation from efficient cause and effect to formal causality, but here it is no longer legitimate to speak of external causality:

> E.g. because there is Fire it does not necessarily follow that there is Fewel for it to feed on, or Smoak sent from it; since History relates that, in Fact, it has happened otherwise. Neither, on the contrary, is the necessary Existence of Fire inferred from Ashes or Smoak. For who doubts but God can immediately create Ashes and Smoak or produce it by other Means? Perhaps you will say that Ashes and Smoak denote an intrinsecal Relation of Fire. I answer briefly, this supposes the Argumentation not to proceed from the efficient Cause or the Effect, but from the formal Cause or Causate.[69]

Barrow uses Zabarella's example to argue for the necessity of a mental examination with the express purpose of reducing efficient causation to a kind of formal cause. Unlike Zabarella, there is no trace of the intermediate stage in Barrow, and this follows from the fact that he is working with mathematical demonstrations in which premises are certain. In Lecture 6, he examines another 'most celebrated and trite' example that has been pivotal in the regressus theory, the lunar eclipse. Thus, from the Earth's interposition between the moon and the sun, it does not follow that there is necessarily a lunar eclipse; indeed, 'for, if God please, the solar Rays may pass through the Body of the Earth, or reach the Moon by an indirect Passage without touching the Earth'.[70] By introducing God's omnipotence into natural philosophy, Barrow diminishes the possibility of an efficient cause having a real necessity in nature; only a formal cause is 'necessary'. For this reason, only mathematical demonstrations can truly be said to be causal, and are deemed to be so. This does not mean that there are no laws of nature, but rather that these are not necessary. Concerning lunar eclipse, he states that

> a Defect of Light then cannot be concluded from the Interposition of an opaque Body, nor this from that. I own according to the Law and Custom of Nature that such Effects do always proceed from such Causes; and therefore we may pardon Aristotle for supposing Nature subject to an intrinsecal Necessity, and not obnoxious to an external superior Power; but in reality it is one thing to happen naturally, and another to exist of Necessity.[71]

Barrow establishes a clear distinction between a logical necessity and natural causation: nature is subject to rule, but sometimes this fails. The effectiveness of agents may change or be frustrated, or the same effect may be caused not only by nature but also by God. No necessary reasoning – and therefore no demonstration – is possible from an efficient cause or from effects. No other conclusion is to be had but that necessary demonstration is based on definition, according to which a thing is such and therefore cannot be otherwise. Demonstration is a 'Syllogism in which one essential Affection is connected with, or inferred from another', and it can be deduced only from a formal cause.[72] Mathematical demonstrations are, according to Barrow, of the most perfect and highest form of demonstration because based on principles and axioms, which are definitions containing essential attributes: they are so much more than any alleged reasoning from any efficient cause or effect. Barrow thus reverses Piccolomini's position. In this way, however, only mathematics, or at least arguments that are like mathematical demonstrations, can provide scientific knowledge.

Lecture 7 considers the principles and axioms themselves, and how they constitute the premises of a demonstration that leads to scientific knowledge. They are such – Barrow claims – because they do not belong to something other, or because they receive immediate assent without requiring any reasoning.[73] But the real issue is the origin of the truth of these principles. Barrow states that the principles of a science cannot be demonstrated from within that science or another particular science, or else they would not be true first principles. The truth of principles belongs to a higher and more universal science, and this is metaphysics:

> But some will ask, How can the Truth of Geometrical Axioms be demonstrated? I will answer, by other Axioms more simple, if any such there are, which are to be drawn from some higher and more universal science, as *Metaphysics*; I say *Metaphysics* which is, or ought to be, the Treasure of the most general and simple Notions; and is therefore by Aristotle named the Mistress of all Sciences.[74]

Metaphysics is that science capable of demonstrating rigorously the proper principles of particular sciences. However, these particular sciences are not subordinate to metaphysics. He explicitly attributes this conception to Aristotle, and in particular to the passage in *Posterior Analytics* 76 a 16–25, which is quoted extensively:

> If this is evident, it is evident too that one cannot demonstrate the proper principles of anything; for those will be principles of everything, and understanding of them will be sovereign over everything. For you understand better if you know from the higher explanations; for you know from what is prior when you know from unexplainable explanations. Hence if you know better and best, that understanding too will be better and best. But demonstration does not apply to another genus – except, as has been said, geometrical demonstrations apply to mechanical or optical demonstrations, and arithmetical to harmonical.

But as is evident from Aristotle's text, there is no mention of metaphysics as a universal science capable of demonstrating principles. However, for Barrow this idea was quintessentially Aristotelian, and indeed to support his interpretation he brings in another passage of *Physics* 184 b 26–185 a 5, that in which Galileo believed himself to have found the origin of principles in metaphysics. But we have seen already that Aristotle is not in favour of this conception of metaphysics, so why is it that Barrow is so comfortable with this conception and in attributing it to Aristotle?

A precious clue is given by Barrow himself in quoting Proclus's *Commentary on the First Book of Euclid's Elements*, where the author nonetheless seems to ignore the existence of a universal science capable of demonstrating the principles of others. He simply claims that 'no science demonstrates its own first principles or presents a reason for them; rather, each holds them as self-evident, that is, as more evident than their consequences'.[75] Looking at the textual tradition and the various interpretations of this passage, however, the situation seems rather different. Alessandro Piccolomini, in his *Commentary on the Certainty of Mathematics*, writes that there is a 'discipline common (*facultatis communis*) to geometry and arithmetic', and he bases his opinion on Proclus, who shows

> in a very evident way … in the first and second book [of his Commentary] that there is a common science to those two [geometry and arithmetic], which claim their own subjects attributes and principles, and which are subordinate to it.[76]

Following the Proclian text, according to Piccolomini there is a science common to the others, capable of its own demonstration and having its own subject, attributes and principles. This common and universal science has a relation of subordination to the other two sciences, but he does not qualify this, and so it is not clear whether this common science or faculty is mathematical or of a different nature. If the relation of subordination is a type of dependence and subservience, then this common science is likely to be a higher form of mathematics, though this is not explicitly stated in the text.[77] However, Barrow did not know Piccolomini – or at least we have no clue to suggest this. Nonetheless, Barrow had in his own library Francesco Barozzi's edition of Proclus's *Commentary on the First Book of Euclid's Elements* (1560), and here, he might have found the idea that all principles derive from a 'divine science (*divina scientia*)' or 'first philosophy (*prima philosophia*)' – 'which Plato calls dialectic in the seventh book of the *Republic*' and 'which Aristotle defines as the mistress of all science in the first book of *Posterior Analytics*'.[78]

The Aristotelian passage in which metaphysics is considered the mistress of all sciences is exactly that quoted by Barrow, yet he drops any reference to Platonic dialectics, opting instead for metaphysics as the only universal discipline. This is a bold solution because – for instance – Petrus Ramus argued against metaphysics as a universal science, and in favour of dialectic in a Platonic fashion.[79] In choosing metaphysics as a universal science, Barrow had perhaps in mind Descartes, though this would not explain the reference to Aristotle. Attributing this conception to Aristotle, Barrow was likely following Zabarella's *Commentary to Posterior Analytics*: Zabarella mentions exactly the Aristotelian passages quoted by Barrow in his seventh lecture. In following Zabarella with this particular argument, Barrow advances the idea of a metaphysical foundation for mathematical principles mathematically demonstrated,[80] an idea that was shared also by Descartes.

This metaphysical foundation for mathematics substantiates the fact – in Barrow's view – that nature is written in numbers and geometrical figures, exactly as Galileo believed, and this allows mathematics to become a powerful descriptive tool for reality, whose description is not merely mathematical hypothesis, but a real and genuine explication of the world – that is, of how things are.

8

Robert Hooke

London. 1667. In reconstructing the early history and mission of The Royal Society of London for Improving Knowledge, Thomas Sprat outlined what for him was the correct method for investigating nature, in opposition to past epistemologies:

> True philosophy must first of all be begun on a scrupulous, and severe examination of particulars: from them, there may be some general Rules, with great caution drawn: But it must not rest there, nor is that the most difficult part of the course: It must advance those Principles, to the finding out of new effects, through all the varieties of Matter: and so both the courses must proceed orderly together, from experimenting to Demonstrating, and from demonstrating to Experimenting again.[1]

He sketches a twofold process. The first was based on experience and experiments, and proceeds towards general rules and axioms. The second used these rules and axioms as principles to determine new effects. This process then should be repeated again and again to be sure of the validity of the scientific explanation. The epistemology that Sprat presents as typical of the fellows of the Royal Society has many similarities and overlaps with Aristotelian regressus.[2] However, in contrast to the Aristotelians, who in Sprat's view had developed a very specific kind of dogmatic philosophy, the new experimental philosophy was based in its methodology for scientific discovery on Francis Bacon's works, which allowed for a more accurate inquiry into natural phenomena.[3]

Sprat's general outline of the Royal Society is somewhat simplistic in picturing the adventures of the institution and in characterizing the nature of the true philosophy which would lead to new scientific discoveries. Several insightful studies have shown how the life in the Royal Society was far from being peaceful.[4] Internal rivalries, underlying conflicts and open controversies were common among the fellows. The Royal Society was united neither in its intent nor in its methodology,[5] even if a generic Baconianism seems to have emerged from the background. Many concurrent epistemological approaches were endorsed by its fellows, and even within a single personality it is possible to trace methodological developments, changes and evolutions. A clue to understanding Sprat's idea of scientific method can be found in Robert Hooke, whose epistemology has been carefully explored by Mary B. Hesse, David R. Oldroy and Michael Hunter.[6] One of the earliest accounts of his approach is a short booklet titled *An Attempt for the Explication of the Phaenomena* (1661), in which he responds to Experiment 35 of Robert Boyle's *New Experiments Physico-Mechanical, Touching the Spring of the Air and Its Effects* (1660). Experiment 35 concerned the behaviour of water or other liquids in small pipes, which gave rise to many conjectures, none of which was particularly satisfying according to Boyle. Hooke's attempt in this writing is to explain the phenomena of nature which are extremely evident in themselves, but on which, for him, no diligent inquiry has been carried out, but only observations with a careless method or none at all.

Hooke starts with a conjecture 'that the unequal height of the surfaces of the water proceeded from the greater pressure made upon the water by the Air without the Pipes A.B.C. than by that within them'.[7]

This conjecture is based on two propositions: (1) 'an unequal pressure of the incumbent Air will cause an unequal height in the waters surfaces'; (2) 'in this Experiment there is such an unequal pressure'.[8] These two propositions are based on experiments which lead to a general conclusion that there is incongruity or inconformity between two different fluids, or between one fluid and one solid. Hooke explains that

> from what cause this congruity or Incongruity of bodies one to another does proceed, whether from the Figure of their constituent Particles, or interspersed pores, or from the differing motions of the parts of the one and the other, as whether circular, undulating, progressive, etc. whether I say from one, or more, or none of these enumerated causes, I shall not here determine; It being an enquiry more proper to be followed and explained among the general Principles of Philosophy, whither at present I shall refer it; as fearing lest it might here seem absurd, without the concatenation of several other Principles to explicate; and knowing it likewise sufficient for this enquiry to shew, that there is such a property, from what cause soever it proceeds.[9]

Experiments lead to a certain general conclusion about a property, the cause of which – Hooke maintains – is not in itself important, at least not in terms of the effectiveness of the property in producing specific effects. Knowledge of the cause remains obscure. This scepticism concerning the cause and this argument directly recalls Newton's conception of gravity, which we will encounter more fully in Chapter 12. From the discovery of this property, one may deduce other effects on other fluids, testifying to the validity of the conjecture but in a more precise manner. Indeed

> from these, and other such trials, it will be found very evident, that to force the air or other fluid into a smaller hole, of an incongruous Solid, there is required a greater force, than there is requisite for the protrusion of it into a bigger. Come we therefore to the explanation of the Phaenomena in the experiment of the rising of water in a smaller Pipe ... That this unequal pressure of the Air caused by its protrusion into unequal holes, is a cause sufficient to produce this effect, without the help of any other concurrent, and therefore is the principal, (if not the only) cause of this Phaenomenon.[10]

According to Hooke, the result of this reasoning is a principle of philosophy. In the 1661 edition of this work, this is not explicit. Indeed, about the conclusion he writes

> I know not also whether it will be requisite to add what things occurr'd to my thoughts, when I had these Phaenomena under consideration; which I shall only propound as Queries, that may serve as hints to some further discovery, and not as Axioms.[11]

Among these hints, we can find the explanation of the phenomenon of gravity as a kind of fluid in which the Earth may be conceived to be immersed. What is important, however, is that this writing will be included in the first major publication under the auspices of the Royal Society, that is, the *Micrographia* (1665). In the 1665 edition of *An Attempt*, the sentence is somewhat different:

> And indeed upon the consideration of this Principle, multitudes of other uses of it occurr'd to me, which I have not yet so well examined and digested as to propound for *Axioms*, but only as *Queries* and *Conjectures* which may serve as *hints* toward some further *discoveries*.[12]

Hooke's epistemology can thus be described as starting from a conjecture, based on previously made observations which one wants to explain, and the requirement is then to test this conjecture through experiments and trials. These will lead to a general conclusion, which can be considered effective if it fits the explanation of not only that particular phenomenon but also other effects. This conclusion, once proven, becomes a principle of philosophy, from which one may argue other conjectures.

In the *Micrographia*, this method is even further elaborated. Starting from an eclectic epistemology mainly based on sensation, memory and reason,[13] Hooke proposes to correct the defects of the senses, the delusion of memory and the rashness of the understanding through a new, real, mechanical, experimental philosophy.[14] This experimental philosophy is founded on a strict examination of the particulars, in which it is not the number of experiences that matters, but rather their weight in establishing deductions and conclusions. This examination is possible through a 'sincere Hand and a faithful Eye'.[15] But these two organs can be assisted by artificial instruments like telescopes and microscopes through which

> the subtilty of the composition of Bodies, the structure of their parts, the various textures of their matter, the instruments and manner of their inward motions and all the other possible appearances of things, may come to be more fully discovered; all which the antient Peripateicks were content to comprehend in two general and useless words of Matter and Form.[16]

Hooke's aim is to amend Aristotelian philosophy and a kind of rationalism of the 'Science of Nature', which was made only by 'a work of the Brain and the Fancy'.[17] He avoids 'any infallible Deduction or certainty of Axioms', and proceeds rather from 'small Conjectures', understood as 'doubtful Problems, and uncertain ghess, and not as unquestionable Conclusions, or matters of inconfutable Science'.[18] However, Hooke does not disdain what he calls the 'superstructure' of the epistemology – that is, the elaboration of scientific theories. His idea is to compile natural and artificial histories for registering particulars 'into Philosophical Tables' from which might be raised 'Axioms and Theories'.[19] The method of 'true philosophy' – the same expression used by Sprat – is

> to begin with Hands and Eyes, and to proceed on through the Memory to be continued by the Reason; nor is it to stop there, but to come about to the Hands and Eyes again, and so, by a continual passage round from Faculty to another.[20]

The difference with the Aristotelian natural philosophy is the conception of a richer type of natural history compiled through an accurate and precise collection of particulars, and the search for 'Schematisms and Textures of Bodies',[21] 'for Motion, Figure and Magnitude',[22] as opposed to matter, form and occult qualities. Hooke's epistemology presented in the *Preface* of the *Micrographia* is that briefly summarized by Sprat in his reconstruction.

*

In the same period, Hooke developed what became famous with the name of *A General Scheme, or Idea of the Present State of Natural Philosophy*, which was published posthumously in 1705. In this text, Hooke makes clear that the main business of experimental philosophy is 'to find out a perfect Knowledge of the Nature and Proprieties of Bodies, and of the Causes of Natural Productions'.[23] The definition reveals an important feature of Hooke's epistemology. First of all, Hooke deals with bodies – that is, with corporeal things. He aims to acquire a perfect knowledge, which means that immediate cognition is not sufficient to grasp the subject of science. The subject of science consists in the nature, attributes and causes of a body, for which in a very Baconian sense he means 'the inward Texture and Constitution' of bodies, 'the Internal Motions, Powers and Energies', and 'how they may be made use of for producing such Changes and Transformations of Bodies from one thing to another'.[24] Hooke's conception is Baconian also in designating the purpose of knowledge

> in order to inabling a Man to understand how by the joyning of fit Agents to Patients according to the Orders, Laws, Times and Methods of Nature, he may be able to produce and bring to pass such Effects, as may very much conduce to his well being in this World.[25]

Knowledge is not for its own sake, but also for the purpose of intervening in nature. The starting point of knowledge for Hooke is experience, and in particular the examination of natural bodies. This examination can be carried out 'by Dissections, Experiments, or Mechanical Tryals',[26] which are the main instruments in writing a reliable natural history. In the later work, *A Discourse of Earthquakes*, written around twenty years later,[27] Hooke writes that mere observations are in large part so superficial, and the Descriptions so ambiguous, that they create a very imperfect Idea of the true Nature and Characteristick of thing described, and such as will be but of very little use without an ocular Inspection and a manual handling, and other sensible examinations of the very things themselves; for there are so many considerable Instances that may by that means be taken notice of ... The use of such a Collection is not for Divertisement, and Wonder and Gazing ... but for the most serious and diligent study of the most able Proficient in Natural Philosophy.[28]

Hooke defends the methodology of the Royal Society regarding the knowledge of nature coming about by means of experiments and trials, as opposed to the naïve naturalism of historians and antiquarians like Conrad Gessner and Ulisse Aldrovandi.[29] Without this new approach to natural history, it would be impossible to find out the material of science. Hooke is extremely critical of the Aristotelians, who claim that their own epistemology is based on sensation and experience, while at the same time proceeding otherwise: moving from the collection of some particulars towards imperfect, generic deductions, which are then used as laws of nature. Furthermore, they adapt observations and experiments to their theories, because they rate their own minds as being the source of all knowledge capable of producing a true image of the world, whereas they should rather be regulating their own thoughts to experience. Moreover, their greatest error lies in their reverence for the *ipse dixit* of the authorities, their repugnance for new discoveries and their fallacious use of syllogism.

Hooke's attack becomes more circumstantial in examining previous experimental philosophy. Though Aristotelians have found some errors and set up new trials, their method of proceeding from observations and experiments has led only to confused conclusions – so much so that the more experiences they collected, and more confused they became. Indeed, for Hooke it is neither the number of experiments nor the subtlety of arguments that transform experience into scientific knowledge: their mistaken way of proceeding produces at best 'new Probabilities and consequently augment[s] Disputes'.[30] The Aristotelian method can never lead to epistemic knowledge. Indeed, the 'Examination and Comparison of so great a Number of Particulars is requisite', but it is not sufficient.[31] Extended activity of the mind is necessary. Hooke states clearly that he does not reject logic, but he is sceptical of a method by which a man

> should be able to leap, from a few particular Informations of his senses, and those very superficial at best, and for the most part fallacious, to the general knowledge of Universals or abstracted Natures, and thence be able, as out of in inexhaustible Fountain, to draw out a perfect Knowledge of all Particulars, to deduce the Causes of all Effects and Actions from this or that Axiome or Sentence, and as it were intuitively, to know what Nature does or is capable of effecting.[32]

Hooke criticizes the foundations of Aristotelian regressus, in particular the first part, which discovers causes from the effects in one of two ways: either (1) on the basis of superficial and fallacious sensation or (2) through starting from uncertain premises of the demonstration. In order to amend Aristotelian epistemology, Hooke promotes a new 'Method or Engine', which assists the intellect in the art of inquiry or discovery.

According to Hooke, no one – 'except the incomparable Verulam' – has come close to the idea of this method. Hooke is referring to Bacon's literate experience and theory of induction, which, as we have seen, assist the mind in determining the material provided by natural history. However, the Baconian method is not sufficient in itself to create the conditions for scientific knowledge to complete this 'Engine'.[33] Only by supplementing induction with the art of algebra does it become possible to proceed with certainty in natural investigations. This algebra is not merely mathematical in Hooke's eyes – as it was not for Descartes either; rather, it is

> a Philosophical Algebra, or an Art of directing the Mind in the search after Philosophical Truths ... I cannot doubt but that if this Art be well prosecuted and made use of, an ordinary Capacity with Industry, will be able to do very much more than has yet been done, and to shew that even Physical and Natural Enquiries as well as Mathematical and Geometrical, will be capable also of Demonstration.[34]

Similarly, Descartes in the *Principles* deals with a kind of logic as the 'art of directing the mind with the view to discovering the truths',[35] which starts from mathematics and proceeds to physics. Hooke reinforces his position by stating that this algebra does not furnish the mind only with numbers, weights and measures for investigating and proving all things, 'but also instructs and accustoms the Mind to a more strict way of Reasoning, to a more nice and exact way of examining, and to a much more accurate way of inquiring into the Nature of things'.[36]

In his manuscript titled *First Algebra Lecture*, dated 1665 – that is, the same year as the publication of the *Micrographia* and a year after Barrow's first series of lectures on mathematics – Hooke explores the idea of elaborating an algebra. He starts to conceive it from mathematics, as a means of extending its dominion over all possible objects. Mathematical sciences are the most certain on account of their subject being something 'abstracted from matter', and by reason of their conclusions which are more universal.[37] There are two kinds of mathematical sciences: geometry and arithmetic. Geometry 'is the knowledge of the properties and affections of continued quantities', and deals with the extension 'as it is considered undivided'.[38] Arithmetic, in contrast, 'is the knowledge of the properties and affections of discontinued quantity', and deals with the extension 'as it is considered divided'.[39] Their main business is to find out, respectively, in the case of the former 'the most abstruse properties' of continued quantities and, in the case of the latter, those of numbers.[40] For the purpose of discovering these properties, according to Hooke, there is only one demonstrative tool – that is, algebra. Algebra is therefore considered comparable to logic, without which no discovery or invention is possible. Indeed, nothing can be known with the 'braine, reason, or the very soule of man' without algebra. Algebra is 'an art, wherein we may see the very grounds and proceedings of ratiocination', through which 'we come to conclusions and axioms, and even to the very foundations of reason itself'.[41]

Algebra, therefore, discovers the axioms and principles of reasoning. It shows how to proceed from 'the most obvious and sensible object' to the highest form of ratiocination, on which 'axioms are built'.[42] Hooke emphasizes – unlike Descartes – that all the axioms are 'derived from the most plain and obvious information of the senses'.[43]

They are not 'innate in us or infused'; rather, they are acquired or deduced from 'a continued series of ratiocinations by comparing, compounding and separating'.[44] These were the three operations used by Aristotelians in the intermediate stage of regressus. Hooke adds also that all these operations of comparison, composition and separation are to be considered ways of 'examining and applying the most sensible properties of bodies'.[45] These examinations are a continuous test and trial between general axioms and particular properties in order to acquire the most evident and certain knowledge. This kind of knowledge will be as certain, obvious and plain as the sum of 'two plus two makes four'. Algebra shows us how to 'proceed to find out the most abstruse mysteries' in any subject whatsoever.[46] It is a process from known to unknown, and encompasses what the Aristotelians believe to be the first part of regressus, up to the establishment of the first principles.[47]

Algebra, in short, is method of reasoning, 'whereby we proceed from the most obvious and known proprieties of a subject, to the invention of the most abstruse enquiry'.[48] But Hooke makes clear that this algebra is not properly speaking part of the mathematical sciences, but rather an instrument applicable to all sciences. Indeed,

> this Algebra is not merely to be bound up to this one only subject of inquiry, viz to the invention of the proprieties of quantities, but can be made use of and does extend itself much further for the seeking and finding out the proprieties of divers other things.[49]

Hooke totally agrees with Barrow on the utility of algebra in every field of knowledge. If it is true that algebra starts from the mathematical sciences, it is not to be restricted only to their field of compass, for its 'plain and easy method' is suitable for any kind of inquiry. In particular, it is most useful if the subject is reduced to a kind of quantity.[50]

Hooke believes that the most perfect algebra is that capable of expressing with the 'fewest characters' possible all kinds of knowledge, and most especially that of the 'first grounds or principles, upon which the ratiocination … is built'.[51] The certainty of algebra is based on the simple and evident nature of every part of the reasoning, starting from the principles or axioms, which depend 'either upon some principle or deduction preceding, or upon some formerly demonstrated and evident axiom'.[52] The simplicity and clearness of the reasoning must be sufficient enough that the entire reasoning, 'the deduction itself and the reason for the deduction are evidently to be seen'.[53] In this manuscript, more than in any other of his works Hooke reveals that the best way of deducing is that advanced by François Viète – that is, his 'method of proceeding by species'.[54] This reference is of particular importance because it sheds light on Hooke's mature epistemology. So what was Viète's method?

In his *Introduction to the Analytic Art*, Viète explains that the way to search for and discover the truth in mathematics is through analysis. He does not, however, follow the common definition of analysis, preferring that advanced by Theon of Smyrna in his *On Mathematics Useful for the Understanding of Plato*. Thus, Viète writes that analysis means 'assuming that which is sought as if it were admitted [and working] through the consequences [of that assumption] to what is admittedly true'. In opposition to analysis there is synthesis, 'which is assuming what is [already] admitted [and working] through the consequences [of that assumption] to arrive at and to understand that which is sought'.⁵⁵ The analytic art then comprises three stages: (1) zetetics, (2) poristics and (3) rhetics or exegetics. Zetetics seems to be the part of analysis that arouses Hooke's chief interest. For Viète, zetetics is the method 'by which one sets up an equation or proportion between a term that is to be found and the given terms'.⁵⁶ The definition in itself is not important; rather, it is remarkable that

> whatever pertains to zetetics begins, in accordance with the art of logic, with syllogisms and enthymemes the premises of which are those fundamental symbols with which equations and proportions are established. These are derived from axioms and from theorems created by analysis itself.⁵⁷

Zetetics, therefore, is a kind of logic, although – Viète adds – it has 'its own method of proceeding',⁵⁸ and this is 'the method of proceeding by species' favoured by Hooke. This method consists in not limiting analysis to numbers, but in working 'with a newly discovered logistic by species (*per logisticem sub specie*), which is far more fruitful and powerful than numerical logistic for comparing magnitudes with one another'.⁵⁹ Species are 'quantities or Magnitudes, denoted by Letters, signifying Numbers, Leneas, Leneats, Figures Geometrical, &c'.⁶⁰

Hooke does not only consider Viète but also takes in various other positions on algebra, as such those of Thomas Harriot, William Oughtred, Lawerence Rook, René Descartes, Pierre Hérigone and John Pell, for each of which he provides a very succinct examination. In the rest of the manuscript, he explains the importance of choosing the right characters for denoting quantities, operations and effects. In another manuscript titled *Mathematicall Language*, he tries to develop his own algebra on the basis of the aforementioned algebrists and to make this compatible with Aristotelian logic.⁶¹ As far as we know, Hooke never fully elaborated his own system of algebra, but what remains interesting, however, is that Viète's method of proceeding in Hooke's view allowed algebra to be extended beyond the realm of numbers in the understanding of reality. Nonetheless, this method requires a mathematization of reality in order to be fully functional. In this powerful tool, Hooke sees the possibility of overcoming traditional mathematics and logic, merging them into a more universal tool which has the capacity to infer with the highest degree of certainty.

Even so, at this stage, Hooke is still not clear whether this philosophical version of algebra encompasses the Baconian method also, or whether it is something that simply assists it, or whether again it is only one instrument besides many others. In his *Of the True Method of Building a Solid Philosophy, or of a Philosophical Algebra*, philosophical algebra is characterized as a universal method which contains 'Baconian part':

> This Method of a Philosophical Algebra, I shall divide into two main Branches. The First shall contain the manner of Preparing the Mind, and Furnishing it with fit Materials to work on. The Second shall contain the Rules and Methods of proceeding or operating with this so collected and qualify'd Supellex.⁶²

There is, therefore, one part that organizes the material provided by experience and collected in natural histories, and this was Bacon's method, and, following this, a second part concerning only rules for directing the mind in how it might acquire scientific knowledge of the same material. In the *General Scheme*, Hooke develops only the first part and provides just a glimpse of the second.

The first part of this philosophical algebra coincides with what the Aristotelians characterized as analysis in the regressus theory – that is, invention – even if Hooke, as we shall see, will never use this name to define it. This first stage has the task of establishing all the possible observations, examinations and experiments which might lead to a complete knowledge of the subject under investigation. The strongly Baconian character of the first part is a clear departure from Descartes, since while for Hooke the material of knowledge comes from experience, for the French philosophers it had its origin in the mind, remaining closer to Proclus's position. Hooke had already criticized this innatist position in his *First Algebra Lecture*.

Once preliminary work on particular cases has been carried out, one should undertake these experiments and trials with care, circumspection and diligence. Hooke suggests that these examinations be performed in the presence of someone who has a different understanding and is without bias in the research.[63] The way of discovering natures and properties of bodies is threefold:

> I. By the Help of Naked Senses; II. By the Senses assisted with Instruments, and arm'd with Engines. III. By Induction, or comparing the collected Observations, by two preceding Helps, and ratiocinating from them.[64]

Hooke is particularly interested in the second way because the first is through simple sensation, while the third entails a combination of the first with the second. Therefore, what is distinctive about Hooke's method lies in the second way, in turn divided into two processes: (1) the reduction of the sensation to a standard and (2) the discovery of sensible properties in bodies which sensation cannot reach. The second way can easily be implemented using instruments like a microscope or telescope to aid the senses. The first way is much more complicated. After all the observations, experiments and trials, Hooke considers it good practice to reduce all sensations to standards, meaning a specific 'Regularity, Certainty, Number, Weight, and Measure'.[65] Reduction was the way used by Galileo to provide mathematical descriptions of phenomena, and it was the first process of Descartes's Rule 5. Like Galileo and Descartes, Hooke is proposing a mathematization of nature. Indeed, in the making of observations, trials and experiments

> great Care and Judgment must be used in exactly determining the Quantity, Quality, Time, Place, Space and several other Circumstances of the Ingredients, Effects, Processes ... that all things may be reduced to some Certainty of Number, Weight, and Measure, and that nothing may be left to doubting ... that no part of these Materials which are for the Foundation may be defective or faulty, which may endanger the whole Superstructure.[66]

In a manuscript note, Hooke explains that reduction is not performed by an immediate operation of sensation, but rather it requires rational activity of the intellect. Indeed, sensation can mislead or misinform the mind, while reduction can provide a more reliable knowledge:

> The Sight doth Represent the Sun and moon as two Small Round & flat bodys of light as if they were noe bigger than a basket or the area of a Round table and as if they were not a miles distance above us in the Sky, but by Ratiocination and procuring other tryalls and Observations we come to be assured that the Bodys of the Sun and Moon are vastly great and of a globular figure, that the nearest of them is many thousand times further Removed from us than we could have Imagined by the common appearance and the other yet vastly further Distance and of a much more prodigious Magnitude.[67]

The observations' reduction into mathematical forms are useful for Hooke, as for Galileo, to minimize the possible deceptions of naked sensation. Hooke provides a useful list on how to produce this reduction:

Reduction of sensations to standards

1. By increasing the power of the senses
2. By making sensible qualities in the object more powerful
    2.1. By activation
    2.2. By assistance
    2.3. By abatement of the quality
    2.4. By concentration of the quality
    2.5. By removing interferences
    2.6. By supplying the desired qualities
3. By inquiring for the effects produced
    3.1. By observing the effects themselves produced
        3.1.1. Intermediate
            3.1.1.1. Always present
            3.1.1.2. Not always present
        3.1.2. Ultimate
            3.1.2.1. Necessary
            3.1.2.2. Accidental
    3.2. By observing the manner of nature's productions or workings
        3.2.1. Secret
            3.2.1.1. No information afforded to the senses
            3.2.1.2. Action at a distance or influxes
            3.2.1.3. Spiritual operations of the mind
        3.2.2. Obvious
            3.2.2.1. How nature proceeds in generation, increasing, decreasing, and destroying a property in a body
            3.2.2.2. How nature distributes the same properties in different bodies

3.2.2.3. How nature proceeds in combining, that is, with what other kinds of propriety the same quality is joined in diverse bodies

3.2.2.4. How nature proceeds in separations, that is, what property seldom or never joins in the same subject

3.2.2.5. To observe the transitions or changes of nature, how from the destruction of one property generates a specific thing

3.2.2.6. To observe the several ways that nature takes in several bodies to produce the same effect

3.2.2.7. To observe how, where, and when nature makes use of the same body or means to produce different effects

3.2.2.8. To distinguish when there seem to be more than one cause conducive to the effect, which of them is most powerful, which less

3.2.2.9. To observe the resemblance and discrepancy of nature

3.2.2.10. To observe the transition of nature in the forms and properties of creatures

3.2.2.11. To observe where and how nature is frustrated, fails or misses in producing its usual effects

3.2.2.12. To observe deviations of nature in framing specific bodies

3.2.2.13. To observe deviations of nature in producing individuals such as monsters

3.2.2.14. To observe concomitant properties of individuals in the greatest and lowest pitch of perfection

3.2.2.15. To observe the various workings of nature in various places and how such circumstances alter its course

3.2.2.16. To observe what things seem most conducive to the perfect or imperfection of productions

3.2.2.17. To observe the workings of nature where it seems to be peculiar in its manner and it seems not anywhere else to follow the like method

3.2.2.18. To inquire after what means is interposed between this peculiar and the more common method

3.2.2.19. To observe how much, and by what degree nature is made to alter its course by art

3.2.2.20. To observe where and by what means art causes nature to deviate

3.2.2.21. To observe the natural and artificial ways of producing the same effect

3.2.2.22. To observe difference between bodies produced by nature and those by art

3.2.2.23. To observe the different ways of producing the same effect by art

3.2.2.24. To observe with what circumstances nature and art do sometime exceed each other
3.2.2.25. To inquire how many mechanical ways there may be of working on, or altering the properties of several bodies
3.2.2.26. To inquire how many mechanical ways there may be of separating bodies
3.2.2.27. To inquire how many mechanical ways there may be of uniting and incorporating bodies into one another
3.2.2.28. To inquire by what means bodies may be changed or transmuted from one thing to another by a real change of all their former properties
3.2.2.29. To observe by what means may be more sensibly explained, that so the nature of insensible operations may be more easily guessed at and imagined[68]

After observations, trials and experiments and the application of the reduction to them, only the most reliable particulars remain in a standardized form and can be outlined in a 'Philosophical Treasury'.[69] This attempt at reduction is completely understandable within Hooke's framework of philosophical algebra, which pointed to scientific knowledge by discovering notions – those philosophical treasures – through which a complete understanding of nature would be made possible. Indeed, algebra worked much better with quantities, as he pointed out in his *First Algebra Lecture*.

However, as is evident from the list, not all these techniques of reduction lead to mathematization. Indeed, this scheme reveals that this operation of reduction to standardization deals with the application of logical and experimental procedures employed by Aristotelians in the second stage of *regressus*, which Galileo also used and which came originally from the *Topics*.[70] It is a detailed list and a rigorization of the rules which were applied in order to establish the connection between causes and effects, and to generate principles and axioms for the demonstration *propter quid*. As we have seen in Viète, this kind of algebra or analysis was perfectly compatible with logic. Hooke, however, transforms the *regressus* theory in at least two aspects. The first is to reduce the relation between causes and effects to quantitative aspects, and in this he was following the secure path of Galileo and Descartes. The second is to introduce the idea of the possibility of changing nature or intervening in it, and in this it is Bacon's methodology that he was refining.

Hooke, however, considers this reduction not sufficient to fulfil the purpose of producing reliable principles for scientific knowledge. In reflecting on his procedure of reduction, Hooke writes that

> these are some of the various ways by which Nature may be trac'd, by which we may be able to find out the material Efficient and Instrumental Causes of divers Effects, not too far removed beyond the reach of our Senses, and which do not very much differ from such Effects as are more material and obvious to our Senses.[71]

Hooke is conscious of the limits of this kind of investigation and of the fact that it provides a kind of knowledge which does not differ significantly from what is acquired through sensation. He adds that

> as for the Discovery of the more internal Texture and Constitution, as also of the Motion, Energy, and operating Principle of Concret Bodies, together with the Method and Course of Nature's proceeding in them: These will require much deeper Researches and Ratiocinations, and very many Vicissitudes of Proceedings from Axiomes to Experiments, and from Experiments to Axiomes, and are indeed the Business of the Philosopher, and not of the Historian.[72]

In order to know the nature and properties of things, further examinations and reasonings are necessary, and these constitute a reciprocal comparison in two steps – first, between axioms and experiments, and then between experiments and axioms.[73] This strategy recalls the continual testing of conjectures and hypothesis with effects that Galileo elaborated in the intermediate stage of regressus, and which was based on the mathematical analysis of Theon of Smyrna and Pappus.

These reasonings from axioms to experiments and from experiments to axioms are the task of the second part of the philosophical algebra, 'which explains the way of making use of the *Penus Analytica*, of raising Axiomes, and more general Deductions from a sufficient Stock of Materials collected according to the Method of the first part, with Integrity, Judgment, and Care'.[74] The second part of the philosophical algebra thus explains how to proceed in a more rigorous manner than the rules prescribed in the scheme for the reduction of sensations to standards to establish axioms and principles – that is, how to make general deductions or inference from effects. This philosophical algebra seems to 'assist the Ratiocination and Invention in detecting the true Causes of things', much more so than any other instrument, and it seems to conclude the inventive part of Hooke's method – that is, his logic of scientific discovery.[75] Extending the concept to philosophy in general, philosophical algebra becomes the inventive part of the entire epistemology as already outlined in the *First Algebra Lecture*.

*

In *A Discourse of Earthquakes*,[76] Hooke implements its epistemology introducing a 'dynamic interplay' between effects and causes.[77] At the same time he abandons the idea of a philosophical algebra as such, transforming it into something different. He writes that in order to attain 'true and certain knowledge of the Works of Nature' – that is, scientific knowledge – one should employ a method composed of two stages: (1) analysis and (2) synthesis.[78] At first glance, this seems to be a resuscitation of the regressus theory, or otherwise an anticipation of the Newtonian methodology that we shall see in Chapter 12, but the epistemological situation is in fact quite different. Indeed, Hooke points out that analysis 'is the proceeding from the Causes to the Effects', while synthesis is the proceeding 'from the Effects to the Causes'.[79] He reverses the definitions of analysis and synthesis as laid out in the Aristotelian method, and this inversion is extremely significant in understanding his epistemology.

In order to grasp the full significance of this inversion and the transformation of the project of philosophical algebra, we should refer to one of the favourite authors of Hooke in his *First Algebra Lecture*, Thomas Harriot. At the beginning of Harriot's treatise *The Practice of the Analytic Art* (1631), he introduces an apparently useless definition for our discourse, that of equation. Equation is 'the clearly determined equality of the sought quantity with some given quantity, when a comparison has been made of one with the other'.[80] Following this definition, Harriot points out that the best method of proof is that which 'proceeds from the principles and elements proper to each subject, composing a chain of logical steps until the proposition is confirmed'.[81] This method is called 'compositive or synthesis'.[82] This is the traditional way of conceiving of synthesis, but according to him it is totally useless for invention in mathematics. Indeed, if a mathematician 'takes the natural route of synthesis from principles and elements of science, he cannot make any progress in finding and demonstrating a solution by logical means'.[83] Therefore, the mathematician necessarily reverses this procedure:

> He makes a start from some unknown and sought quantity which occurs in the problem, as if it were a known and given quantity; he then proceeds by resolving it in a chain of logical steps, until he reaches an equality between the quantity which was taken as given ... he then proceeds by resolving it in a chain of logical steps, until he reaches an equality between the quantity which was taken as given ... and some quantity which really is given. If this equality can be found and properly established by such a technique, then, from this equality, the sought quantity itself either will be self-evident, or can be elicited through a further application of the technique and so the problem can, at length, be solved.[84]

Harriot characterizes this method as 'resolution' or 'analysis' and explicitly absorbs Viète's definition of analysis coming from Theon of Smyrna and Pappus. He is sufficiently aware of this appropriation and inversion as to write in Definition 5:

> The words 'composition' and 'resolution,' which were introduced in these two definitions are the usual ones Mathematicians employ, by which they signify explicitly, when necessary, the two contrary paths of reasoning used in constructing demonstrations.[85]

The two paths for constructing demonstrations are those of composition 'from the prior to the posterior', and analysis 'from the posterior to the prior'.[86] The *a priori* or *a posteriori* character is in relation to the nature of the knowledge, and not the order of the things. Indeed, Harriot conceives these two methods as being the same logical process as that contained in the form of the classical regressus. He points out that

> for, the elements and axioms of the sciences, if they are legitimately constituted, ought to be convertible; hence it comes about that those which in nature are antecedents can, in reasoning, be consequents. It is for this reason that logical progress through deductive steps is, of necessity, equally firm and demonstrative in both directions.[87]

Harriot characterizes analysis as the method proceeding from what is prior to what is posterior – that is, from axioms to elements – while synthesis is the method proceeding from what is posterior to what is prior – that is, from elements to axioms. He is aware that he is reversing the traditional logic of regressus, but in both cases the inference is demonstrative because of the possibility of convertibility. Convertibility between elements and axioms becomes an essential issue for Harriot to the extent that, in spite of the different notions of analysis and synthesis, mathematical method and regressus can proceed universally and necessary in a demonstrative way. This is something that we have already found in Zabarella, and which constitutes a break with Aristotle, who maintained that convertibility was proper only to the mathematical sciences because their reasonings (1) are based on definitions in which there is identity between subject and predicate, (2) are equations and (3) employ propositions in which the predicate pertains necessarily and essentially to the subject. This was not obvious for physics and ethics for Aristotle, but it was possible for the regressus theorists.

In the case of logical regressus, convertibility is made possible by finding an essential connection – that is a definition – through which we can say that something is a set of properties, or that a certain set of properties constitutes something. In the case of mathematics, there is an equation of $x = y$, which may be stated alternatively, but with the same value, as $y = x$. Between these two kinds of reasonings – that of mathematics and regressus – algebrists did not recognize a real difference because it was possible to formalize a definition in the regressus whereby what we obtain is that $x = (a, b, c, d)$, which is the same as saying that $(a, b, c, d) = x$. The most important part was to find the definition or the essential connection, but once this was established – for Aristotelians through the negotiation of the intellect – the entire system of philosophy could be algebrized or mathematized. This is what Hooke tries to develop in his philosophical algebra – that is, to formalize the natural investigations. But in order to do so, he had either to reduce all natural phenomena to mathematical terms in which convertibility is in most cases granted – and this was the role of reduction – or to identify the relation of the properties with the subject, and this was the role of the twenty-nine rules listed in the *General Scheme*. In one way or another, both processes should lead to a standardization, which constituted the basis of this algebra. But for unknown reasons, he was unable to carry out a project that would come to occupy the most brilliant minds of the eighteenth century.

As already indicated, in *A Discourse*, Hooke abandons the project of philosophical algebra, but he does not change his mode of thinking. He clearly suggests an equivalence between analysis and synthesis with processes that progress from 'Axiomes to Experiments, and from Experiments to Axiomes',[88] outlined in the *General Scheme*. Indeed, the parallelism from causes to effects and from axioms to experiments, and then, vice versa, from effects to causes and from experiments to axioms, works perfectly also in considering the problem of convertibility where the transition is made after the mathematization of the subject under investigation. It also explains the obsessive request to reduce experience to number, weight and measure.

This twofold method is essential because it is not always possible to begin with certain premises from which 'the conclusion necessarily follows'.[89] Indeed now tho' in Physical Inquiries, by reason of the abstruseness of Causes, and the limited Power of the Senses we cannot thus reason, and without many Inductions we may arrive to so great an assurance and limitation of Propositions as will at least be sufficient to ground Conjectures upon, which may serve for making *Hypotheses* fit to be enquired into by the *Analytick* method, and thence to find out what other Experiments or Observations are necessary to be procured for the further progress in the *Synthetick*, which will questionless so far inform us of the general and universal progress of the Operations of Nature, that nothing but what is really the truth shall be proposed but the absurdity and insufficiency thereof will presently be detected and proved.[90]

In particular, for Hooke the analytic method 'supposes the thing to be already done and known, which is the thing sought and to be found out; this begins from the highest, most general and universal Principles or Causes of Things'.[91] This is Viète and Harriot's definition of analysis, taken from Theon of Smyrna and Pappus, and applied, however, not only to mathematics but also to natural philosophy. But what is the function of analysis? Hooke provides a striking example of the application of the analytic method in natural investigations after the discovery of the hypotheses. This method should be that followed by 'Mr. *Newton* in a Treatise of his now in the Press', and the reference is most likely to *The Mathematical Principles of Natural Philosophy*. In this case, as in many others concerning the study of nature

> from an *Hypothesis* being supposed or a premeditated Design, all the *Phaenomena* of the Subject will be *a Priori* foretold, and the Effects naturally follow as proceeding from a Cause so and so qualified and limited. And in truth the *Synthetick* way by Experiments Observations &c will be very slow if it be not often assisted by the *Analytick*, which proves of excellent use, even tho' it proceed by a false position; for that the discovery of a *Negative* is one way of restraining and limiting an *Affirmative*.[92]

The analytic method has the role of determining and assessing hypotheses through comparisons of the causes and principles with the particulars in order to check whether that principle is a hypothesis to be refuted or to be maintained.[93] The role of synthesis, in contrast, is much less distinct. Indeed, he characterizes it rather ambiguously as the method

> more proper for experimental Inquiry, which from a true information of the Effect by a due process, finds not the immediate Cause thereof, and so proceeds gradually to higher and more remote Causes and Powers effective, founding its Steps upon the lowest and more immediate Conclusions.[94]

The ambiguity lies in the fact that synthesis seems to encompass, or at least overlap with, the first part of the philosophical algebra – that is, the stage based on experience and on the collection of phenomena. But in this specific case, synthesis follows analysis. The ambiguity partially disappears if we consider that in *A Discourse* Hooke is no longer dealing with a philosophical algebra and perhaps the two parts – one more Baconian and the other more logical – are no longer useful for characterizing his method. Still, synthesis follows analysis.

In *A Remonstrance Concerning Some of Hooke's Discoveries* (26 June 1689) – written some two years after *A Discourse*[95] – the situation changes once again, and becomes even more complex. This short lecture provides a summarized version of Hooke's epistemology as composed of seven stages. The first stage is the reduction to geometrical certainty of the powers and effects of bodies, limiting their description to number, weight and measure. This is perhaps the most important way adopted by Hooke in order to aspire to mathematical certainty in knowledge of nature. The second way consists of the discovery of a new property by means of which it is possible to find out the true essence of a body. The third is to invent artificial ways or instruments to penetrate the nature of bodies more effectively.

The fourth stage is perhaps the most complicated. Indeed, it prescribes searching by a proper method of reasoning from effects to causes for such theories that seem capable of directing the mind in a further process of examination by means of experiments. Theories here assume the connotation of hypotheses. Hooke adds as a second thought 'Synthetick' above the words 'proper method', as is evident in the manuscript. With synthesis, the connection between effects and causes is not yet necessary. This necessity comes with the second part of the fourth stage – that is, 'Analyticall Resolution' – which shows how effects are 'necessary Results' of the theories. Therefore, in this manuscript, synthesis precedes analysis, unlike in *A Discourse*. The fifth moment prescribes the need

> to proceed with the further examination of such Experiments in order to the confirmation of the Doctrine propounded if they answered in all particulars to the effects that were expected or to the Amending and Limiting and further Restraining thereof if somewhat new and not expected occurred thereupon.[96]

In other words, axioms and theories are to be proved from effects and experiments, and this was synthesis in *A Discourse*. During this phase, a new particular might be found that would confute the results of the previous stages. The sixth stage requires that these new phenomena be collected, constituting the material for a new investigation. Once new phenomena are discovered through the theories then the seventh stage claims 'to produce such geometricall Demonstrations of asserted proprietys as put the Doctrine beyond further Dispute',[97] following 'the best method of proof' advanced by Harriot. There are, therefore, many more stages of composition according to this manuscript version of Hooke's epistemology.

The reading of this manuscript leads us to various conclusions. The epistemology outlined in *A Remonstrance* is more in line with the preceding methodological tradition. The first three moments are mainly based on observations and experience, which provide better certainty if they are reduced to mathematical standards. After these three steps, there is an inference that leads to the formulation of general principles. Then these principles are tested on other particulars and effects until the relation between causes and effects becomes certain and necessary. Once the principles are established as such, then it is possible to discover new effects and phenomena by applying a mathematical or geometrical demonstration.[98] In all these stages, Hooke's process does not differ significantly from Galileo's epistemology.[99] For the Pisan scientist knowledge starts from sensate experiences acquired through the use of the telescope. The experiences were reduced to mathematical terms in order to provide mathematical descriptions, which led to the general definition of a law. This law was tested and, once confirmed, it was applied for the discovery of phenomena.

The examination of all of Hooke's attempts shows how he tries to reconcile Baconianism, the Aristotelian logical tradition and mathematical analysis. He develops an original epistemology capable of providing a reliable account of new experience and experiments. His idea of a philosophical algebra was soon abandoned, or at least it did not receive further elaboration. In his later work, Hooke returns to a classical methodological distinction between analysis and synthesis, but he conceives it in a very peculiar and original manner – through a mathematical lens as a method for testing hypotheses and proving new phenomena. The constant preoccupation in his methodology is the reduction of natural objects to quantities – that is, the transformation of natural philosophy into mathematics – but this mathematization happens at the expense of the knowledge of the causes that characterized Aristotelian theory of scientific knowledge.

9

John Locke

Oxford. 1652. After five years at Westminster School, John Locke was admitted to Christ Church. Ramism never took root in Oxford, at least not to the extent that it did elsewhere in England. The university curriculum was mainly Aristotelian and Scholastic, with a strong emphasis on Zabarella's work and Jesuit texts. This kind of university philosophy was unattractive for the young Locke. He gained little satisfaction from his studies: he took against the captious logical discussions of his professors; he complained of losing a great deal of time 'because the only philosophy then known at Oxford was the Peripatetic', full of obscure terms and useless questions.[1] He preferred to study medicine with its concreteness and strong attachment to experience.

Though long regarded the founder of empiricism, Locke is no longer credited in the scholarship with being in all respects an exponent.[2] Indeed, a large part of his epistemology aims to establish the criteria for certain knowledge and to recognize the boundaries that divide science from mere opinion on the basis of rational processes.

Locke's detachment from the previous philosophical tradition is clear from the time of the so-called *Draft A* (1671) of *An Essay Concerning Human Understanding* (1689). *Draft A* is radically different from the definitive version of the published book, so much so that we may speak of two diverse works. The first draft clearly shows how Locke's ideas depart from the epistemological approaches we have considered so far in this book. The foundation of Lockean epistemology is explicit already from the very paragraph, sect. 1, in which he establishes that true knowledge is based on sensation. In particular, knowledge is generated by the senses in connection with particular objects, which provides simple ideas of things. In *Draft A*, there is no definition of what he means by 'idea', which we can find in the *Essay* as 'The Object of the Understanding When a Man Thinks'.[3] Such a notion is in perfect agreement with the Cartesian spirit; indeed, in the *Replies* to the *Third Objections* Descartes wrote that he took the 'word "idea" to refer to whatever is immediately perceived by the mind'.[4] Unlike Descartes, Locke seems to extend the notion of 'idea' to every representation of the object in the understanding when the understanding thinks. Indeed, Locke employs this notion 'to express whatever is meant by *Phantasm, Notion, Species*, or whatever it is, which the Mind can be employ'd about in thinking'.[5]

In the Aristotelian-Scholastic tradition, albeit referring to the same thing, phantasms, notions and species constituted three different mental objects, each pertaining to different faculties of the mind with different characteristics. There is in Locke a sharp detachment from the previous epistemological tradition in this sense. What remains unchanged, however, is the relation between ideas and objects – that is, the former can be considered mental representations of the latter, deriving from a passive affection of the mind. Locke thus remains Aristotelian in saying that the mind cannot deny having or receiving ideas that come from sensation. Mind is passive in relation to sensation.

Purely because of this relation with sensation, these ideas can be understood as sensible qualities, and they constitute the first object of the mind, what the Aristotelians called *primum cognitum*. Locke adds that the continuous exercise of sensation in connection with some objects is called experience or observation, and this has led to the discovery that a certain number of simple ideas are always conjoined. For this reason, the understanding supposes them to pertain to a single subject, usually called 'substance' by the philosophical tradition.

Locke emphasizes that the understanding 'supposes' the existence of this substance, but it has no certain knowledge, to the extent that the knowledge of these substances is better in one 'who having oftenest & with greatest care examind all the simple sensible qualities of any subject, findes such a number of them certainly & constantly united together'.[6] Knowledge of these substances is not certain because the 'collection of simple Ideas into one compound specific Idea' depends 'upon us', because 'enquiry attention & observation is different in one man from another, & different in the same man at different times'.[7] On the contrary, knowledge of sensible ideas is not affected by such accidental and contingent conditions; indeed, only these ideas are 'soe cleare & destinct & perfect in the understanding that it never mistakes one for an other'.[8] To sum up, it is possible to acquire certain knowledge of sensible qualities, but never of substance. In this way Locke undermines the foundations of Aristotelian epistemology, delivering a fatal blow also to Bacon, Hobbes and Descartes, who conceived knowledge of essence and of essential relations the ground of every science.

This epistemological difference between sensible direct knowledge and that of substances, which is indirect, led Locke to a more detailed investigation of the various degrees of knowledge relating to complex ideas. If a complex idea is constituted only by some simple ideas that pertain to it, then the complex idea will be imperfect. If a complex idea is constituted of simple ideas, which pertain only to that object and not to others, then the complex idea will be distinct. If a complex idea is constituted of a set of simple ideas, which are specific to the object, then knowledge of this complex idea will be perfect.

Perfect knowledge is essentially that of mathematics, which deals with extensions and numbers. For Locke, its certainty is indubitable

> that by constant observation of our senses espetialy our eys we come to finde that such & such quantitys have such & such proportions compard with other v.g. that the 3 angles of a triangle are equal to two right ones, or that one side of a triangle being produced the exterior angle is equal to the two interior opposite angles, which being tried in several tryangles & by noe body found in any one triangle otherwise, passes into an universal acknowledgd truth, & is received as undoubted axiom ... The evidence then of demonstration, or the certainty of knowledg we atteine by it is thus, That having from our senses or sensation got

the Ideas of Extension & number & by repeated observations about tem atteind certeine knowledg of aequality or inequality of them compard to one another, which always reteine the same proportion when ever we compare them together, we collect from hence such propositions which having found to be true we call Maximes ... it is to be observed that all this knowledg which we atteine soe perfect & fully is of proportions of numbers & extensions which are soe ex necessitate rei, i.e. are inseperable propertys of the angles or figures where we finde them.[9]

In other words, knowledge is perfect and certain because the connection between numbers and extensions acquired by comparisons is necessary. Indeed, 'the connection of Ideas within us & their existing soe without us doe always exactly agree'.[10] This is exactly the necessary connection that the theorists of regressus looked for in natural philosophy by means of comparisons. However – Locke immediately points out – mathematical knowledge does not equate with the Scholastic definition of knowledge as knowledge of causes, because demonstration provides only 'the equality or inequality of quantitys compard & their several proportions, but not the cause that produced them'.[11] It is merely knowledge coming from a process of comparison, and not knowledge of causes and effects.[12] Nevertheless, mathematical knowledge is the only kind of knowledge that can be called 'demonstrative' in the Aristotelian sense of being certain, necessary and true.

All other kinds of knowledge that are not concerned with extensions and numbers – that is, they deal with a 'substantial thing' – possess only that particular kind of certainty coming from sense, which is always particular, never universal, because of the epistemological boundaries of experience. In sect. 11 Locke affirms that this depends on the fact that there are no 'measures & standards whereby accurately to measure' other sensible qualities.[13] It follows that if there were a way to mathematize all that comes from sensation, then all varieties of knowledge would become demonstrative – that is, true and certain.[14]

Yet for Locke, it is not possible to have mathematical knowledge of all phenomena. For this reason, true and certain knowledge of substantial things is possible if, and only if, a given simple idea is contained in the definition of the thing itself. In such a case, however, knowledge would be only verbal or nominal, and not real. Indeed, only rarely through sensible knowledge does one know 'the specific essentiall Ideas that belong to every particular individual of any species'.[15] As with the Aristotelians, Galileo, Hobbes and Descartes, Locke's idea of necessary, universal and certain knowledge depends on knowledge of the essence. Still, Locke rejects the idea of knowing essence, and therefore denies that such a kind of knowledge is achievable. Indeed, when we 'have observd this simple Idea to belong to all the individuals of that species that my senses have met with & then predicate it of but doe not include it in the definition of the name of that species', we have not 'certaine knowledg of it'.[16] In sensible knowledge, one cannot find that certainty, convertibility and perfect correspondence between ideas. For instance, in the idea of 'human being' we

> have put toegeather these following ideas v.g. 1° A face usual to that species which thus consisting of many particular Iedas for brevitys sake & the thing being

obvious I take for one. 2° Two hands with five fingers on each. 3° Two legs. 4° Upright posture. 5° Liveing. 6° A power of laughing. 7° A power of speakeing. 8° Of reasoning i.e knowing the consequence of words or propositions one to an other. 9° of judging i.e guesse at the truth of words being of things. I say when I have put all these 8 Ideas togeather & thereof in my minde framd the Idea which I call a man I cannot thereby certainly know that wheere any 7 of those 8 doe realy exist togeather that there the 8th is necessarily also.[17]

Knowledge of complex ideas is similar to that of the connection between causes and effects. The only certainty of this causal knowledge is provided by the senses, yet from the senses it is impossible to derive universal and true propositions, unless – Locke points outs – 'it be of those powers which I include in the Idea of that subject or definition of that name'.[18] He clarifies this conception using an example typical of the Aristotelian tradition, that of the magnet

> if I say I know this universal proposition to be true that all load stones wil draw iron it is certainly true when I include this power of drawing iron in the very Idea of the thing I call a load stone or make that quality or power part of the definition of the word load stone. But if I say that a stone of such colour weight hardnesse & those other simple Ideas which usually meet in a load stone will certainly draw iron, of this I have noe certain knowledg, & therefor in universal propositions connecting causes & effects I cannot be assured that they are true.[19]

Before Hume, Locke establishes that even if one has habitual knowledge of an effect produced by a cause, from this knowledge one cannot infer with certainty that the cause will produce the same effect, unless the experience is complete. In short, what Locke requires is a complete induction that determines the certainty of the extension of the knowledge in question to all possible agents which can generate this particular effect. Locke is aware that it is only possible to infer contingent and never necessary conclusions from experience, because the number of experiences is always limited. For Locke, a complete induction on the basis of experience – that is, a comprehensive knowledge of the causes and effects – surpasses human understanding, which is dependent on experience for all its ideas. By 'causes' he means, in Hobbes's wake, exclusively efficient causes,[20] detaching himself from the Aristotelians, who considered formal causes the preeminent explanatory element. A universal, certain, and necessary knowledge of efficient causes by induction is impossible, indeed it would be only 'a grosse kinde of knowledge', which testifies the existence of the phenomenon, but not of its cause.[21]

What is lacking – and what, for instance, Hobbes looked for in Aristotelian fashion – is the way of generating the effect that Locke calls *modus operandi*. The reason for this lack, according to Locke, is that in the latent process of these *modi operandi* there are alterations of particles, which are impossible for sensation to see or perceive. Locke's epistemology is influenced by his corpuscular conception of matter – that is, by his natural philosophy.[22] However, if the senses were able to know these particles, which act as causes, then a perfect and certain knowledge of natural phenomena would become possible.[23] Locke seems to allude to this in the *Essay* when he states that if we had senses so accurate as to be able to discern the particles of bodies, as is the case in the use of microscopes, then our knowledge and ideas would be qualitatively different.[24] It is mainly a problem of epistemology of perception, which tackles directly the question of transduction – that is, the inference from macroscopic to microscopic worlds – and which will occupy Locke's investigation in all phases of the preparation of his masterpiece.

In *Draft B*, Locke does not change his position. He points out that 'the transactions of our owne thoughts' on ideas does not lead beyond the clear and distinct knowledge provided by simple ideas, which was what the deniers of the theory of regressus affirmed when they established that no knowledge could be qualitatively superior than that provided by sensation.[25] However, 'laborious & exact scrutiny into the nature of the things & a searching out all their qualitys & propertys' would lead to a more distinct and clearer knowledge of complex ideas.[26] This process of transition from a more confused and obscure notion to one that is clearer and more distinct is for Locke – in contrast to the Aristotelians, Bacon and Descartes – not characterized by a mental process, but rather regarded as an operation that refers continuously to experience and observation, the only tools by means of which certainty of knowledge is achievable.

*

In the final version of the *Essay*, Locke takes this position to an extreme, also emphasizing, however, the importance of the operation of the understanding in generating knowledge, even if this knowledge is not as certain as that coming from simple ideas acquired by sensation. In this case too, as we will see better in the following pages, there is an inversion of Aristotelian epistemology. Indeed, the operations of senses fall under the domain of certain knowledge, while operations of the understanding come under probable knowledge. Now, the proper senses do not deceive also for the Aristotelians, but in dealing with matter they provide a merely accidental and contingent quality of knowledge, which is not scientific, unlike the intellect, which deals with universals and generates science. The Lockean epistemological shift is thus evident. In looking for the criteria of certainty, Locke aims to assess which objects can be known and which are superior to the understanding[27] in such a way as to obtain clear and distinct knowledge, avoiding the danger of straying from the correct path of reasoning.[28]

With this broad context, in the *Essay*, Locke focuses on reason, which was not the subject of the previous two drafts. Reason is the faculty for inferring unknown truths from first principles, or from already known and established propositions: it is a process from the known to the unknown just like the Aristotelian conception of method.[29] Reason is thus the proper faculty for epistemology as understood within the Aristotelian tradition: it relies on deductive reasonings like demonstration, as we shall see.

The chapter 'Of Reason' shows the greatest debt to the Aristotelian tradition, and indeed it is in this chapter that we find the harshest criticism against Aristotelian syllogism. Reason, as we have mentioned previously, is not only the faculty capable of generating 'clear and fair deductions' from true principles[30] but also the specific faculty for extending knowledge and for determining assent. It is characterized by two intellectual faculties: (1) sagacity and (2) illation. Sagacity is concerned with the logic of finding intermediate ideas, which are used for comparisons. Illation, instead, arranges and orders ideas in such a way that their connections become evident. Without reason it would be impossible to recognize the 'certain Agreement or Disagreement of any two *Ideas*, as in Demonstration, in which it arrives at Knowledge; or their probable connexion, on which it gives or with-holds its Assent, as in Opinion'.[31] Locke makes clear the twofold way through which logical inferences, and in particular Aristotelian syllogism, may be employed. In the case of demonstration, reason leads to certain knowledge, in the case of hypothetical syllogism it leads to opinion.

Reason is particularly useful in discovering and examining the grounds for the probability of propositions, which are considered true without full certainty. In other words, reason is necessary for establishing the foundations of scientific knowledge because it perceives the necessary or probable connection among the ideas. Such a task for determining the foundations of scientific knowledge, examining in particular the connections between ideas, was typical in the Aristotelian tradition of reasoning called mental consideration or negotiation. Locke seems to refer implicitly to a joint working between the empirical and rational parts, which brings to mind a kind of evolution of the Aristotelian negotiation and which we have already examined in Bacon and in Galileo. In reason, Locke states, we may consider four stages:

> The first and highest, is the discovering, and finding out of Proofs; the second, the regular and methodical Disposition of them, and laying them in a clear and fit Order, to make their Connexion and Force be plainly and easily perceived; the third is the perceiving their Connexion; and the fourth, the making a right conclusion.[32]

The first step is the determination of proof. In Locke, 'proof' refers to that intermediate idea necessary to prove agreement between two other ideas, that is, the middle term.[33] The discovery of this middle term pertains to sagacity, which is 'a quickness in the Mind to find out these intermediate Ideas, (that shall discover the Agreement or Disagreement of any other,) and to apply them'.[34] The first part of Locke's method is thus that of the discovery of an intermediate idea, which connects two other ideas. Since all ideas come from sensation, this proof mainly characterizes the empirical moment. These ideas, however, cannot be collected haphazardly and randomly, otherwise any kind of inference would be impossible. For this reason, it is necessary to arrange them in an orderly manner with the effect that in a third moment one might establish the connections that potentially emerge from these cognitions. Having established these connections, a correctly inferred conclusion is possible.

Locke makes the additional comment that these four stages are typical of mathematical demonstration. The epistemic ideal of certainty that is the rightful goal of reasoning is that of mathematics, and indeed it is not by chance that 'most of Locke's examples of demonstrations, or demonstrative knowledge are mathematical'.[35] In this fourfold process, Locke goes against the preeminent epistemological position that syllogism held for the Aristotelians;[36] indeed he is persuaded that it is sufficient to establish a connection between proofs – that is, how different intermediate ideas are related to the extremes – to understand the validity of a logical conclusion.

Locke's entire epistemology is based on the possibility of establishing these agreements or disagreements between ideas. It is necessary, therefore, to determine: (1) by what way ideas are acquired and (2) by what way ideas are compared.

Locke is clear in stating that without reason knowledge of general and abstract truths is not possible.[37] The formation of these general and abstract truths follows for Locke a process that recalls Aristotle's *Posterior Analytics* II.19:

> The Senses at first let in particular Ideas, and furnish the yet empty Cabinet: And the Mind by degrees growing familiar with some of them, they are lodged in the Memory, and Names got to them. Afterwards the Mind proceeding farther, abstracts them, and by Degrees learns the use of general Names. In this manner the Mind comes to be furnish'd with Ideas and Language, the Materials about which to exercise its discursive Faculty: And the use of Reason becomes daily more visible, as these Materials, that give it Employment, increase.[38]

These truths are composed of ideas acquired first of all from impressions of external things,[39] and indeed the understanding is more familiar with particular things, which are by themselves evident than with general and abstract ideas.[40] For Locke, all the material of reason comes from experience, which constitutes the foundation of knowledge.[41] Experience is both internal to the mind and its various operations, and external in relation to the objects of the world. The source of the former is reflection, the source of the latter sensation. These, according to Locke, are 'the only Originals, from whence all our Ideas take their beginnings'.[42]

This sensation or perception, which is considered as the first faculty of the mind, is, as already pointed out, merely passive and provides the ideas of objects.[43] There is another faculty called retention, through which the mind preserves and collects these ideas.[44] This capacity to remember ideas is a different kind of perception, not purely passive: indeed, 'the appearance of those dormant Pictures' depends 'on the Will'.[45] A third faculty of the mind is called discernment, that is, the faculty to distinguish between several ideas.[46]

In order to understand the complex faculty of discerning, we should consider that all general and more abstract truths are composed of ideas, which can be either simple or complex. Simple ideas are prior to every composition and produce a uniform appearance or conception in the mind. Such perceptions could be clear or obscure, confused or distinct. A confused perception of something is not sufficient for scientific knowledge: indeed, if the 'Mind had a distinct Perception of different Objects, and their Qualities, it would be capable of very little Knowledge'.[47] For acquiring scientific knowledge, therefore, it is necessary that all ideas or perceptions of the mind are 'clear and determinate', that is distinct.[48] According to Locke, a perception is clear and distinct when it cannot be further analysed[49] – that is, when the perception is associated with specific qualities identified by analysis, and without such qualities the perception of that object ceases to be. In fact, Locke characterizes the quality of an external object as 'the Power to produce any Idea in our mind'.[50]

He next distinguishes two kinds of qualities: (1) primary qualities, which are inseparable from objects and to be found in all their particles, and (2) secondary qualities, which are those that have the power to produce sensation in the mind by means of primary qualities. If a primary quality is removed, then the object of knowledge vanishes and one may argue therefore that that quality is what specifically characterises the object. The mind has a clear and distinct perception when all the qualities that produce the ideas of the object are identified, and when the absence of any of these might lead to the loss of the object. This was one of the objectives of the negotiation of the intellect in all its different variations. To achieve this objective, the faculty of discerning divides the various ideas 'wherein can be found the least difference, thereby to avoid being misled by Similitude, and by affinity to take one thing for another'.[51] Locke adds that when the ideas are fixed in memory, 'mind makes the particular Ideas, received from particular Objects, to become general': this process is called abstraction.[52] Sensation, memory, discerning and abstraction are the pillars of Aristotelian epistemology grounded on *Posterior Analytics* II.19, as we have already seen, and this is what Locke in his *Essays*, in spite of the differences, seems to follow.

Up to this point, we have seen how Locke explained the origin of ideas, but he has not yet determined the process that leads to the clarity and distinctness of scientific knowledge. This problem is tackled at the end of the second book where Locke characterizes how an idea that is clear or obscure, distinct or confused, relates to the sense of the sight. An idea is clear when the mind 'has such a full and evident perception, as it does receive from an outward Object operating duly on a well-disposed Organ'.[53] A distinct idea, in contrast, 'is that wherein the Mind perceives a difference from all other' – that is, it has at least one specific characteristic without which it would be confused with other ideas. Therefore, a confused idea 'is not sufficiently distinguishable from another' or, more specifically, it is confused 'when it is such, that it may as well be called by another name'.[54] Confusion leads to the difficulty of discerning two things which should be considered separate – that is, it becomes problematic when the faculty of discerning does not find a specific characteristic that identifies one idea set against another. For this reason, in order to examine whether an idea is confused or not, it should be compared with other similar ideas to determine in a better way what distinguishes them.[55] The operation of the mind in separating or collecting ideas not neatly perceived as distinct is called judgement.[56]

Thus Locke characterizes knowledge in general as 'the perception of the connexion and agreement, or disagreement and repugnancy of any of our Ideas'.[57] This was a significant departure from the Aristotelians, who considered this agreement in relation to the essence of the thing; as it was from Bacon, who in the agreement of forms saw the constitution of the object; and also Hobbes, who conceived this agreement as the relation generated by the efficient cause, as we shall see later in this book. So, for Locke, there are four different possible modalities: (1) identity or diversity, (2) relation, (3) co-existence, or necessary connection, and (4) real existence.

Unlike the philosophers encountered in previous chapters – who believed as necessary a certain work of the intellect, typically comparison, to determine agreement and disagreement concerning identity and diversity – for Locke this operation is 'without any pains, labour, or deduction, but at first view', almost intuitively by means of the natural power of the mind to perceive and distinguish.[58] Indeed, through the first operation of the mind – that is, of perceiving ideas – mind apprehends that every idea has existence and is also distinct from others. By this process, the mind clearly and immediately understands 'each Idea to agree with itself, and to be what it is; and all distinct Ideas to disagree, i.e. the one not to be the other'.[59] It is an immediate and not-inferential perception that shifts the epistemological process from a discursive comparison between two or more ideas to the immediate intuition of this comparison. Only in this way can we avoid inconsistencies in Lockean epistemology between sensible knowledge and the definition of knowledge in general,[60] at least regarding what concerns the modality of identity and diversity. If agreement or disagreement between ideas were conceived as a discursive process, then it would be impossible to understand how knowledge may be based on sensation, which provides content and material, rather than on purely mental ideas detached from experience.[61]

The second modality is that of relation. If the agreement or disagreement were based only on identity and diversity, which shows only that an idea exists and is not another, no other kind of knowledge would be possible beyond the negative statement that a thing is not another thing. For Locke, there must be positive knowledge that determines the relations between ideas. This knowledge is neither immediate nor intuitive indeed, only that which determines identity or diversity is intuitive. Rather is it discursive, and this means that from the very beginning the agreement or disagreement is simply supposed and only subsequently is established as certain through looking for positive relations – that is, similarities – between ideas.

The third modality of agreement and disagreement leading to knowledge is reminiscent of the necessary connection in the search for demonstrability by the authors considered in this book. It concerns the co-existence or non-co-existence in the substance of particular qualities that always and necessarily inhere in them. Locke gives the example of the complex idea of gold, which is always conjoined with the quality of having a certain weight, or of being yellow, malleable, fusible and so on. These qualities are necessarily coexistent, and without them knowledge of gold would be impossible. Locke does not explain how these qualities are known, other than being discerned naturally by the mind, producing the idea of the object by means of sensation.[62] Unlike the Aristotelians, Bacon, Hobbes and Descartes, Locke sees no other explanation as possible, on the grounds that knowledge of substances is primarily knowledge of secondary qualities, which depend on the primary qualities of particles not perceivable by sensation. Without sensation of these particles, it is impossible for the mind to determine their necessary connection or incompatibility. Furthermore, Locke maintains, it is also impossible to discover a necessary connection between primary and secondary qualities. For this reason, certain and scientific knowledge of substance is not possible, as had been previously believed. Knowledge of substances is merely probable or conjectural. Locke's corpuscularism leads him to a position of epistemological pessimism, which would eventually be overtaken by the progress of chemistry.[63]

The last modality is the most problematic, but Locke deals with it in a rather off-hand fashion. Its tenet is that the idea in the mind should be in agreement with something existent, which fits with Lockean realism.

After establishing that knowledge in general is the agreement or disagreement between two ideas, Locke points out that there are two ways in which the mind possesses truth. The first way is through actual knowledge, which is also characterized as the 'present view' that mind has of the agreement and disagreement of ideas and their connections.[64] This kind of knowledge is immediate and refers directly to sensation. There is then habitual knowledge, which is constituted by the relations between all the actual, clear and distinct perceptions collected in the memory. This kind of knowledge is fundamental, otherwise every kind of knowledge would be extemporaneous.

Locke identifies two degrees of habitual knowledge. The first relates to intuitive knowledge that comes from the memory of an agreement or disagreement between ideas. The second degree, of equal certainty, comes from the memory of the conviction and persuasion of a previous demonstration – that is, when the agreement or disagreement between two ideas has been perceived by means of another idea, entirely different from an idea that generates perception. Locke offers the example of the proposition 'three angles of a triangle are equal to right angles', in relation to which the process of demonstration can be forgotten, although its truth is evident and certain as soon as one recollects this proposition. Mathematics is specifically based on this kind of knowledge, because once a specific thing is known as certain – given the immutable character of mathematics – this thing will always be true. In general, this kind of knowledge can be defined as demonstrative, and being based on the memory of a chain of reasoning, it is less perfect than intuitive knowledge, which immediately recognizes agreement or disagreement.

The various kinds of knowledge are thus characterized by various degrees of evidentiality. Intuitive knowledge is the most evident, because the agreement or disagreement between ideas is immediately perceived. The truth is perceived as 'they Eye doth light'.[65] It is also the most certain knowledge. The second degree of evidentiality is when the agreement or disagreement between ideas is not perceived in an immediate manner and this kind of knowledge is 'reasoning'.[66] It is made possible through a proof, which is, as we have already shown, the intermediate idea, leading the mind to perceive agreement or disagreement between ideas. This kind of perception is specifically called demonstration. Demonstration is not as evident as intuitive knowledge. Indeed, demonstration requires 'steady application and pursuit', which is not 'without pains and attention'.[67] It is based on a preceding doubt, coming from the fact that the perception of the agreement or disagreement was not immediate. According to Locke, demonstrative knowledge is based on intuitive knowledge: 'In every step Reason makes in demonstrative Knowledge, there is an intuitive Knowledge of that Agreement or Disagreement, it seeks, with the next intermediate Idea, which it uses as a Proof'.[68]

The necessity of this intuitive knowledge gave origin to the first principles of scientific knowledge, or maxims, *ex praecognitis et praeconcessis*. They are considered true in and of themselves, establishing a priori agreement or disagreement between ideas. Such principles for Locke are useless for the discovery of new knowledge; rather, they are the result of knowledge. Locke warns that

> those who have this Traditional Admiration of these Propositions, that they think no Step can be made in Knowledge without the support of an *Axiom*, no Stone

laid in the building of the Sciences without a general *Maxim*, but distinguish between the Method of acquiring Knowledge, and of communicating it; between the Method of raising any Science, and that of teaching it to others as far as it is advanced, they would see that those general *Maxims* were not the Foundations on which the first Discoverers raised their admirable Structures, nor the Keys that unlocked and opened those Secrets of Knowledge.[69]

Locke distinguishes neatly the method of discovery from the order of teaching, as Renaissance logicians did, and he believes that only the first stage of the discovery is fundamental for acquiring new knowledge. There is no regressus according to Locke because everything is resolved within experience. Certainty of knowledge is based only on the stage of the acquisition of knowledge and not on its order or arrangement.

In particular, demonstrative certainty is for Locke equivalent to that of mathematics, even if he does not deny that it is possible also in the case of other kinds of knowledge. The epistemic ideal is once again mathematical, and the fact that this degree of certainty is only attributed to mathematics according to Locke can be imputed to the lack of a method or of the application of the mind within the discovery. The main reason, however, lies in the fact that with mathematics 'in comparing their Equality or Excess, the Modes of Numbers have every the least difference very clear and perceivable'.[70] While in with respect to other ideas, which are not mathematical, differences 'are mad and counted by degrees', and 'we have not so nice and accurate a distinction of their differences, as to perceive, or find ways to measure just Equality or the least Differences'.[71] For instance, it is difficult to distinguish between the various shades of the colour 'white'. Such differences come from the particles, which are invisible to the senses and thus cannot be subject to immediate perception. For this reason, it is difficult to determine the cause of differences between physical objects. These causes are never perceived directly, rather they must be sought.

The logical consequence of this examination of different levels of evidentiality in human knowledge is that it is impossible to aspire to intuitive and immediately evident knowledge of all things. Furthermore, rational knowledge does not cover all ideas because only rarely is one able to find intermediate ideas for demonstration that can point out agreement or disagreement between ideas. In other words, the extension of human knowledge is limited and less comprehensive than both reality and ideas, because often the mind cannot find agreement or disagreement between them. These limits of scientific knowledge are also the boundaries of probability, because where the former ceases the latter begins. For this reason, in contrast to the views of the Aristotelians Locke regarded knowledge of substances as no more than probable, hypothetical and conjectural. In order to overcome this hypothetical character of the knowledge of substance coming from experience, the Aristotelians introduced the negotiation and a certain mental meditation of the intellect Locke, however, stops at experience, at what the Aristotelians, Bacon, Galileo, Hobbes and Descartes would call the analytical part of the method. A negotiation of the intellect does not and could not exist because the clarity and distinctness of knowledge depends only on sensation.

Human knowledge cannot be improved by the appraisal of principles or hypotheses or the contemplation of ideas. There exists no rational or intellectual moment, which can extend knowledge. Only experience 'must teach me, what Reason cannot'.[72] Indeed, mere reasoning

> will carry us but a little way in the certain discovery of the other Properties in those Masses of Matter, wherein all these are to be found. Because the other Properties of such Bodies, depending not on these, but on that unknown real Essence, on which these also depend, we cannot by them discover the rest.[73]

Locke does not deny that a 'Man accustomed to rational and regular Experiments shall be able to see farther into the Nature of Bodies, and guess righter at their yet unknown Properties'; however, these are only opinions. The only manner of getting and improving knowledge is 'by experience and history'.[74] Even if for Locke the possibility of a real science of natural phenomena is in doubt, the Baconian methodological approach of collecting experiences and doing experiments might be fruitful,[75] at least for refuting hypotheses too hastily accepted.[76] Hypothesis should be based always on 'matter of fact' and on 'sensible experience', while it is not appropriate to 'presume on matter of fact' on the basis of hypothesis or theories. The unique task of hypothesis is to aid memory after a thing is discovered, but, Locke adds, hypotheses 'very seldom are sound & sure enough without experience to warrant our practise or lead us into the right way of operation'.[77] These Lockean reflections, which are directly opposed to Cartesians philosophy, will have an impact on the elaboration of Isaac Newton's second edition of the *Principia*, especially in reaction to the positions of those who charged him with tinkering with the occult in his notion of gravity.

In Locke, clearness and distinctness remain within the firm boundaries of direct sensation, and no mental or intellectual process can make knowledge scientific, certain and evident. There is a mental process, which Locke calls 'rational experiment', which helps to make probable conjectures about causes, but without achieving scientific knowledge of the same. It allows for the radical act whereby the observational inaccessibility of microscopic things like corpuscles may be overcome. This process is *analogy* and by means of it causes – even if not known perfectly and scientifically – 'can appear more or less probable, only as they more or less agree to Truths that are established in our Minds, and as they hold proportion to other parts of our Knowledge and Observation'.[78] The effectiveness of analogy depends on other truths having already been established by sensation, and where its conclusions are coherent with the entire corpus of knowledge and observations:

> Observing that the bare rubbing of two Bodies violently one upon another, produces heat, and very often fire itself, we have reason to think, that what we call Heat and Fire, consists in a violent agitation of the imperceptible minute parts of the burning matter.[79]

This 'we have reason to think' is a supposition or an argument by analogy, which establishes the possibility of an inference from what is observable to what in principle is unobservable. But how is it possible that transduction of this kind may be compatible with an epistemology based on sensation like that of Locke? While for Aristotelians,

Bacon and Descartes, analogical reasoning by means of comparison led to certain and scientific knowledge of causes, for Locke it led only to probable knowledge. Analogy is therefore 'the only help we have' to infer from macroscopic to microscopic, 'it is from that alone we draw all ground of probability'.[80] Probability, rather than certainty, makes of analogy an epistemological inference for Locke. But there is much more than this. Indeed, the very idea of this kind of reasoning is based on a broader philosophical view based on the conception of the harmony, continuity and uniformity of nature as God's creation, which Locke shares with Newton:

> Finding in all parts of the Creation, that fall under human Observation, that there is a gradual connexion of one with another, without any great or discernible gaps between, in all that great variety of Things we see in the World, which are so closely linked together, that in the several ranks of beings, it is not easy to discover the bounds betwixt them, we have reason to be persuaded that, in such gentle steps Things ascend upwards in degrees of perfection ... Observing, I say, such gradual and gentle descents downwards in those parts of the Creation that are beneath man, the Rule of Analogy may make it probable, that it is so also in things above us and our observation.[81]

The limits of observations are the potentialities of the analogy. Locke does not dismiss the central role of sensation but is unable to deny that something may also exist at some level of probability beyond the power of observation. This is justifiable on the basis of a very specific vision of nature based on extra-epistemological assumptions like that of harmony or continuity. Or, to be more precise, there is the introduction of these principles of nature from metaphysics into epistemology. This introduction constitutes the ground for probable knowledge.

Therefore, in Locke's epistemology – as Peter R. Anstey has pointed out – analogy, on the basis of resemblances, 'consists in reasoning from effects to causes or in reasoning from relations between qualities in observable objects to relations between qualities in unobservable objects',[82] but it leads only to probable knowledge. In sect. 40 *Of the Conduct of the Understanding*, Locke writes

> Analogy is of great use to the mind in many cases, especially in natural philosophy; and that part of it chiefly consists in happy and successful experiments. But here we must take care that we keep ourselves within that wherein the analogy consists. For example, the acid oil of vitriol is found to be good in such a case, therefore the spirit of nitre or vinegar may be used in the like case. If the good effect of it be owing wholly to the acidity of it, the trial may be justified; but if there be something else besides the acidity in the oil of vitriol which produces the good we desire in the case, we mistake for that analogy which is not, and suffer our understanding to be misguided by a wrong supposition of analogy where there is none.[83]

The limits of analogy are the limits of rational reasoning in providing certain and scientific knowledge beyond the contribution already furnished by sensation. For this reason, being based on resemblance, analogy can picture only an approximation of reality, which is far from the epistemic certainty required in scientific knowledge, even if sometimes 'a wary Reasoning from Analogy' can lead 'into the discovery of Truths, and useful Productions'.[84] The possibility of having certain knowledge of the physical world seems to fall, with the exception of what is directly, but passively perceived by sensation. Only mathematics can aspire to true scientific knowledge.

10

Gottfried Wilhelm Leibniz

Hannover. 1696. It was approaching the end of the year, and Gottfried Wilhelm Leibniz was collecting material for his history of the House of Braunschweig when, in one of the rare moments of pause, he wrote to his friend Gabriel Wagner on the value of Aristotelian logic imbibed during his university years – exactly the opposite of Descartes' attitude sixty years before in the *Discourse on the Method*. He described as childish the dismissal of what is learnt in youth as completely inappropriate and useless. Indeed, in his mature years he discovered the importance of what he had studied:

> Before I entered a class in which it was taught, I was steeped in the historians and poets, for I had begun to read history as soon as I could read at all, and I found great pleasure and value in verse. But as soon as I began to learn logic, I was greatly stirred by the classification and order which I perceived in its principles. I came at once to notice that there must be something great in it, as far as a lad of thirteen years could notice such a thing. My greatest pleasure lay in the categories … I soon made the amusing discovery of a method of guessing or of recalling to mind, by means of the categories … I took great pleasure in such matters and wrote out all kinds of stuff, but then forgot it and let it be lost. Many years later, however, I found some of it and discovered that it did not entirely displease me. Later on I found the value of these exercises when I came to work out certain problems … This is what I wanted to say about the great, and for the most part tested, advantages of the known logic when it is rightly used.[1]

On this logic, Leibniz built the foundations of his epistemology. Leibniz's epistemological reflections are not confined to a single writing, but are scattered across several works, now considered canonical by the scholarship for reconstructing his theory of knowledge: *Meditations on Knowledge, Truth and Ideas* (1684), *Discourse of Metaphysics* (1686), *New Essays on Human Understanding* (1704) and *Monadology* (1714).[2]

Generally, it would be appropriate to create an outline of Leibniz's theory of knowledge starting from his earlier works, but in this specific case a better understanding comes at the very beginning of the *Monadology*, which may be considered his last epistemological work. In fact, this short treatise seems to exclude any kind of contribution from experience and to characterize Leibniz as rationalist. Leibniz starts with his idea of monad as simple substance, as atom of nature, which has no windows, 'through which anything could come in or go out'.[3] This would indicate that for Leibniz no contribution from experience is possible and that all knowledge comes from within the monad. Indeed, among monads there is no mutual influence, and contact only occurs among them through the intervention of God. Leibniz, therefore, comes to the conclusion that each monad 'represents the whole universe'.[4] The very fact that monads represent the same universe explains the a priori reasons why things are necessary. In other words, by Leibniz's thinking this guarantees a scientific knowledge of the universe. However, monads do not represent the whole universe in a uniform way, but only from their own perspective. The differences among monads consist in the fact that their representations of the details of the universe are generically confused and 'can only be distinct with respect to a small part of things, namely those which are either closest or largest in relation to each monad'.[5] Leibniz seems to propose a modified version of Plato's epistemology, according to which all that is knowable is always already known by the mind, but by means of a process of anamnesis a new portion of knowledge is discovered, and this portion differs from mind to mind. Unlike Plato, Leibniz does not conceive of this knowledge as a kind of recalling, but rather he tries to develop a more articulated epistemological framework.

For Leibniz, in a monad, 'there is a great multiplicity of small perceptions, in which there is nothing distinct',[6] and we are sure to have at least some kind of distinct perception. Otherwise, the mind would be in a permanent state of stupor and nature would have in vain provided animals with

> organs which bring together a number of rays of light or of undulations in the air, thus making them more effective by combining them. There is something approaching this in the case of scent, taste, and touch, and perhaps in numerous other senses which we are not aware of.[7]

It is a matter of fact, therefore, that monads have distinct perceptions and that in monads there is a transition between indistinct and distinct perceptions. At the beginning of the *Monadology*, Leibniz writes that a perception is 'the transitory state which incorporates and represents a multitude within a unity or within a simple substance'.[8] This perception must be distinguished from apperception, which involves conscious perceptions. Indeed, as we have seen, there are perceptions which are not apperceived, and this was the main mistake which Cartesians made when they asserted that souls alone were simple substances. Leibniz introduces a refinement into the Cartesian conception, establishing the existence of small or indistinct perceptions for all simple substances. Souls according to Leibniz are not only simple substances but 'those which have perceptions which are more distinct and accompanied by memory'.[9] Memory is crucial because – albeit different from reason – it provides a kind of sequencing according to which animals

> which are struck by the perception of something of which they have had a similar perception before are led to expect whatever was connected to that previous

perception, and are therefore taken with feelings similar to those which they had on the previous occasion.[10]

If human beings only use memory for sequencing their perceptions, they are like empirics with history, but without theory, basing their ideas on the collection of particulars. Baconian history is therefore important for acquiring scientific knowledge, but not in itself sufficient. Here we find a first glimpse of the idea that Leibniz's epistemology is more than rationalist, and that experience, experiments and observations play a crucial role.

Memory and history are not sufficient because what distinguishes human beings is science, which is the knowledge of necessary and eternal truths.[11] Leibniz, therefore, seems to exclude any contribution of experience, if by experience we conceive of something coming from the external world of the monad and which can extend knowledge. What is external must be understood as what is different from the 'I' to which all the conscious perceptions refer. In a letter to Simon Foucher in 1675, Leibniz writes that

> There are two absolute general truths, that is, which speak of the actual existence of things: the first, that we think, and the second, that there is a great variety in our thoughts. From the former it follows that we exist and from the latter it follows that there is something else besides us, that is, something else besides that which thinks, something which is the cause of the variety of our appearances.[12]

The first truth is obviously taken from Descartes, but the latter, however, is instrumental for Leibniz in explaining how in the unity and uniqueness of substance there are multiple representations rather than one. Leibniz establishes a causal relation between object and perception, which G. H. R. Parkinson calls 'causal theory of sense perception'.[13] There is not a conscious perception that is not caused by an object, and this object is first of all external. In a letter to Queen Sophie Charlotte, Leibniz states that 'external senses are necessary for us to think and, if we had none of them, we would not think', simply because external objects stimulate the transition from little perceptions and conscious perceptions.[14] Thus 'senses provide the content for reasoning and we never have thoughts so abstract that something sensible is not mixed with them'.[15] Unlike Locke, however, for Leibniz knowledge is not limited to sensation – that is, sensation is not the only source of cognition.

Leibniz now broaches the problem of establishing the objectivity of perception, and how from the epistemological standpoint it is possible to effect the transition from little unconscious perceptions and distinct perceptions. Both these questions were without solutions in the *Monadology*.

Leibniz answers the first problem in a letter to Bartholomew Des Bosses and in *On the Method of Distinguishing Real from Imaginary Phenomena*. In order to establish if what happens in the mind is in agreement with what happens out of it, it is sufficient 'that those things that happen in one single mind are in agreement among them and with what happens in any other mind'.[16] Leibniz adds that the most valid clue is the agreement of our experience with the experiences of other minds.[17] For Leibniz, such agreement, however, is granted only by God according to the model of pre-established harmony, limiting thus his epistemology to an opaque dimension:

> Thus there is a perfect harmony between the perceptions of the monad and the motions of the body, pre-established from the beginning between the system of efficient causes and that of final causes. It is in this that the accord and the physical union of soul and body consist, without either one being able to change the laws of the other.[18]

The only solution, even if not demonstrative, is that the reality of phenomena is assessed by 'success in predicting future, past and present phenomena'.[19]

The problem of the process of distinction, in contrast, was to be the subject of the *Meditations*. The context of its publication was the discussion around Antonie Arnauld's book *Of True and False Ideas* (1683), which criticized Nicolas Malebranche's representationalism in *The Search after Truth*.[20] Malebranche gave his immediate response with a *Reply* to Arnauld, who in turn defended his position in the *Defence*. However, in a letter of autumn of 1684 to Ehrenfried Walhter von Tschirnhaus, Leibniz wrote that he had not yet read the works of the controversy between Malebranche and Arnauld.[21] And nor do the *Meditations* attempt to thematize Lockean epistemology. This is reserved for the *New Essays* and is itself remarkable because many of the unpublished writings, – written before the reading of Locke's *Essay* – seem to reflect on or respond to Lockean doctrines.

The *Meditations* remains a work perfectly inscribed within Cartesian epistemology, which, according to Leibniz, did not explain sufficiently and following necessary criteria the ways of distinguishing the various kinds of ideas and of knowledge. Leibniz writes that

> knowledge is either obscure or clear, and again, clear knowledge is either confused or distinct, and distinct knowledge either inadequate or adequate, and adequate knowledge either symbolic or intuitive: and, indeed, if knowledge were, at the same time, both adequate and intuitive, it would be absolutely prefect.[22]

And so he provides an overview of all the fundamental concepts for understanding the way in which the human mind acquires scientific knowledge.

Leibniz characterizes as obscure that notion 'which is not sufficient for recognizing the thing represented',[23] in other words, where the mind has not the capacity to distinguish it from other ideas. Conversely, knowledge is clear when the mind admits recognition of the thing represented. Though clear, however, this knowledge can still be either confused or distinct. Knowledge is clear and confused when one 'cannot sufficiently enumerate marks one by one in order to differentiate one thing from others, even though the thing does indeed have such marks and requisites into which its notion can be resolved'.[24] The mind recognizes a diversity of ideas without, however, being able to identify the marks that characterize that thing specifically. This kind of knowledge is typical of sensation. Indeed, the difference between two colours is clear at first sight, but the mind does not recognize distinctly the marks that determine this difference. On the contrary, a distinct notion is that 'notion connected with marks and tests sufficiently to distinguish a thing from all other similar'.[25] Knowledge is distinct, therefore, when it is possible to enumerate all marks sufficiently to determine that thing and distinguish it from another. Distinct knowledge is also adequate when 'every mark that enters into a distinct notion is, again, distinctly known, or when analysis has been carried to completion'.[26] Indeed, distinct knowledge is inadequate when the mind's knowledge of the enumerated marks is confused and indistinct. When it is possible to conceive all marks that pertain to a thing in a distinct way, then the mind has adequate and intuitive knowledge. Sometimes in instances of complex knowledge, when analysis is required, 'we do not usually grasp the entire nature of a thing all at once' and 'in place of the things themselves we make use of signs'.[27] This kind of knowledge is called symbolic.

According to Leibniz, we can have scientific knowledge of the ideas of things only when knowledge is intuitive. Indeed, Leibniz states, the term 'idea' has often been used inappropriately, as in the case of the Cartesians. Furthermore, an idea is true when a notion is possible – that is, it stands without contradiction. This possibility is knowable either a priori or a posteriori. A posteriori knowledge of truth occurs when the mind has experience of something existing, wherefore if it exists then it is possible. A truth can be known a priori only when a notion is resolved into its elements, of which the mind knows the possibility. Therefore, 'whenever we have adequate knowledge, we also have a priori knowledge of possibility, by virtue of having carried an analysis to completion'.[28] This kind of knowledge is specific to 'when we understand the way in which a thing can be produced',[29] – that is, when the mind has a causal definition. For Leibniz – who rejects Descartes's definition of clearness and distinctness – the method that leads to scientific knowledge, and which allows the transition from little perceptions, through distinct perception, to adequate ideas, is analysis. However, Leibniz is sceptical as to 'whether people can ever produce a perfect analysis of their notions or whether they can ever reduce their thoughts to primitive possibilities or to irresolvable notions'.[30]

All that we have seen up to now seems to argue in favour of a strict rationalism that excludes any kind of empiricism or appeal to experience, since every representation is already in the monad. Leibniz seems at the opposite pole to Locke, as a number of scholars suggest. However, as we have also seen, experience in the Leibnizian sense plays a fundamental role in transforming what is unconscious into something scientifically knowable. Scientific knowledge is possible only by means of analysis. A complete analysis, especially of physical phenomena, seems hard to realize for Leibniz, while he is more open to the possibility of such in the case of knowledge depending only on reason.

In *On Universal Synthesis and Analysis, or the Art of Discovery and Judgment*, Leibniz states that in matters of fact or in contingent things, which depend on observations and experiments, 'primary truths (for us) are those that are perceived immediately within us or those of which we are conscious within ourselves'.[31] Leibniz re-iterates the notion already aired to Des Bosses, and *On the Method of Distinguishing* establishes that the existence of thinking beings and the fact that there are many different thoughts caused by multiple things depend on these primary truths. These things are only contingent and not metaphysically necessary; therefore,

> we must regard the agreement of phenomena as truth, since such agreement does not occur by chance but has a cause. Certainly it is only through this agreement among phenomena that we distinguish dreams from waking, and we predict that the sun will rise tomorrow only because it has fulfilled our faith so often.[32]

Once again, the connection between perceptions and phenomena is based on the predictive power of the causal knowledge of the mind. Purely on the grounds that experience plays such a fundamental role, Leibniz desires a history of phenomena. If we add to this the Baconian history of phenomena, we add truths abstracted from experience, and from here it is possible to constitute the possibility of mixed sciences. Leibniz, more than Locke therefore, seems to have developed the original Baconian idea.

A history of phenomena is not sufficient in itself for the emergence of scientific knowledge – indeed, it would be like a 'store well provided with all kinds of wares but without any order or inventory'.[33] The allusion to the epistemology of empirics is quite explicit. In the *Monadology*, as I have already shown, Leibniz criticizes the attitude of the empirics, who do no more than collect data. Leibniz is aware that 'we need a particular art for arranging as well as for ordering and combining our experiments, so that useful inductions can be made from them, causes discovered, and aphorisms and anticipations'.[34] Leibniz's terminology directly recalls Bacon, with the use of induction, aphorisms and anticipations for acquiring scientific knowledge after the compilation of the natural history of phenomena. Leibniz points out that this art is constituted of synthesis and analysis. Synthesis is achieved 'when we begin from principles and run through truths in good order, thus discovering certain progressions and setting up tables, or sometimes general formulas'.[35] In the Aristotelian tradition, this kind of synthesis is the demonstration *propter quid*, from causes to effects. This kind of art would be satisfactory enough if we were to know the truth of all principles, many of which have been, however, previously discovered by others. If the certainty of these principles is in doubt, then analysis should be performed. Indeed, 'analysis goes back to the principles … just as if neither we nor others had discovered anything before'.[36] Analysis is not sufficient, and synthesis should follow. Leibniz re-proposes the Aristotelian theory of regressus. Leibniz points out that there are two kinds of analysis. One is used in algebra and advances by leaps. This is clearly not the kind of analysis helpful for understanding natural phenomena. The second kind of analysis is 'special and far more elegant but less well known'.[37] This is known as reductive analysis, and it should be responsible for providing the primary elements of scientific knowledge in relation to natural phenomena. However, in *On Universal Synthesis and Analysis*, Leibniz does not explain how this reductive analysis works.

We can find a precious clue to what Leibniz means in *The Secret Analysis in Physics*. The secret analysis of natural phenomena consists in 'the reduction of the confused qualities of the senses … to the distinct qualities that accompany them'.[38] According to Leibniz, confused qualities for the senses are those like colours for sight, flavours for taste, odours in the case of smell and so on, while distinct qualities are 'number, size, shape, motion, and consistency'.[39] The last two distinct qualities in particular pertain to natural phenomena. The reduction involves finding out whether distinct qualities accompany certain confused ones – that is, whether confused qualities can be understood in a necessary way through distinct qualities which allow us to determine the nature and the essence of the phenomenon. However, there is no attempt to explain how to perform this reduction.

*

We can find a key to Leibniz's art and his conception of mixed sciences in the *On Universal Synthesis and Analysis*, in the short paper on *Physical Truths*, which constitutes a pivotal text according to Anne-Lise Rey for understanding Leibnizian provisional empiricism.[40] In this work, Leibniz distinguishes three kinds of physical truth: (1) intellectual, (2) sensible and (3) mixed.

Intellectual physical truths are known only by reason and are divided into axioms and theorems. Axioms are self-evident by nature and understandable by everyone, while theorems require a further demonstration. Sensible physical truths are known only by sensation and are divided into histories or inductions. Histories are collections of singular truths, while inductions infer a universal truth from the agreement of many singular cases. Inductions are of two kinds: (1) observations and (2) experiments. Observations are those inductions for which we are the mere spectator, as in celestial observations, while with experiments we are not merely spectator but ourselves manipulate nature, as with chemical reactions. Not all inductions have the same degree of certainty according to Leibniz: 'for instance nobody can doubt that the sun will rise tomorrow, while that rhubarb purges is not equally certain'.[41] For people, in general, it is definitely a much clearer understanding that the sun will rise tomorrow than that rhubarb helps to restore health, without introducing any further experiment or reasoning. Mixed physical truths are sensible and intellectual at the same time. For instance, that a concave mirror collects rays is known by observation – that is, the angles of reflection and of incident are equal and also by geometrical reasoning. Therefore, according to Leibniz, natural investigations should combine observations with experiments and demonstrative reasoning in order to acquire certainty and scientific knowledge.[42]

Leibniz explores the combination of these two stages in *An Introduction on the Value and Method of Natural Science*. The work was written between the summer of 1678 and the winter of 1679, but, as we shall see, it seems to tackle some important issues of Lockean epistemology in a very direct way. Leibniz starts from a clear Aristotelian position by establishing that the proper power of the mind is understanding and that 'we will be the happier the clearer our comprehension of things and the more we act in accordance with proper nature, namely, reason'.[43] The activity of reason is impeded by the passions of bodies, which are impossible to evade entirely. Therefore, Leibniz concludes, knowledge of bodies is important 'to perfect our mind through the understanding of the purposes and causes of things' and 'to conserve and nurture our body, which is the organ of the soul', reducing the potential for what is harmful for reason.[44] Leibniz particularly focuses on knowledge of the causes and purposes of things, which entails learning the 'mode of operation', to borrow an expression employed by Locke in the drafts of his *Essay*.

Such knowledge is possible for Leibniz, first of all, by 'setting up a history of nature in which experiments will be brought together in a catalogue'.[45] As we have seen, however, experiments 'are to be combined with accurate and thoroughly extended reasonings after the manner of geometry, for only in this way can causes be discovered'.[46] Experiment results are particularly useful for bringing 'to light true principles from effect to causes'.[47] This kind of reasoning is analytic and consists of a long 'chain of reasons and involves a kind of geometry or calculus and cannot be understood without much thought'.[48]

Leibniz's epistemology moves on from the ashes of the Aristotelian theory of regressus. He makes the point that the discovery of causes, without which no advances in physical sciences would be possible, cannot be obtained through quick flights and leaps – as was Bacon's view also – but only through a process of profound and *quasi* geometrical reasonings. Leibniz, however, is sceptical about the possibility of applying analysis to natural phenomena. Indeed, 'unless principles are advanced from geometry and mechanics which can be applied with equal ease to sensible and insensible things alike, nature in its subtlety will escape us'.[49] Like Locke, Leibniz believes that the nature of things can be beyond our power of knowledge because 'a corpuscle hundreds of thousands times smaller than any bit of dust which flies through the air, together with other corpuscles of the same subtlety' cannot be the subject of sensation.[50] Because of this, 'reason must supply this most important lack in experience' for reason deals with these corpuscles 'as easily as can a ball by the hand of player'.[51] As we have seen, Leibniz admits what Locke denies – that is, an intervention or a working of the mind to supply the deficiency of knowledge acquired only by experience, just as the Aristotelians did with the negotiation of the intellect. This work of the mind in a perfect world would be the a priori act of contemplation of God, and this method of discovery of the nature of bodies is difficult and 'not everyone should undertake it'.[52] Indeed, Leibniz claims that 'sensible effects are too greatly compounded to be readily reduced to their first causes'.[53] Leibniz evinces a kind of pessimism, reminiscent of Locke, in his sense of the acquisition of knowledge in this way.

There is another possible method of discovering causes which is conjectural and proceeds a priori 'by hypothesis, assuming certain causes, perhaps, without proof, and showing that the things which now happen would follow from these assumption'.[54] However, Leibniz adds that 'no firm demonstration can be made from the success of hypothesis', and in this way we can acquire only probable, not scientific knowledge. Hypotheses may fit perfectly with phenomena but cannot provide scientific knowledge. As with Locke, Leibniz believes that hypotheses mainly serve memory because 'phenomena are virtually contained in the hypothesis from which they can be deduced, so that anyone who remembers the hypothesis will more easily recall these phenomena'.[55]

There is a further conjectural method a posteriori, which proceeds from experiments and 'rests for the most part upon analogies', like a person who 'seeing that lye fights against acids, reduce all corporeal conflicts to those of acid and alkali'.[56] In particular, analogy is helpful in investigating (but not finding) 'the true causes of things',

> for it is always easier to discover the cause of a phenomenon which several things have in common ... the cause of the same phenomenon can be investigated more easily in one subject than in another, as anatomists who dissect different animals well know.[57]

Leibniz's analogy works not only via resemblances, but also by means of comparisons analysing the common features of phenomena. Again like Locke, Leibniz uses the example of acids and alkalis to explain analogy. Furthermore, once more like Locke, he warns 'against the abuse of analogies' because they do not lead to certain conclusions, even if they can be extremely useful when making predictions 'about matters of which we as yet have little experience'.[58]

The only viable reasoning method for human beings for acquiring scientific knowledge is that which proceeds from experiments to the discovery of causes capable of resolving a phenomenon into its various attributes, and also of seeking the causes and effects of each attribute:

> We must resolve every phenomenon into all its circumstances by considering separately colour, odour, taste, heat, and cold, and other tactile qualities, and finally the common attributes of magnitude, figure, and motion. Now if we have discovered the cause of each of these attributes in itself we will certainly have the cause of the whole phenomenon. But if by chance we do not come upon the reciprocal and permanent cause of certain attributes, but only several possible causes, we can exclude those which are not pertinent here.[59]

The method Leibniz proposes in this passage is not dissimilar to that process suggested by the Aristotelians with the negotiation of the intellect. Looking for the essential elements of a phenomenon and for their causes is the first step towards understanding the cause of the phenomenon. The way in which Leibniz characterizes the process is an evolution of the negotiation of the intellect:

> Assume two attributes, A and L, of the same phenomenon and assume that there are two possible causes of A, namely, b and c, and two of L, namely, m and n. Now, if we establish that cause b cannot exist along with either m or n, it follows necessarily that the cause of A is c. If we can further establish that m cannot exist along with c, then the cause of L must be n.[60]

Like Locke, Leibniz is aware that perhaps the complete enumeration of possible causes is not always in the power of the human mind, but, unlike Locke, he believes that this method at least makes exclusion certain and can provide the foundation for scientific knowledge. What is hard to perform is the complete analysis of the attributes. Indeed, 'of the attributes which are presented to the senses, some are simple, others are compounded of simple ones'.[61] Composite attributes must be resolved into simple ones. Simple attributes can be simple on account of their 'nature and intellectual reasons' or 'with respect to our senses'.[62] Simple attributes by nature are those such as 'to endure' or 'to be itself', while simple attributes with respect to the senses are those such as 'coldness' or 'heat'. The problem with simple attributes with respect to the senses is that sensation does not show

> by what mechanism the state of a body is produced which brings about the sensation of warmth in us, yet the mind properly perceives that warmth is not

something absolute which is understood in itself but that it will only then be adequately understood when we explain of what it consists or distinctly describe its proximate cause.⁶³

In most cases, attributes which are simple to the senses cannot be explained or described, but rather their existence simply comes to light through the senses. No further reasoning can provide knowledge about them, only experience. Leibniz explains this idea with the example of

> a land where men do not know the sun and fire and have blood which is cold, not warm; surely they cannot be made to understand what heat is merely by describing it, for even if someone were to explain to them the innermost secrets of nature and even interpret perfectly the cause of heat, they would still not recognize heat from this description if it were presented to them, for they could not know that this peculiar sensation which they perceived in their minds is excited by this particular motion, since we cannot notice distinctly what arises in our mind and what in our organs. But if someone kindles a fire near them, they would at length learn what heat is.⁶⁴

These attributes are confused with respect to the intellect and a reference to sensation is always necessary. So Leibniz draws a clear distinction between distinguished, confused attributes and distinct attributes, which pertain properly to the intellect and whose resolution is known even if they are not primary notions. On the one side, there are simple distinct attributes – attributes such as 'to be itself' or 'to endure' – which do not require a resolution because they are understood in themselves. On the other side, there are distinct attributes that are characterized by a definition such as the definition of gravity as the striving towards the centre of the Earth. These definitions require a further resolution, according to which some attributes may be classified as either confused or distinct. Confused attributes in a definition must be resolved in order to have a perfect understanding of definition. However, the resolution of these attributes does not necessarily lead to distinct attributes. Indeed, Leibniz writes quite explicitly that 'there is a way of resolving confused attributes experimentally into other attributes … which does not make them cease to be confused'.⁶⁵ Resolution is indeed the instrument for investigating the causes of confused attributes. The process of this kind of resolution looks familiar if we look back to the Aristotelian tradition. Leibniz states that in order to understand confused attributes, they must be related with other attributes and with the subject that contains them. But these subjects are knowable only through attributes, and therefore to relate confused attributes with subjects means relating them with a collection of other attributes which constitute the subjects. The attributes that concur in the constitution of the subject can be either confused or distinct, and therefore a confused attribute can be related with confused or distinct attributes. What happens for Leibniz is that the relating of an attribute to others consists 'in making apparent their concurrence in the same subject, their connection with each other, their compatibility, and on the other hand, how one can be changed into another or can be produced out of several others'.⁶⁶ So assessing the relation between the various attributes in a subject with a confused attribute means to understand whether the confused attribute can co-exist in the subject with other attributes, whether the confused attribute concurs in the existence of the subject and whether an attribute can replace another. Leibniz calls this kind of resolution experimental in opposition to intellectual resolution, and makes two compelling examples:

Experimental resolution	Intellectual resolution
The colour green arises from a mixture of blue and yellow, no change taking place in the coloured object but only the eye. Furthermore, the separate ingredients can sometimes be distinguished with a microscope, each with its own colour, yellow or blue. We cannot yet say with any certainty, however, that blue and yellow are prior to or simpler in nature than green, for we do not understand, but merely experience, that green arises out of yellow and blue.[67]	We understand though we may not experience that a square is made by two right isosceles triangles joined by a common hypotenuse and lying in the same plane or that from two odd numbers there arise an even number.[68]

Experimentally, Leibniz writes, we can show that two colours form a third colour, but we cannot explain it with mere use of the intellect, we need sensation. This kind of knowledge remains probable because 'we do not grasp how the third colour is given us through the confused appearance of these two colours'[69] – that is, certain knowledge of the cause is not possible. In opposition to this experimental resolution, which does not lead to scientific knowledge, Leibniz clearly compares the intellectual resolution to definition, as the Aristotelians did in discussing the negotiation of the intellect. Indeed, Leibniz writes that 'in intellectual resolution or in definition' – which means they are two interchangeable operations leading to the same result – the mind 'understands that which is described when the ingredients of the description are understood'.[70] By contrast, in experimental resolution made by sensation alone, what is resolved does not cease to be confused. There is a kind of scepticism about knowing physical phenomena perfectly, which is due to the limited capacity of the human mind.

Leibniz suggests a possible way out by stating that in resolving a subject, the mind should first consider distinct attributes such as 'duration, magnitude, motion, figure, angle, and other circumstances'.[71] These attributes allow for 'the application of mathematics to physical science' which leads to scientific knowledge. Scientific knowledge of physical phenomena is therefore possible, not absolutely, but only from the perspective of those attributes which characterize the nature of things mathematically. According to Leibniz, human beings know from a position of certainty and rigour only what is mathematizable, while everything relating to matter evades the possibility of scientific knowledge:

> Since everything confused is by its nature resolvable into the distinct, even though it may not always be in our power to do this, it follows that all qualities and mutations of bodies can, according to their nature, at length be reduced to certain distinct concepts. But in a body viewed as matter only, or as that which fills space, nothing can be conceived distinctly beyond magnitude and figure … Thus material things can be explained through magnitude, figure, and motion.[72]

This is a bold Aristotelian statement, which quite explicitly recalls the second book of *Metaphysics*. Leibniz upholds the idea that 'the truth of the matter is that one must give the reason for such qualities and explain how they arise in a body'.[73]

So how do Leibniz and Locke differ? Underlying Leibniz's epistemological pessimism – which can be compared to Locke's position – there is the conviction that there are two kinds of truths, that of reason and matters of fact. Already in the early *Dissertation* (1666), Leibniz established that there were two eternal truths, 'which exist, not by the will of God, but by their own nature'.[74] Next, there are truths that have their 'basis in existence, not in essence, and which are true as if by chance, i.e. by the will of God'.[75] Primary truths, which are eternal, are those of mathematics and are analysable. Truths concerning existence, in contrast, are termed historical and are collections of observations. Historical truths are not known through analysis or demonstration, but by means of mere induction. This distinction establishes a break with the previous epistemological model – indeed, it awards the characteristics of the truth to mere contingent knowledge, whereas all the other philosophers, Locke included, believed that truth was yielded by necessity alone. In other words, there is a science also of contingent things.

In the *New Essays*, Leibniz states that there are propositions of fact and propositions of reason. In this case, he does not use the term 'truth' but prefers the more generic cognitive word 'proposition'. He establishes that the difference between these two kinds of propositions does not consist in universality – indeed, both may be considered universal. Where these propositions differ is in this respect: the universality of propositions of fact depends on induction and observations – that is, a contingent universality (what Aristotelians called generality) – while propositions of reason are universals by nature. Leibniz adds that there are also mixed propositions

> that derive from premises some of which come from facts and observations while others are necessary propositions. These include a great many of the findings of geography and astronomy about the sphere of the earth and the paths of the stars, arrived at by combining the observations of travellers and astronomers with the theorems of geometry and arithmetic. But logicians have a principle which states that a conclusion cannot be more certain than the least certain of the premises; so these mixed propositions have only the level of certainty and generality that observations or propositions of fact have.[76]

Leibniz reduces these three typologies of propositions to those two kinds of truth already identified in the *Dissertation*, stating that mixed propositions have the same epistemic value as those of fact. It would seem a superficial remark following on from what Leibniz had previously said on mixed sciences, but in truth it deals a hard blow to Aristotelian epistemology. Indeed, in his comments on the Lockean chapter *On the Reason*, Leibniz explains the peculiarity of these mixed proposition in the analytic process:

> According to Pappus, analysis as practised by the ancients was the procedure of taking the proposition which is to be proved and deriving consequences from it until something given or known is reached. I have noted that for this to be effective the propositions must be reciprocal ones, so that a synthetic demonstration can move backwards along the path which the analysis has followed; but it is still a matter of drawing inferences. It is worth noting here, though, that in the hypotheses of astronomy and natural science the return journey cannot occur; but neither does a success demonstrate the truth of the hypothesis. It does indeed make it probable; but as this probability appears to sin against the rule of logic which, tells us that truths can be derived from falsehoods, it may be thought that logical rules are not entirely applicable to matters of probability. My reply to this is that although it is possible that a truth should follow from a falsehood, it is not always probable, especially when a single hypothesis explains many truths – something which is rare and not easily encountered.[77]

Albeit Pappus's conception of analysis, as we know it, is different from that of the Aristotelians, Leibniz seems to deny in every way the possibility of regressus in the fields of natural sciences and astronomy. Leibniz rejects regressus for the same reasons maintained by Aristotelians, who denied the validity of synthesis for acquiring new knowledge: (1) knowledge based on weak premises cannot be epistemologically stronger than the premises themselves and (2) knowledge derived from induction and observation is always probable. Indeed, regressus seems to be possible for Leibniz only if the propositions are reciprocal,[78] as also the Aristotelian supporters of regressus argued. While it was their belief that with the negotiation of the intellect the propositions of natural science could also become reciprocal, for Leibniz this is not possible. Thus even 'when this reverse process is not demonstrative – in natural science, for instance – it still sometimes yields great likelihood in situations where the hypothesis readily explains many phenomena which would otherwise be puzzling, and which are quite independent of one another'.[79]

Leibniz does not dismiss the possibility of attaining truth in the natural sciences, and thus in the *Monadology* he provides a justification for how they must be considered as true, albeit with aspects of their content remaining not perfectly analysable. Leibniz characterizes the truths pertaining to reason as necessary – that is, their opposites are impossible – while truths of fact are held to be accidental and contingent – that is, their opposites are possible. Whereas the philosophical tradition termed these truths of fact probable knowledge or opinion, Leibniz elevates this kind of cognition to the realm of truth. The common root of these two versions of truth rests on two epistemological principles: that of contradiction, according to which that which is true is the opposite to that which is the false; and that of sufficient reason, according to which it is impossible that a fact is true or exists without a sufficient reason, and the latter determines that the fact is necessary in this way and not otherwise. Truths of reason are based on the principle of contradiction, while truths of fact on that of sufficient reason.

With regard to these two principles, Leibniz transforms an attitude of epistemological pessimism concerning the analysis into a moderate optimism: what is contingent is also known to be true. According to Leibniz – as already anticipated in the *Dissertation* – it is possible to find the cause of a necessary truth by means of analysis, chasing the pathway back until one reaches primary simple truths, as in the case of mathematical truths. Conversely, with respect to contingent things it is necessary to refer to the principle of sufficient reason, which guarantees their truth. Indeed, even if the proximate causes of contingent things are not perfectly known in most instances, for the reason that the particular causes of effects can be infinite – and they should thus be analysed in an infinite process – the mind establishes that the ultimate cause is outside of the infinite causal chain of contingent things. Being outside of this chain, ultimate reason is necessary and this is God. Even if the infinite analysis of contingent things is not possible, and the mind must be content with the most proximate causes, God as the ultimate cause guarantees the validity of this knowledge. Therefore, knowledge of contingent things is also true.

The reference point of certainty and truth remains for Leibniz that of mathematics. The tool for finding this truth is analysis, and this is the tool that must be applied to both mathematics and natural science. Even if in the natural sciences a complete analysis is impossible, it is nonetheless possible to achieve truth. The epistemological model, which at the very beginning seemed to exclude a collaborative epistemology between experience and mental process, ends up being closer to that developed by the Aristotelians than Lockean philosophy. If perhaps what Isaac Newton said is true – that 'Mr Leibniz never found but a new experiment in all his life'[80] – it is equally true that Leibniz was perfectly aware of the importance of the experiment for the foundation of scientific knowledge.

11

Isaac Newton

London. 1688. The *Mathematical Principles of Natural Philosophy* had been published the previous year in 1687.[1] Its author was one of the most eminent members of The Royal Society of London and was soon to become the most influential figure in England, if not the single most important philosopher and scientist in the eighteenth century. The individual in question was Isaac Newton. His masterpiece was an immediate success, thanks to the propaganda exercised by the Royal Society's fellows all over Europe. It has been called the 'Newtonian moment'. His success was paralleled only to the obscurity of the text.[2]

The *Principles* was received in four different reviews: the *Philosophical Transactions*, the *Bibliothèque Universelle*, the *Journal des Sçavans* and the *Acta Eruditorum*.[3] The reviewers were, respectively, Edmund Halley, John Locke, Pierre-Sylvain Régis and Christoph Pfautz, though in the latter three cases the reviews appeared anonymously. All of these reviewers were eminent scholars, who but nonetheless with different philosophical and scientific interests and backgrounds. Their reviews were generally positive. Halley was not an impartial judge: he saw and revised the drafts of Newton's work before publication. Pfautz made the most competent critical assessment, while Locke was Newton's friend and probably he had some difficulty understanding the mathematical apparatus developed in the text.

Régis's review, by contrast, was more troublesome for Newton not because it dismantled some of his conclusions, but because it made him reflect on his global philosophical enterprise. Indeed, Régis was mainly concerned with philosophical issues and did not apply himself to explaining Newton's mechanics. His main critique was that Newton's demonstrations were merely mathematical: they provided a wonderful description of the world in numbers, but did not give a physical account of how that world was in reality. Newton's mechanics, according to Régis, had no bearing on metaphysics or on natural philosophy because it did not explain the true nature of 'force', 'gravity', 'matter' or other similar concepts. Régis emphasized how Newton's argument proceeded 'only by hypotheses which are, for the most part, arbitrary, and which therefore can serve as foundation only for a treatise in the realm of pure mechanics'.[4] In other words, Régis was describing Newton's lifetime achievement as Andreas Osiander had done for Copernicus *De revolutionibus orbium coelestium* (1543): merely a mathematical hypothesis capable of providing calculus consistent with observation, but useless for attaining the true causes of things. The review ended with an appeal to Newton to write an exact physics, leaving behind the hypothetical conceptualizing of his mechanics.

Though Newton never openly confessed it, this criticism struck him deep inside. The idea that his *Principles* elaborated only a simple theory that was not metaphysically well-founded began to circulate via this review among his opponents, especially the Cartesians and Leibnizians.[5] Leibniz explicitly writes that Newton's attempts to explain the motions of heavenly bodies through astronomical laws is worthwhile, but the introduction of gravitation was slipping back into barbarism, a return to occult qualities or Scholastic tendencies, without a metaphysical basis. Indeed, it is true that 'everything happens mechanically in nature, but the principles of mechanism are metaphysical'.[6]

Newton's answer was not immediate. He was thinking carefully about how to respond to such criticism, and there is evidence of his troubled reflections in the manuscript notes preparatory to the *Opticks*, datable back to 1701–4:[7]

> These Principles I consider not as occult Qualities supposed to resulting from the specific forms of things but as general Laws of Nature by wch ye things themselves are formed: Their Truth appearing to us by phænomena, though their causes be n^t y^t explained. To tell us that every species of things is endowed wth an occult specificly Quality by wch it acts, is to tell us nothing; but to derive two or three general Principles of motion from Phænomena, & after wards to tell us how the properties & actions of all corporeal things follow from those manifest Principles, would be a very great step in Philosophy, tho the causes of those Principles were not yet discovered: & therefore I scruple not to propose the Principles of motion above mentioned, they being of very general extent.[8]

Newton's reaction is quite subtle, but clearly entails the employment of an elaborated and highly original form of regressus. First of all, he deals with principles. These principles are not occult qualities, but something that derives from phenomena. Newton uses the verb 'appear', which means that they are manifest through experience. This is the first analytic stage of regressus. However, he maintains that these principles are not sufficient to provide the perfect cause of the effect. Better, they may form the explanatory causes of the effects, but their necessity and universality are not yet discovered. In order to assess their scientific character – wherein lies their necessity and universality – Newton points out that it is from this that properties common to all corporeal things are derived. This is the synthetic stage of regressus. Recapitulating, therefore, from some phenomena and specific cases, some general principles are extracted. These general principles can then be tested to see whether they say something about all things and not only about the samples considered at the beginning. This procedure, Newton adds, represents a great step in natural philosophy.

Newton had a wide range of means of accessing regressus theory – from Aristotelian textbooks or from early modern philosophers like Thomas Hobbes, especially in his formative years as his Trinity Notebook evinces.[9] But variants of regressus were endorsed by some eminent scientists at the Royal Society.[10] Even if Newton's absorption of these epistemological ideas was superficial,[11] there is good reason to suspect that there is much more beyond the manuscript notes pointing to a personal re-elaboration of an epistemological tradition.[12] As Alan E. Shapiro has remarked, it is in this period that Newton starts to pay concerted attention to logical doctrines typical of the Aristotelian tradition – the method of analysis and synthesis, the process of induction, deduction from effects – all of which will come together to elaborate what Newton calls experimental philosophy.[13]

There is good reason too to believe that Newton was inspired not only by the Aristotelian tradition, but more immediately by the polemics against the Cartesians and he aimed to confront them on the same methodological level. Newton's epistemological and methodological reflections and interests are triggered in reaction to his critics.[14] The long period of his optical theory's gestation, concluding with the publication of his *Opticks* in 1704, is characterized by the publication of Descartes's *Posthumous Works on Physics and Mathematics* in 1701. There is no explicit evidence that Newton read this book, as there is for other Cartesian works present on the shelves of his library. However, it is also well-known that Newton's reading was not confined to the books he owned. Descartes's posthumous publication contained the *Rules for the Direction of the Mind*, which was previously unknown to the larger public, and which contained – as we have seen in Chapter 7 – a peculiar interpretation of the Aristotelian theory of regressus. The study of these methodological doctrines in a period of intense and lively debate with the Cartesians allowed Newton to clarify his logic of scientific discovery and had a profound influence on the genesis of the second edition of the *Principles*.

In the same series of manuscript notes, Newton writes a long passage that was not included in the first English edition of the *Opticks*, but it will find a place in the Latin edition, the *Optice*, published in 1706. In these notes, Newton writes as follows:

> Many experiments are wanting for completing the Analysis of this part of Nature & coming to a clear & distinct knowledge of all ye causes of these things, many more for perfecting the Analysis of all Nature & making a full & clear discovery of all the first Principles of Natural Philosophy. To compass this is a work wch requires many heads & hands & a long time & yet this ought to be done before we proceed from the first Principles by Composition to explain Nature.[15]

Here the use of the expression 'clear & distinct knowledge' seems to refer to that kind of cognition envisaged in the old regressus theory. Only through experiments, according to Newton, is it possible to discover all the first principles of natural philosophy, which are at the very beginning offer only a very tentative possibility of explaining effects from causes. Indeed, many heads and hands, and a long time – which directly recalls the hard work of the negotiation – are necessary in order to pass on to the synthetic stage of regressus, that is, 'composition' in Newtonian terms. Newton is particularly concerned with the analytic stage because he is aware that it represents the foundation of all his method:

> The business of Experimental Philosophy is only to find out by experience & Observation not how things were created but/what is the present frame of nature ... This inquiry must proceed first by Analysis in arguing from effects to causes &

from compositions to ingredients. And when we have found the principles, the causes & ingredients of things we may proceed by Synthesis from those Principles to explain the things of this method ... This Analysis consists in arguing from compositions to ingredients, & from motions to the forces producing them & in general from phænomena to their causes, & from particular causes to more general ones till the argument end[s] in the most general. The Synthesis in assuming the causes discovered & established as principles of Philosophy & from them explaining the Phænomena proceeding from them & proving the Explanations.[16]

This very elaborate manuscript annotation is again a reflection on the regressus theory, but even more complicated in comparison to what we have previously considered. First of all, Newton states that the method of experimental philosophy is that of regressus, which seems to be composed of at least of two stages: analysis and synthesis. Analysis is a form of reasoning from phenomena which leads to their causes, where the phenomena are motions and the causes their forces. These causes do not explain how things come into being, but how nature works. Next, Newton shifts from this merely epistemological standpoint to physics, relating his investigations on gravitation to his logic of discovery. This is a pivotal step because here Newton explains that out of some experimental cases he has derived some particular causes. From these particular causes, he then moves on to establish more general and universal causes, which will serve as principles of philosophy to be used in explaining all other possible cases. The process that supersedes the transition from particular causes to general causes, however, remains quite obscure in this note.

An even more detailed explanation of the regressus theory occurs in another manuscript annotation, where Newton for the first time characterizes what was for the Aristotelians the intermediate stage of regressus. Here I offer a tripartite division of the explanation in order to make the three stages explicit:

[1] The method of Resolution consists in trying experiments & considering all the Phænomena of nature relating to the subject in hand & drawing conclusions from them

[2] & examining the truth of those conclusions by new experiments & so proceeding drawing new conclusions (if it may be) from those experiments & so proceeding alternately from experiments to conclusions & from conclusions to experiments untill you have come to the general properties of things

[3] And when you & by experiments & phænomena have established the truth of those properties, then assuming those properties as Principles of Philosophy you may by them explain the causes of such Phænomena as follow from them: wch is the method of Composition.[17]

The first and the third stages are the classical inference from effects to causes and from causes to effects. The intermediate stage is a continuous work of comparison between causes and effects, between logical conclusions and observations – that is, a series of tests and trials in order to assess the most universal principle that it is possible to find. Newton's characterization is rather vague here, but it is reminiscent of Aristotelian activity of the negotiation of the intellect. However, he had not yet established the ultimate rules for the experimental philosophy that will control these comparisons.

While this discussion is entirely absent in the *Opticks*, it occupies the final pages of the *Optice*.[18] Again, writing very much against the Cartesians in *Quaestio* 20, Newton states that the ancient philosophers were on his side in supporting the theory of gravitation, but that recent natural philosophers (*Physici recentiores*) in their speculations on natural things offer no explanatory cause, and therefore 'feign hypotheses' (*hypothesium commenta confingentes*) to provide reasons for all phenomena. In this way, they build castles in the air with their metaphysics.

Newton is clearly distinguishing between two approaches to natural philosophy, and one of these was that of speculative and rationalist thinkers who failed to consider experience as a yardstick.[19] The other was that of experimental natural philosophy, whose 'only business and goal is that of proceeding from effect to causes up to we reach the first cause'.[20] Newton was opposing his own method of investigating natural things to what Descartes had developed in his Rule 14 with his 'entire business' of the human mind. Here were two businesses based on phenomena, but each completely different in their procedure, at least in the eyes of the Cartesians of the time.

Equally interesting is Newton's *Quaestio* 23, which will become – albeit with a different wording – the famous Query 31 in the final edition of the *Opticks* (1730). Newton argues by analogy for the existence of attractive forces between particles being comparable to those between planets. He refrains from giving an ontological explanation for the nature and origin of these forces, but his aim is to investigate the laws and the properties of attraction from the starting point of natural phenomena. His discussion operates almost as a dialogue with Locke's *Of the Conduct of the Understanding*, published posthumously in 1706, the same year as the *Optice*. Indeed, Newton mentions some of Locke's own examples, such as oil of vitriol, nitre or vinegar, which may have been some of the many experiments performed at the Royal Society. What is interesting, however, is the relation established by logical analysis and physical separation, which is a clear echo of the Baconian project. Furthermore, in contrast with the final version of Query 31, *Quaestio* 23 makes explicit that his polemical target was the Cartesians.[21] And Newton adds a final claim, which will cause him a lot of trouble. He declares that God at the very beginning created matter with specific properties, among which were 'active principles, such as that of gravity'.[22] These principles are general laws of nature and not occult qualities that arise from the forms of things, and the reality of them is clear from their effects.

These principles, according to Newton, are discovered by analysis and synthesis both in mathematics and in natural philosophy.[23] The equivalence of method in mathematics and natural philosophy seems an innocent and naïve tenet but – as Niccolò Guicciardini has shown – it is full of disruptive implications which were far from obvious in Newton's times.[24] The first and most important of these was that, for Newton, investigations in natural philosophy – if conducted with a proper method could reach the same level of certainty and necessity as mathematical conclusions. This was a hard statement for some members of the Royal Society to digest, and no less so for their probabilism to absorb.[25] If they all agreed that mathematics represents the ideal of epistemic certainty – this is the case with Hooke and Locke and many others – they did not agree that natural philosophy, based on sensation and observation, could reach the same degree of certainty and necessity. It was the old vexed question that tormented the Aristotelians from the reading of the second book of the *Metaphysics*, and which led philosophers like Alessandro Piccolomini to deny that mathematics and natural philosophy provide the same level of epistemic certainty.[26]

However, the most important consequence is not the idea that both mathematics and natural philosophy have the same epistemic validity, but that they share the same method.[27] The exact phrasing translated from the Latin runs as follows:

> As in mathematics, so in natural philosophy (*physica*), the investigation of difficult things by the method that is named analytic (*Analytica*) ought ever to precede the method of that is called (*Synthetica*). Analytic method consists in making experiments and observations of phenomena, proceeding from compounds to ingredients, from motions (*motibus*) to forces (*vires moventes*), and in general, from effects to causes, and from particular causes to general causes, till they become the most general. Synthetic method consists in assuming the causes discovered and established as principles, and by them explaining the phaenomena proceeding from them, and proving explanation.[28]

Newton is reformulating the regressus theory here, and it is applied in two different realms, that of physics and that of mathematics. But it is well known that mathematical analysis is not the kind of analysis employed in the regressus, as should have been made abundantly clear by the many examples in this book. How should we interpret this? Guicciardini brilliantly highlights this difficulty by showing that the problems tackled in the first two books of the *Principles* have to do with analysis in the strict mathematical sense, and therefore the problem can be overcome 'by noting that the pronouncements on the method of analysis and synthesis in natural philosophy do not refer to the mathematical propositions of the first two books ... but to the deduction of universal gravitation in the third book'.[29] But still the difficulty remains because the two methods are different. For this reason, for the sake of consistency, the expression 'as in mathematics' (*quemadmodum in mathematica*) should be read in its weakest sense, conveying that just as there are two methods in mathematics, there are also two methods in natural philosophy. Also pointing in this direction is the manuscript note on optics where Newton writes as follows:

> As Mathematicians have two Methods of doing things wch they call Composition and Resolution & in all difficulties have recourse to their method of resolution before they compound so in explaining the Phænomena of nature the like methods are to be used.[30]

If this is the case, therefore, there would only be an analogy between these methods. Furthermore, it is possible to add that this *Quaestio* 21 deals with optical problems and not with gravitation, and therefore the methods can change. But again, in the same annotation Newton makes clear that this twofold method is what he employed in the *Principles*.

Thus in Mathematical Principles of Philosophy I first shewed from Phænomena that the all bodies endeavoured by a certain force proportional to their matter to approach one another, that this force in receding from ye body grows less & less in reciprocal proportion to ye square of the distance from it & that it is equal to gravity & therefore is one & the same force with gravity. Then using this force as a Principle of Philosophy I derived from it all the motions of the heavenly bodies & the flux & reflux of ye sea, shewing by mathematical demonstrations that this force alone was sufficient to produce all those Phænomena, & deriving from it (a priori) some new motions wch Astronomers had not then observed but since appear to be true, as that Saturn & Iupiter draw one another, that ye Variation of ye Moon is bigger in winter then in summer, that there is an equation of ye Moons meane motion amounting to almost 5 minutes wch depends upon the position of her Apoge of ye Sun.[31]

Hence, there is in fact a relation which goes beyond mere analogy regarding the methods of analysis and synthesis practised by mathematicians and natural philosophers. A possible solution might be to consider measurable physical quantities as exclusive objects of investigation in the *Principles*, as Andrew Janiak has suggested.[32] In this way, analysis from effects to causes may be carried out as mathematical analysis if the objects are measurable physical quantities, and their measurements will thus become the experiments that are instrumental in testing and proving the possible cause established at the end of the first stage of regressus. Once causes which may be conceived as laws are identified – as in the case of gravity – then it is possible to apply them in the synthetic stage of regressus to other phenomena, and so to explain new effects. According to Newton, this method will allow natural philosophy (*Philosophia naturalis*) to advance in its discoveries, and to achieve the credentials of a perfect science (*perfecta Scientia*). In this way, the *Optice* can be interpreted as a revision of the regressus theory.

In a draft of a letter to Roger Cotes – written before the second edition of the *Principles*, but after the *Optice* – Newton makes it clear that he did not consider hypotheses 'in explaining the Phenomena of nature' as holding a position in opposition to arguments 'founded upon Phaenomena by Induction or to Principles settled upon such arguments'.[33] Somewhat more explicitly, he writes that in experimental philosophy it is necessary to distinguish propositions as principles from propositions as hypotheses. Propositions as principles 'are deduced from Phaenomena by proper Arguments & made general by Induction' and this is 'the best way of arguing in Philosophy for a general Proposition'. Hypotheses, by contrast, 'are not deduced from Phaenomena by proper arguments'.[34]

Newton had in mind this distinction already in the 1670s, when he was fresh from his university studies. In 1672, in his *Queries to Be Determined by Experiments, Positively and Directly Concluding His New Theory of Light and Colours*, Newton writes as follows:

The proper Method for inquiring after the properties of things is to deduce them from Experiments. And I told you, that the Theory, which I propounded, was

> evinced to me, not by inferring 'tis thus because not otherwise, that is, not by deducing it from a confutation of contrary suppositions, but by deriving it from Experiments concluding positively and directly.[35]

Newton here is arguing that his method is to proceed in a positive way experiment by experiment, and so avoiding hasty generalization. He is openly criticizing the Aristotelian demonstrative induction, as expressed in Smith and in Sanderson, and which involved the traditional enumeration of the particulars, but where a correlation in nature was evident, it was possible to introduce the clause *et sic de caeteris* and to bring the argument to conclusion.[36] The reference to the Aristotelians is clear in the letter to Henry Oldenburg, Secretary of the Royal Society, dated 6 February 1672:

> For what I shall tell concerning them [colours] is not an Hypothesis but most rigid consequence, not conjectured by barely inferring 'tis thus because not otherwise or because it satisfies all phaenomena (the Philosophers universall Topick), but evinced by ye mediation of experiments concluding directly & wthout suspicion of doubt.[37]

Newton contrasts his inference from experiments – which leads to necessary conclusions – with hypothesis, which is not based on solid grounds and arguments. In this set of hypotheses, he includes suppositions coming from dialectical enumeration and demonstrative induction, Cartesian hypotheses that should fit all effects and even universal Topics, already rejected by Bacon. Newton is clearly rejecting a conjectural and probabilistic idea of scientific knowledge, in favour of an epistemic ideal of necessary knowledge and irrefutable conclusions similar to that of Aristotle and Bacon.[38] This passage was expunged from the version published in the *Royal Transactions* because it contravened the epistemological ideas of important members of the Royal Society like Robert Boyle and Robert Hooke.[39] The paper from which it was deleted is pivotal for understanding Newton's early logic of scientific discovery in practical terms. Indeed, judging from what he writes retrospectively on analysis and synthesis in his *Optice*, Newton is struggling to find a regressus theory of his own which would lead to epistemic knowledge.

The paper, titled *New Theory about Light and Colors* (1672), meticulously describes the method through which Newton reaches the disruptive conclusion that light consists of a heterogeneous mixture of rays that are refrangible to different degrees. The first stage is constituted of experience. Newton uses a glass prism to experience in various way 'the celebrated *Phaenomena* of *Colours*'.[40] He starts, therefore, from the analysis of phenomena and effects, what he calls a circumspect consideration based on observation. From this early consideration, he understands that the experience gained does not match what he had expected according to the cause established by Descartes's laws of refraction. The cause was clearly wrong.

This discovery leads him to a second moment in which through a series of observations and trials Newton tries to understand the discrepancy between effects and causes. First of all, he examines various material circumstances – the thickness of the glass or the termination with shadow or darkness, for example – to see whether they had an influence, but none in fact did. Newton, therefore, makes various conjectures, for which the term 'hypothesis' is never used, although they are consistently introduced with the words 'I suspect'. After several attempts at speculation, which are critically examined and then additionally scrutinized by means of calculation, he achieves no results. The reason for this repeated failure is that the 'computation was founded on the Hypothesis of the proportionality of the *sines* of Incidence, and refraction'.[41] Newton criticizes the use of a hypothesis not grounded on experience, abandons it and starts to look for another cause. 'The gradual removal' of all his suppositions led to an '*experimentum crucis*'[42] – to borrow a typical expression from Robert Hooke through which he discovers the 'true cause' of a phenomenon, and the second stage is then complete.

There follows a third moment in which other phenomena and effects are explained by means of the cause. Newton examines the origin of colours, why a rainbow appears in falling drops of rain, and the reason for the unexpected experiment that Hook made in his *Micrographia*.[43] These three stages reflect the procedure of the regressus: (1) general experience from effects, (2) examination of the cause and (3) explanation of other effects from the cause.

Regarding the methodology for his new theory of light and colours, in a reply to a question posed by Ignace-Gaston Pardies, Newton states as follows:

> The doctrine which I explained concerning refraction and colours, consists only in certain properties of light, without regarding any hypothesis, by which those properties might be explained. For the best and safest method of philosophizing seems to be, first to inquire diligently into the properties of things, and establishing properties by experiments and then to proceed more slowly to hypotheses for the explanation of them. For hypothesis should be subservient only in explaining the properties of things, but not assumed in determining them.[44]

It is clear that what Newton can concede is a kind of hypothesis as principles of knowledge and explanation (*principia cognoscendi*) but not as an ontological cause (*principia essendi*).

In the letter to Cotes, therefore, Newton distinguishes – as in the *Optice* – two ways of investigating nature, that of hypothetical philosophy and that of experimental philosophy:

> Experimental Philosophy reduces Phaenomena to general Rules & looks upon the Rules to be general when they hold generally in Phaenomena ... Hypothetical Philosophy consists in imaginary explications of things & imaginary arguments for or against such explications, or against arguments of Experimental Philosophers founded upon Induction. The first sort Philosophy is followed by me, the latter too much by Cartes, Leibnitz & some others.[45]

The period in which Newton was writing these reflections to Cotes coincided with Leibniz's attack on his approach. The attack appeared in the form of a letter to Nicolaas Hartsoeker published in three journals – *Memoires de Trévoux* (March 1712), *Memoris* of Literature (May 1712) and *Journal des Sçavans* (December 1712) – all of which circulated widely in Europe, discrediting Newton and his greatest achievement – that is, the theory of gravitation:

> If you only claim the will of God for his [explanation of gravitation], you are referring to a miracle, and even a perpetual miracle; for the will of God operates by miracle, whenever we cannot make account of this will and of its effect on the nature of objects. For example, if someone says that it is a will of God for a planet to go around its orb, without causing anything & keeping its movement, I say it will be a perpetual miracle ... it is the same in the suppositions of your atoms ... if you look for this reason only in the will of God, you will find it only in a miracle. Ancients and moderns, who avow gravity is an *occult quality*, are right if they understand by that that there is a certain mechanism unknown to them, by which bodies are impelled toward the center of the Earth. However, if their view is that this is done without any mechanism, by a simple *primitive quality*, or by a law of God which brings about this effect without using any intelligible means, then it is unintelligible occult quality, which is so very occult that it can never become intelligible, even if an angel, not to say God himself, should try to explain it.[46]

As Alan Shapiro has suggested, Leibniz was attacking the last three *Quaestiones* of Newton's *Optice*.[47] The reference to Newton's Latin text is obvious in the allusion to the ancients and the moderns, and to the properties and attribution of the particles' properties to God. In this letter, the philosopher of the pre-established harmony, Leibniz, was charging Newton with introducing miracles and not natural explanations into its system. Leibniz was not alone, for Cartesians in general reacted in the same way. Among others, perhaps the most interesting response was that of Joseph Saurin, who attacks Newton in the final paragraph of his *Examen d'une difficulté considerable proposée par M. Hughens contre le Systeme sur la cause de la Pesanteur* (1709):

> He likes to see gravity as an inherent quality in bodies, and to bring back the much disparaged ideas of occult quality, and attraction. We must not tell ourselves that in our physics research we can never put ourselves above all difficulties; but let us not stop philosophizing always on clear principles of mechanics; if we abandon them, all the light we can have is extinguished, and here we are again plunged back into the ancient darkness of Peripathetism, from which heaven wants to preserve us.[48]

All this criticism led Newton to revise the *Principles* and the *Opticks*. Before examining the revision of the *Principles*, which was published in 1713, it will be useful to highlight the differences between, on the one hand, *Queastio* 20 and *Quaestio* 23, and, on the other, Query 29 and Query 31, which appeared in the second English edition of the *Opticks* (1718). In relation to the passage of the *Quaestio* 20 in which Newton wrote that the 'main business of natural philosophy' was to proceed from phenomena and to discover causes from effects, he adds in Query 29 that all this must be done 'without feigning Hypotheses',[49] in clear opposition to the Cartesian model of demonstration. And while in *Quaestio* 23 he stated that, like clauses, the laws and properties of attraction are deduced from phenomena without knowing their nature and origin, in Query 31 he writes that 'it is the business of experimental philosophy to find them [laws and properties] out'.[50] Moreover, in the passage in which he defends himself against the charge of introducing occult qualities, he adds a long paragraph in order to distinguish his position from that of the Aristotelians:

> The Aristotelians gave the Name of occult Qualities not to manifest Qualities, but to such Qualities only as they supposed to lie hid in Bodies, and to be the unknown Causes of manifest Effects: such as would be the Causes of Gravity, and of magnetick and electrick Attractions, and of Fermentations, if we should suppose that these Forces or Actions arose from Qualities unknown to us, and uncapable of being discovered and manifest. Such occult Qualities put a stop to the Improvement of natural Philosophy, and therefore of late Years have been rejected.[51]

Newton is afraid that his position might be confused – as was the case with Saurin in 1709 – with an occult philosophy of the old style. The main and pivotal difference – representing a great rupture with the Aristotelian tradition – is that he conceived the result of the analytic stage of regressus in a totally different way. As we have seen, analysis proceeds from effects to causes, and this is very much in the vein of Aristotle. However, in proceeding from effects to causes, for Newton what is discovered is not the 'cause' in the Aristotelian sense, but rather the principles or laws of nature. These are causes of the phenomena in the sense that they provide a functional description of *how* the effects work, but do not explain *why* the effects are such. Thus the concept of cause is radically transformed. Laws of nature are causes in an improper sense, looking at them from the traditional – that is, metaphysical – perspectives of the Aristotelians, Cartesians and Leibnizians. Even if, therefore, Newton endorses the epistemology of regressus – sometimes using the same old terminology to characterize analysis as proceeding from effects to causes – he breaks with the Aristotelian tradition in signifying something very different: there is a de-ontologization of the cause.[52] That Newton was re-working the Aristotelian tradition emerges from a manuscript annotation probably written for the preparation of the second edition of the *Principles*, but closely reflecting what he writes in Query 31 against the Aristotelians. Newton believes that the study of laws and effects of the electric forces represents a great advancement in natural philosophy even if 'the cause of these forces is ignored'.[53] If the cause is not known, what is it exactly that *is* known?

Newton provides an answer by critically revisiting once again the old Aristotelian terminology involved in the regressus theory. First of all, one should observe the various phenomena. This observation is preliminary to the discovery of 'proximate causes' (*causae proximae*) – that is, those causes which were found at the end of the first stage of regressus. These causes are then further investigated, with a view to establishing the most proper cause for the effect. This passage is remarkable because it effectively characterizes the intermediate stage of regressus. According to Newton, the passage from proximate causes to the most proper, or ultimate, or supreme cause, is made only by reference to experience – that is, by investigating phenomena. He does not take long to explain this intermediate stage and how to work with phenomena in order to attain the ultimate cause. And once this cause is established, it is possible for Newton, 'by arguing a priori'[54] (*argumentando a priori*) – that is, with a demonstration *propter quid* – to work systematically down from the cause towards its effects.

In this process – which is the true method of natural philosophy – nothing is left for metaphysical enquiry – that is, no hypotheses are admitted. The causes must be considered exclusively as principles *cognoscendi*, and not *essendi*. In this sense, once again Newton is breaking with the Renaissance Aristotelians, who believe that in the first stage of regressus what was found were causes and principles *cognoscendi*, and only in the second stage, once these causes and principles have been demonstrated to be the true reasons of the effects, do they become causes and principles *essendi*.[55] Newton never argues this – that is, that there is a passage from *cognoscendi* to *essendi*: he is only willing to admit that synthesis proves that principles work and are real, but not that they are essential or constituent elements of phenomena. So the focus is on the epistemic grounds, rather than on the constitutive.

In its final form, Query 31 expands Newton's explanation of the analytic method. There are two major additions: (1) the process of analysis is characterized as a form of induction and (2) hypotheses are excluded from natural explanations. The passage varies quite significantly from the Latin version, shedding light on the difficulties that we have found in *Quaestio* 21. Query 31's variants read thus:

> This Analysis consists in making Experiments and Observations, and in drawing general Conclusions from them by Induction, and admitting no Objections against the conclusions, but such as are taken for Experiments, or other certain Truths. For Hypotheses are not to be regarded in Experimental philosophy. And although the arguing from Experiments and Observations by Induction be no demonstration of general conclusions; yet it is the best way of arguing which the Nature of Things admits of, and may be looked upon as so much the stronger, by how much the Induction is more general. And if no Exception occur from Phaenomena, the Conclusion may be pronounced generally. But if at any time afterwards any Exception shall occur from Experiments, it may then begin to be pronounced with such Exceptions as occur.[56]

From induction, according to Newton, one derives general conclusions, which must be considered true until possible objections emerge through experiments. At this stage no hypothesis – that is, any fantastic explanation is admitted. Newton is aware, as all the Aristotelians were, that analysis does not lead to demonstration. Knowledge acquired by analysis is only provisional, but it is the best possible way of arguing and establishing a logic of scientific discovery.[57] For this reason, the method of natural philosophy cannot be that of mathematics, and this suggests that we should read 'as in mathematics' in *Quaestio* 21 in the weakest sense. Furthermore – and most notably – it is clear that Newton has revised his idea that the certainty of natural philosophy can be the same as mathematics, and he is aligning himself more in favour of provisionalism in comparison to his earlier conceptions, which had provoked the 1672 reaction of Royal Society members – first and foremost, Hook. This reading would be consistent with what Newton had already written in the *Preface* of 1687 edition of the *Principles*:

> The basic problem of philosophy seems to be to discover the forces of nature from the phenomena of motions and then to demonstrate the other phenomena from these forces ... in book 3, by means of propositions demonstrated mathematically in books 1 and 2, we derive from celestial phenomena the gravitational forces by which bodies tend toward the sun and toward the individual planets. Then the motions of the plants, the comets, the moon, and the sea are deduced from these forces by propositions that are also mathematical.[58]

Once again, Newton emphasizes the epistemological priority of the regressus theory. But in this case the first stage of regressus is encompassed mainly by mathematical analysis, the conclusions of which are applied to finding new physical phenomena. By doing this, Newton would claim that there is an overlap or analogy between the analysis of trajectories followed in mathematics and that from motions to forces in physics. But this analogy is not explained in the 1687 edition, opening it up, indeed, to the possible interpretations or criticisms that what he had written in the first two books was merely a mathematical hypothesis for how nature works. This is not Newton's strategy, however, at least from what it is possible to learn from his optical investigations. Since 1672, his research on optics shows that before the transition to the synthetic stage, mathematical analysis is of great help – exactly as we have seen in the case of Galileo. If what I have argued is correct, the alleged 'transition' from mathematics to natural philosophy is not a passage between two distinct realms of knowledge, two different conceptual worlds – that of the mathematicians and that of natural philosophers – but they are simply two moments of the same epistemological process – what Newton will call experimental philosophy – which leads to knowledge of the system of the universe and all its effects. The mathematical apparatus of the first two books is 'only' an epistemological phase and in no way can be read as a mathematical hypothesis, as Newton's opponents claimed. However, until the second edition of the *Principles* Newton did little to resolve this ambiguity.

*

In 1687, the epistemological core of the *Principles* was briefly exposed in the author's preface to the reader. After the condemnation of the status of mechanics as a science, Newton points out that the whole difficulty of natural philosophy lies in investigating 'the forces of nature from phenomena of motion' and then demonstrating 'other phenomena from these forces'.[59] Newton makes immediately explicit his twofold method of investigation and any reader of his time would at once understand that his way of proceeding was that of regressus. This twofold process is what led Newton to explain the celestial phenomena concerning effects on the Sun, on Earth, on the Moon, on comets and on the sea. However, Newton's method is not confined to celestial mechanics. He believes that it is possible to derive other phenomena from the principles of mechanics through the same kind of reasoning:

> For many things lead me to have suspicion that all phenomena may depend on certain forces by which the particles of bodies, by causes not yet known, either are impelled toward one another and cohere in regular figures, or are repelled from one another and recede.[60]

Newton is clearly introducing an argument by analogy – that is, transduction – supposing that what happens to observable bodies may be applied to unobservable particles. This claim is based on an assumption concerning the homogeneity and uniformity of nature, which is not explicable at the epistemological level and whose foundation lies in Newton's endorsement of a peculiar form of natural philosophy – shared also by Locke, as we have seen in Chapter 8. This represents an opaque zone of his theory, which Newton constantly re-elaborates in the various editions of the *Principles* and in which he continues to keep faith.

In the *Preface*, Newton makes clear two other important issues in connection with his epistemology. First of all, there is the mathematical nature of his investigation, which aims to 'reduce the phenomena of nature to mathematical laws'.[61] This means that a significant part of his method will employ the instruments of mathematics. Second, he declares his disinterest in the causes of the forces, emphasizing that his only aim is to deal with forces as causes or principles of phenomena. He detaches his position from any ontological interpretation of the cause, and dismisses any kind of investigation into the nature of the cause of forces, which is considered only as a *causa inferendi* or *cognoscendi*, rather than a *causa essendi*. As already mentioned, this represents a peculiar transformation of Aristotelian regressus theory, implying – as pointed out in Chapter 2 – that the causes were both *causa inferendi* and *causa essendi*. His idea of cause is not that of essential or formal cause. His epistemology is based on the abandonment of 'substantial forms and occult qualities'.[62] While this aspect of the thinking was completely understandable to his contemporaries, who no longer believed that natural philosophy was based on the study of substantial forms, much less acceptable was discarding the idea of a mechanical cause for the explanation of natural philosophy. This was particularly true for Cartesians, who charged him with introducing occult qualities, though he denied this from the very beginning of his work. So there remained a lack of clarity in his epistemology, and it needed to be better explained if he was to make his natural philosophy more understandable and more generally acceptable also. This is what Newton intended to do in the second edition.

In the 1713 edition, we find important elements of Newtonian epistemology in Roger Cotes's preface. These traces are particularly relevant because they reveal Newton's redefinition of his epistemology in light of the controversies that the first edition of the *Principles* and the *Optice* triggered. Clearly, Cotes's words cannot reflect *in toto* Newton's position, but given the nature of their epistolary exchange, we can safely say that they share the basic ideas contained in this long preface. Identifying Cotes's main epistemological tenets is therefore relevant to understanding Newton and the evolution of his thinking. Cotes's strategy as editor of the second edition is clear: he wants to detach Newton from any other conceivable philosophical schools from the past, by emphasizing his original contribution.

Cotes identifies three epistemological approaches in the study of nature. The first comes from the Scholastic philosophers and its origin is attributed to Aristotle. Their main idea was that 'individual effects arise from the specific natures of bodies'.[63] The main problem with this conception is that it does not make 'the causes of those natures' explicit, and therefore the reasoning takes place mainly in words. Of course, the issue for Cotes is not that Scholastics fail to investigate the nature of causes, because Newton did not consider them either. The point is rather that is that every effect has its own particular cause and no generalization is possible. Whether this was Aristotle's real position is neither here nor there. What Cotes is pursuing is that this epistemology is unfeasible because it endows 'the individual species of things with specific occult qualities, on which the operations of individual bodies depend in some unknown way'.[64] Cotes wants clearly to defend Newton from the charge of 'occultism', by attacking this idea and attributing it to the Aristotelians.

The second epistemological approach is that of the Cartesians, who Cotes claims are guilty of having introduced 'occult fluids' as mere hypothesis: 'those who take the foundation of their speculations from hypotheses, even if they then proceed most rigorously according to mechanical laws, are merely putting together a romance, elegant perhaps and charming, but nevertheless a romance'.[65] And the third approach – that of Newton and of other members of the Royal Society – is called experimental philosophy. These experimental philosophers do not feign hypotheses, but rather 'they proceed by a twofold method, analytic and synthetic':

> From certain selected phenomena they deduce by analysis the forces of nature and the simpler laws of those forces, from which they then five the constitution of the rest of the phenomena by synthesis.[66]

In other words, Cotes identifies the method of experimental philosophy with regressus. If so – that is, Newton's method is that of regressus – why had not previous philosophers, who had the same logical instruments, not discovered gravity? Cotes's answer is revealing: they had, but only through guesswork, unlike Newton, who 'was the first and only one who was able to demonstrate it from phenomena and to make it a solid foundation for his brilliant theories'.[67] There was something in Newton's regressus theory that was missing in previous epistemologies.

We are able from this distance to safely say that this was his mastery of mathematics, but at that time his calculus was far from being clear, and in fact it was long disputed. Cotes obviously knew Newton's excellence in mathematics, but in his preface he prefers to focus the attention of the reader on another aspect that Newton introduced only in the second edition of the *Principles* – that is, one of the *Rules for Philosophizing*. We will see in a moment the importance of this new conceptual introduction. But before that, it is important to understand why Cotes did not first of all emphasize Newton's greatest achievement in mathematics. The reason is that an exclusive emphasis on mathematics would have led to the Newtonian system being interpreted as a kind of hypothesis, while Cotes's intention was the exact opposite – that is, to show that mathematics was an instrument for reaching far more revolutionary conclusions in the description of world.

Newton had been certain about this since 1672, as we have seen, but to substantiate his belief he had to show why mathematics could lead to a description of physical reality. In his mind, the third book of the *Principles* was functional in showing the fruitfulness of his mathematical apparatus, but his statement is somewhat ambiguous:

> In the preceding books I have taught principles of philosophy that are not, however, philosophical but strictly mathematical – that is, those on philosophical things can be disputed ... It still remains for us to exhibit the system of the world from these same principles.[68]

Newton states that in the first two books he had developed principles of philosophy, which are mathematical in the sense that they do not explain the nature of things, but how things work. In Definition 8 in the first book, he asserts that in dealing with 'attractions', 'impulses', or 'propensities toward a center', he considers 'these forces not from a physical but only from a mathematical point of view'. Then he adds that by words of this kind he is not 'defining a species or mode of action or a physical cause or reason', and nor he is 'attributing forces in a true and physical sense to centres (which are mathematical points)'.[69] In order to prove that these notions provide a description of reality, Newton has to show that (1) they are the only principles that can explain phenomena – that is, the effect cannot be explained in any another way, and (2) they are principles which may also be used for other phenomena. Methodologically speaking, this means that it is necessary the transition from Stage 1 to Stages 2 and 3 of his method. There was no such transition in the first edition, or at least this transition caused serious problems because the third book of the 1687 edition started with a series of nine hypotheses, which rendered Newton's work even more hypothetical in the eyes of his opponents. All of this was of course against the author's own opinion. Newton was working on a solution from the time of the 1687 edition to that of 1713, and thus throughout the elaboration and publication of the *Opticks* and that of the *Optice*.

The major change undergone by the *Principles* between the two editions is no doubt the transformation of the *Hypotheses* into the *Rules of Philosophizing*. This transformation was triggered by the will to respond to the Cartesians, and by Newton's awareness that the transition between the first two books on mathematics to a third book that was much more focussed on natural philosophy required further explanation. The idea to change the hypotheses into *Rules* is a reaction to the Cartesians on at least two grounds. First, after all the polemics and controversies with Cartesians and Leibnizians, Newton does not want to give the impression of introducing hypotheses into his natural philosophy. He believes that his principles are genuine laws of nature and not merely fictional explanatory devices. Second, he conceives the *Rules* as superseding Descartes's *Rules*, which in the meantime had become available in the *Posthumous Works*. Newton did the same with the *Principles*, which had the original intention of supplanting Descartes's *Principles philosophiae*, free of all the metaphysical apparatus that the Cartesian implied. We are confronted by two very different ideas of 'principles'.

The scholarly debate around these *Rules* is vast,[70] but the best contribution for understanding their purpose comes from Zvi Biener.[71] He suggests that the role of the *Rules* 'was similar to that ascribed by some late scholastics to a third stage of investigation, the *negotiatio intellectus* (or *contemplatio mentalis*)'.[72] In particular, he claims,

> The rules are principles for, first, winnowing down the multiplicity of possible causes of celestial phenomena to a single true cause – gravity – and, second, for allowing gravity to be taken universally, despite being inferred from a limited range of phenomena. Like the *negotiatio*, the rules are a guide for thinking, not for direct empirical investigation. And like the *negotiatio*, they provide the necessary link between the *Principia*'s analysis ... and synthesis ... Newton's repeated changes to the rules were meant to clarify how they can execute *this* function, and thus to answer the charge that the *Principia* had not, perhaps could not, provide genuine causal knowledge.[73]

Biener's working hypothesis is very suggestive. In light of what we have seen in the previous chapters, I want to go a little bit further, bringing out what the *Rules* leave in the background, that is the central role of mathematics developed in the first two books in determining the principles. Furthermore, if the *Rules* are the procedure that characterizes the negotiation of the intellect in the regressus theory, they are the principles for thinking about and directing what is in Newton's mind across the whole series of tests, observations and experiments that are necessary for finally extracting the real cause or law of the effects. This process is evident, as we have seen, in his early optical investigations, but at that time he never systematized it. In 1713, the situation is different. The *Rules* are placed at the front of the third book titled *The System of the World* and serve to formalize this process, defining thereby how to establish the true cause or law of phenomena. In this sense, the *Rules* are not simply *Rules for Philosophizing*, but rather the *Rules* of experimental philosophy, without which Newton's epistemology would be without foundation. The *Rules* explain why the principles developed in the first two books are indeed principles, and how they may help to illuminate other phenomena.

Newton conceives five different rules, three of which appear in 1713, one in 1726, and one he discarded immediately after writing it. Their gestation was quite complicated for Newton, because he tended to develop his thinking by engaging with criticism.[74] For this reason, the *Rules* are a good indicator for understanding Newton in the making. Rule 1 corresponds to what was Hypothesis 1, with a short addition:

Hypothesis 1. *No more causes of natural things should be admitted than are both true and sufficient to explicate their phenomena.* For nature is simple and does not indulge in the luxury of superfluous causes.

Rule 1. *No more causes of natural things should be admitted than are both true and sufficient to explicate their phenomena.* As the philosophers say: Nature does nothing in vain, and more causes are in vain when fewer suffice. For nature is simple and does not indulge in the luxury of superfluous causes.[75]

This is a rule of epistemological parsimony, which comes from Aristotle and had developed over time in various forms like Ockham's razor. Beyond its simplicity, it is the most quintessentially 'metaphysical rule', and it entails a conception of reality strictly bounded by the cause–effect relation. If there are two causes that separately cause the same effect, either the causes are the same, or the two causes are not the real and true cause. If the latter is the case, then one should look for what is common to the two causes in order to identify this real and true cause this singular and fundamental cause. This was one of the first rules to be applied in the Aristotelian regressus in order to identify whether the cause identified in the first stage was the real cause of the effect. This was the univocal co-implication of cause and effect, which was essential for the transition from analysis to synthesis in Zabarella, and which was possible for Aristotelians before him only in the field of mathematics. Zabarella made this co-implication possible in natural philosophy by applying a strictly demonstrative induction. Would Newton have supported this position?

There are good reasons to believe that the answer is positive if we conceive the cause in a de-ontologized way – that is, if the cause is not the form of the effect, but a law through which the effect may be. In other words, it is so when the co-implication is between a cause and a kind of effect. If other causes can explain the effect, the effect is not properly speaking explicated. Going beyond the Aristotelian tradition and proposing an epistemology with only a minimum of metaphysics, Newton understands the connection between cause and effect in a strictly mathematical way, rather than in an essential or substantial form. This was what Aristotle, Nifo, Balduino and other Aristotelians had hoped for, but they had not believed in the possibility of achieving it, because they were not willing to abandon their metaphysics and the primacy of essence as the main kind of cause.

Newton's attachment to the parsimony rule is evident in Rule 2, which is a clarification and an addendum to Rule 1, so much so that the two rules can be read seamlessly:

> Rule 2. *Therefore, the causes assigned to natural effects of the same kind must be the same.* Examples are the cause of respiration in man and beast, or of the falling of stones in Europe and America, or of the light of a kitchen fire and the sun, or of the reflection on our earth and the planets.[76]

In the third edition published in 1726, Rule 2 is slightly, but significantly, different. In the first part, there is the addition of the clause 'so far as possible' (*quatenus fieri potest*), which means that in its definitive form the rule reads as follows:

> Rule 2. *Therefore, the causes assigned to natural effects of the same kind must be, so far as possible, the same.*[77]

The introduction of this clause is extremely significant telling. Rule 2 is the foundation of Newton's method. Indeed, without it, the transition from analysis to synthesis would be impossible – that is, it would be impossible to improve and extend knowledge after the discovery of the cause. In fact, in the analysis, once the cause is established, no more is known than the relation between that specific effect and that specific cause. Only when that specific cause is related to other effects of the same kind can we then acquire new knowledge through synthesis. Otherwise, what has been discovered in the analysis is only the cause of an effect which cannot be generalized. This would mean that each effect had its own cause. But if it were so, then there would have been a return to the Scholastic conception that Roger Cotes criticizes in the 'Preface' to the 1713 edition: 'individual effects arise from the specific natures of bodies'.[78] However, these natures cannot be considered principles because they are not more universal and general – that is, they do not explain more than the effects of which they are cause. In other words, no principles of natural philosophy can be established without Rule 2, which introduced the requirement of convertibility between cause and effect in a way that Zabarella had already conceived as being the only way of breaking a conception of regressus that involved to a circular argument. Synthesis adds knowledge to analysis, otherwise authors such Neritonensis would be right in supporting the idea that the first stage of regressus was sufficient to acquire scientific knowledge.

Newton is perfectly aware of this problem, especially in natural philosophy and in relation to mathematics, where the cause in his understanding lies not in the concrete nature of the body but a general law, on the simple basis that otherwise there would be a law for each effect. The addition of the clause 'so far as possible' is important because in experimental philosophy the co-implication between cause and effect is not essential – it is not between an effect and its substantial form, as it was for Balduino and Nifo. Being based on experience and experiment, the co-implication or the convertibility is possible in so far as observation attests that the effects are of the same kind. If they are of the same kind, then it is possible to attribute the same cause to them. With the introduction of the extra clause, Newton shows his awareness of the limits of regressus, in its being based on experience, which cannot be otherwise than limited and therefore open to subsequent emendation, rejection or confutation. This is the core of Newton's provisionalism. Without Rule 2, he would not have been able to extend his discovery of the law of gravitation as a special attractive force to the explanation of other phenomena, as he successfully did in the third book of the *Principles*.

It is not surprising that Cotes believes that 'in this Rule there is the foundation (*fundatur*) of all philosophy'.[79] He introduces two formulations of the same Rule 2:

> Effects of the same kind – that is, effects whose known properties are the same – have the same causes, and their properties which are not yet known are also the same.[80]

> It is in accord with reason that the same effect can arise from several causes somewhat different from one another; but the true cause will be the one from which the effect truly and actually does arise.[81]

Rule 2 has two significant characterizations. First, it aims to expand knowledge towards something that is not already known, and therefore it becomes possible to infer the unknown from the known because of effects that are of the same kind. The fact that effects are 'of the same kind' and are not the same is crucial, and this distinguishes it significantly the early Aristotelian formulation of the regressus. The applicability of the regressus is possible because the formulation of the cause is de-ontologized and mathematical – that is, universal in such a way that it can explain not only that specific effect from which the cause is discovered. Furthermore, Rule 2 – by means of the clause 'so far as possible' – recognizes that the same effect can be generated by multiple causes. However, if Rule 2 is related to Rule 1, as Newton indeed believes, then it is clear that the aim is to find the only true cause of the effect, which lies in the common aspects of the possible causes of an effect.

The nature of the research in pursuit of the true and first cause is attested by Cotes, who writes that 'causes generally proceed in a continuous chain from compound to more simple; when you reach the simple cause, you will not be able to proceed any further'.[82] This a typical Aristotelian response to the regress to the infinite in searching for the cause. If one looks to the infinite for a cause, then no explanation of phenomena would be possible. The phrase 'so far as possible' provides Newton with the possibility of not explicating the grounds for deciding what the effects 'of the same kind' are, because knowledge of these effects comes from experience. Experience can always testify otherwise. Therefore, the connection between cause and effects of the same kind is both epistemic and provisional on account of the very nature of Rule 2, which lies in experience. Rule 2 is thus fundamental because once one has established, for instance, that the law of gravitation works in one case – that is, in the relation between Earth and Moon – it can be extended to other effects of the same kind. As Cotes states,

> The revolutions of the primary planets about the Sun and of the secondary planets about Jupiter and Saturn are phenomena of the same kind as the revolution of the Moon about the Earth; furthermore, it has been demonstrated that the centripetal forces of the primary planets are directed toward the center of the Sun, and those of the secondary planets toward the centers of Jupiter and of Saturn, just as the Moon's centripetal force is directed to the center of the Earth; and, additionally, all these forces are inversely as the squares of distances from the centers, just as the force of the Moon is inversely as the square of the distance from the Earth; therefore it must be concluded that the nature is the same for all things (*universis*).[83]

Newton's Rule 2 directly recalls – as well for the examples he provides – Bacon's *instances of alliance*. Bacon writes that these instances reveal

> that the operations and effects which are classed as belonging to one heterogeneous nature also belong to other heterogeneous natures; and this proves that the supposed heterogeneity is not genuine or essential, but simply a modification of a common nature.[84]

This principle shows that even if there are multiple causes for an effect, one may argue that these causes have something in common, which is the true cause of the effect. Therefore – as Bacon puts it – this argument is of the 'highest value in lifting and raising the mind from differences to kinds'.[85] The most compelling example is that of heat. Usually, Bacon states, philosophers identify three kinds of heat, that of animals, that of fire, and that of heavenly bodies. It is believed that they are different in nature and essence, and for this reason that they are absolutely heterogeneous. However, experiments lead to the conclusion that the cause of the heat of the Sun is the same as that of the heat of a kitchen fire. Newton does not deal with heat, but with the light of the Sun and that of the kitchen fire, but the inference is the same. Rather interestingly, Bacon provides many examples of the kind of rule that Newton tackles in the *Principles* – for example, that of circular and rectilinear motion, that of ebbing and flowing or relating to weight, and one relating to light. The importance of Rule 2 was clear both to Aristotelians for their regressus theory and also to members of the Royal Society for the experimental methodology. Rule 2, according to David Hume, was 'Newton's chief rule of philosophizing'. Indeed, he believed that it was 'entirely agreeable to the Rules of Philosophy, and even of common Reason, where any principle has been found to have a great Force and Energy in one instance, to ascribe to it a like Energy in all similar instances'.[86]

According to Cotes, Rule 2 is fundamental in two key aspects: (1) 'if it is taken away, there is then nothing we can affirm about things universally', that is, as we have said, there would only be individual explanations; (2) 'the constitution of individual things can be found by observations and experiments; and proceeding from there, it is only by this rule that we make judgements about the nature of things universally'.[87] The second raises a different aspect in comparison to the first one. There is not only the universality of science at stake. Indeed, it is concerned with the constitution of individual things (*constitutio rerum singularium*) – that is, how each thing is made. This takes Newton almost to the point of Rule 3.

In the 1713 edition, Rule 3 replaces Hypothesis 3 of the first edition, which stated that 'every body can be transformed into a body of any other kind and successively take on all the intermediate degrees of qualities'.[88] This hypothesis regarding the transformation of one body into another body has been discussed at length by the scholarship, giving rise to a number of controversies, since it entails a peculiar concept of matter which relates to Newton's alchemical activities. From the epistemic standpoint, however, this hypothesis finds an antecedent in Bacon. In the second book of the *Novum Organum*, Bacon states that the axiom or rule of transformation of bodies (*praeceptum sive axioma de transformatione corporum*) is of two kinds. The first considers the body as a combination of simple natures. For instance, gold is made up of the following qualities: tawny colouredness, heaviness, extension, malleability and so on. Therefore, for 'he who knows the forms and methods of superinducing tawny colour, weight, ductility, stability, melting, solution and so on, and their degrees and manners, will take pains to try to unite them in some body, and from this follows the transformation into gold'.[89] He adds, furthermore, that 'this mode of operation proceeds from what is constant, eternal and universal in nature' – in other words, what might be termed simple natures. The second kind of transformation does not proceed by means of these simple natures, but through investigating concrete bodies as they naturally behave in nature, as in the example where 'one is investigating the origins, means and process by which gold or any other metal or stone is generated from their base substances or elements to the perfect mineral'.[90] According to Bacon, this kind of inquiry is not only concerned with the generation of bodies, but also with the other motions and operations of nature.

This background is helpful for understanding the various passages from Hypothesis 3 to Rule 3, which James E. McGuire has carefully examined in all its various facets. He has identified at least three stages of development before the final version was reached.[91] Rule 3 in its 1713 version reads:

> Rule 3. *Those qualities of bodies that cannot be intended and remitted and that belong to all bodies on which experiments can be made should be taken as qualities of all bodies universally.* For qualities of bodies can be known only through experiments; and therefore qualities that square with experiments universally are to be regarded as universal qualities; and qualities that cannot be diminished cannot be taken away from bodies. Certainly idle fancies ought not to be fabricated recklessly against the evidence of experiments, nor should we depart from the analogy of nature, since nature is always simple and ever consonant with itself. The extension of bodies is known to us only through our senses, and yet there are bodies beyond the range of these senses; but because extension is found in all sensible bodies, it is ascribed to all bodies' universalities. We know by experience that some bodies are hard ... That all bodies are impenetrable we gather not by reason but by our senses ... That all bodies are movable and persevere in motion or in rest by means of certain forces (which we call forces of inertia) we deduce from the properties of observable bodies. The extension, hardness, impenetrability, mobility, and force of inertia of the whole arise from the extension, hardness, impenetrability, mobility, and force of inertia of each of the parts; and thus we conclude that every one of the least parts of all bodies is extended, hard, impenetrable, movable, and endowed with a force of inertia. And this is the foundation of all philosophy. ...
>
> Finally, if it is universally established by experiments and astronomical observations that all bodies on or near the Earth gravitate toward the Earth, and do so in proportion to quantity of matter in each body, and that the Moon gravitates toward the Earth in proportion to the quantity of its matter, and that our sea in turn gravitates toward the moon, and that all planets gravitate toward one another, and that there is a similar gravity of comets toward the sun, it will have to be concluded by this third rule that all bodies gravitate toward one another. Indeed, the argument from phenomena will be even stronger for universal gravity than for impenetrability of bodies, for which, of course, we have not a single experiment, and not even an observation in the case of the heavenly bodies. Yet I am by no means affirming that gravity is essential to bodies. By inherent force I mean only the force of inertia. This is immutable. Gravity is diminished as bodies recede from the Earth.[92]

Rule 3 presents a long explanation, most likely on account of its difficulty. The first problem relates to the question of how qualities that are to be taken as pertaining to all bodies universally are able to fulfil both the condition of being qualities that cannot be intended or remitted *and* qualities on which experiments can be undertaken.[93] This interpretation seems to find support in a handwritten annotation to Newton's own copy of the 1687 edition:

> Gravity (in Earth) is a quality of all bodies which are round the Earth (in so far as they are related to the Earth), and on which are allowed to make experiments, and is proportional to the quantity of matter in each thing and cannot be intended and remitted and therefore according to Hypothesis III is a property of all bodies.[94]

For this exposition, it is obvious that Hypothesis 3 is no longer the hypothesis of the first edition but has moved to some extent in the direction of Rule 3. Gravity seems to satisfy both conditions of Rule 3. This interpretation faces the problem that in the 1727 edition at the end of Rule 3, Newton identifies at least one quality that can be taken as a quality of all bodies, which, however, seems still to have the capacity to be remitted and intended. This quality is famously gravity. So, how may we read the note in Newton's own copy? Here, in fact, Newton restricts gravity to the Earth and to bodies on the Earth, where gravity neither decreases nor increases, and hence it is always the same for all bodies on the Earth. Therefore, in this particular note gravity can satisfy both conditions, but only because it is not generalized. Once generalized – that is, when gravity is not considered only in relation to Earth but also as it recedes from it – it diminishes and therefore cannot satisfy the first condition.

The only really consistent way of reading Rule 3 is to view Newton as grouping two classes of objects under one umbrella.[95] The first class considers qualities of bodies that cannot be intended or remitted, using a terminology that recalls Oxford calculators – still popular in Newton's time.[96] The second class, in contrast, concerns qualities that pertain to all bodies on which experiments can be done. The common umbrella in this context is that all the qualities of these two classes can be considered qualities of all bodies universally.

In his explanation, Newton does not waste any time expounding why qualities that cannot be intended or remitted may also be the qualities of all bodies universally, yet he spends several paragraphs on the second class of objects. Why so? This is no doubt a particularly remarkable aspect of the writing, because it is as if Newton is taking it for granted that qualities that cannot be intended or remitted are qualities of all bodies. With regard to Bacon's axiom of transformation of bodies of the first kind – considering in the case of simple natures extension, heaviness, and so on – we have come to understand that from these simple natures it is possible to generate bodies, and learned that this generation proceeds 'from what is constant, eternal and universal in nature'. In Newtonian terms, 'what is constant, eternal and universal in nature' is what cannot be intended or remitted. Qualities in Newton play the same role as simple natures in Bacon. Simple natures in Bacon – as Lisa Jardine has pointed out – are determinate physical properties which produce the effects of a body.[97] According to Bacon, in order to determine what is constant and universal – as we have seen in Chapter 4 – we should investigate all things that in Newtonian terms are intended and remitted.[98] Once we have taken these away, we are left with something that pertains to this kind of body. The Aristotelians adopted a similar process in the negotiation of the intellect in order to remove all the accidents and find the essential qualities or properties that constituted the nature of the cause. This was the strategy used by Descartes in his mental inspection to find the essential qualities of wax. Now the question is whether – for Newton – hardness, impenetrability, mobility and force of inertia were qualities that could not be intended and remitted, and therefore were able to be considered essential, as they were with the Aristotelians and Cartesians. Looking at the manuscript, the answer to this seems positive:

> Axiom 4. Qualities which are intended and remitted such as motion and rest, heat and cold, wet and dry, light and darkness, colour and blackness, opacity and transparency, a good smell and a bad smell, acidity, bitterness and sweetness, volatility and immovability, liveliness, health and sickness, do not come in for consideration here. What cannot really be intended and remitted such as magnitude, impenetrability, solidity, mobility and that inertia which resist to motion and mutations are usually properties of all bodies. And the reason is because a quality cannot be remitted cannot be taken away from the whole; instead, that which can be taken away, if it were to be taken away from some parts of the whole, it could be remitted in the whole. Impenetrability indeed is usually described as the essence of bodies and hence attributed to all of them; but essential properties of bodies do not become known by the natural light [of the mind]. We gather only from sensation that bodies that we touch are impenetrable because this quality in them is not intended or remitted, we attribute to all bodies alike, no less to heavenly bodies and to bodies imperceptible to the senses than to those bodies which we touch: and that too with an argument which we think so strong that this quality is referred to the essence of bodies and is therefore considered to be the firmest foundation of all philosophy. Therefore the axiom on strength of which we gather this property of bodies must not be repudiated.[99] ... Therefore this axiom is the foundation of all philosophy. At all events nature would not be sufficiently simple and consistent with itself [unless] the qualities of bodies on which it is allowed to make experiments, as many of them as are found to be immutable, apply in common to all bodies.[100]

Hardness, impenetrability, mobility, inertia are all, according to this annotation, qualities that cannot be remitted or intended, and hence – Newton deduces – they are also essential. But they are not the only qualities that cannot be remitted or intended. Indeed, as we have already seen, Newton deals at length with the reason why there are also qualities common to all bodies on which experiments can be made.[101] This seems to be for Newton the pivotal aspect. It is important for his argument because it enables him to emphasize the experimental procedure through which he has established that hardness, impenetrability, mobility and force of inertia are qualities that pertain to all bodies universally. And by the same experimental means, he has discovered another kind of quality which may be ascribed to bodies universally, because on those bodies it is possible to undertake experiments. Of this kind of quality is gravity. However, gravity is not essential – indeed, it is not a quality that cannot be intended or remitted since it increases or decreases in relation to distance from the Earth. That there are qualities that satisfy only one of the two conditions is clear from the observation in the above-mentioned annotation that only some of the qualities of bodies on which experiments are done are found to be immutable, not all of them. Gravity is not immutable, while force of inertia is immutable and inheres to the body.

We should note that the main argument of Rule 3 works even without the first part on qualities that cannot be intended and remitted. Indeed, in another manuscript note, Newton offers this conclusion quite emphatically, here not considering the first class of objects: 'The laws (and properties) of all bodies on which it is possible to institute experiments, are laws (and properties) of all bodies whatsoever.'[102] In the 1713 edition, however, Newton did include the first class of objects. One might ask why Newton needs to introduce qualities that cannot be intended or remitted if he has discovered through experience all the other qualities like hardness, impenetrability and so on just as he has done for gravity.

Newton's strategy is brilliant. All qualities such as hardness, impenetrability and force of inertia are found through experiments. Experiments alone are not sufficient to guarantee such qualities' universal and necessary character, which is fundamental for Rule 2, otherwise causes or laws cannot be generalized, but would pertain only to individual substances. Yet this, as we have seen, would amount to the return of a certain kind of substantial form, which Newton himself denies. So these qualities may be considered universal and necessary only if they are included in the first class of objects. Indeed, it is impossible to conceive of a body without extension. This is exactly what Descartes said, but according to Newton the knowledge of extension comes from experiments. What is important for Newton is to show that what comes from experiments *can* be attributed to all bodies universally, and this is proven because none can doubt that an essential attribute like extension is not universal. Therefore, something that is known by experiments can be attributed universally. This provides Newton with the second step of his argument: gravity, like extension, hardness and mobility, is known by experiments, and therefore it *can* be attributed to all bodies universally. Newton is not obliged to demonstrate that gravity is also an essential quality, but only that it can be universally attributable. For this reason, he has no problem asserting that gravity is not essential to a body. That gravity is a quality that pertains to all bodies universally, however, is evident from what Cotes writes in his preface paraphrasing Rule 3:

> Since all terrestrial and celestial bodies on which we can make experiments or observations are heavy, it must be acknowledge without exception that gravity belongs to all bodies ... The extension, mobility, and impenetrability of bodies are known only through experiments; it is in exactly the same way that the gravity of bodies is known. All bodies for which we have observations are extended and mobile and impenetrable; and from this we conclude that all bodies universally are extended and mobile and impenetrable, even those for which we do not have observations. Thus all bodies for which we have observations are heavy; and from this we conclude that all bodies universally are heavy, even those for which we do not have observation.[103]

In this passage, it is even more evident that there is a second crucial aspect for Newton in Rule 3 – that is, the transition from observable to unobservable things – in other words, the necessity to extend these qualities to *all* bodies, and not only those on which experiments have been done and which are generally observable. This is possible only thanks to Rule 2. If there are the same kind of effects in observable and unobservable bodies, then these effects are attributable to the same quality. This leads Newton to conclude that if two things, no matter whether they are observable or unobservable, coincide as to all their known qualities, these things are of the same kind.[104] As indicated above, the existence of effects of the same kind is extracted only inductively.[105] Rule 3 develops, therefore, a very complex methodology in which occur various degrees of induction, analogy and transduction.

Newton was perfectly aware of the difficulty of Rule 3. In a manuscript note, probably written immediately after 1715, discussing Leibniz, Newton writes as follows:

> He [Leibniz] denys none of Mr Newtons experiments. He denys not theis third Rule of Philosophy. And yet from the Experim^{ts} & that Rule universal gravity necessarily follows. But he denys the conclusion. And indeed he has a very good faculty at denying conclusions. That third Rule is the Rule of Induction. And without it no Proposition can become general in Naturall Philosophy. Without it we cannot affirm that all bodies are impenetrable. And the argument by Induction for universal gravity is as strong as the argument for universal impenetrability. Yet Arguments from Induction are not Demonstrations. They are only to take place till some experimental exception can be found.[106]

Rule 3 is called the 'rule of induction', and according to Newton it is the only conceptual tool that can be used to reach universal conclusions in natural philosophy, otherwise there would be a return to the Scholastic occult quality. Newton had not yet conceived his Rule 4, which will be included only in 1726 edition of the *Principles* and arose as an addendum of Rule 3:

> Rule 4. *In experimental philosophy, propositions gathered from phenomena by induction should be considered either exactly or very nearly true notwithstanding any contrary hypotheses, until yet other phenomena make such propositions either more exact or liable to exceptions.* This rule should be followed so that arguments based on induction may be nullified by hypotheses.[107]

Rule 4 serves to armour Rule 3 from the attack of any who, like Leibniz, though accepting the experimental process, did not share Newton's conclusion. Induction becomes the primary instrument for the logic of discovery. But it is not simple enumeration – which Newton rejected from his early youth – but a form of demonstrative induction, supported by the firm principle of harmony, continuity and uniformity inherent in nature as God's creation.[108]

In Rule 4, Newton admits that the conclusions are only provisional, yet he believes that they should be considered as almost (*quamproxime*) true and accurate so that they can lay the foundations for epistemic knowledge. Provisional knowledge can function as epistemic knowledge, until a phenomenon confutes it. Conclusions can become principles, which then serve the third stage of regressus.[109] The importance of rejecting hypotheses and defending true principles deduced from phenomena by induction does not consist only in the fact that Newton was afraid that his law of gravity might be considered as having an occult quality; rather, it lies in the validity of the experimental method and of the logic of the scientific discovery. In the *Scholium Generale*, after refuting Descartes's theory of vortexes, Newton shows his awareness of the provisional character of his knowledge. He explicitly states that he was not able to yet assign 'a cause to gravity', or 'to deduce from phenomena the reason for these properties'.[110] But he is not interested at this stage in metaphysical investigations, and nor does he feign hypothesis. Rather, according to Rule 4, any metaphysical or physical or occult or mechanical cause must be rejected in experimental philosophy until a further effect can contradict the general conclusion, either inferred by deduction from the phenomena or from the generalization of induction. In Newton's specific case, it is not necessary to know what gravity is – the main research question of the Aristotelians and Cartesians – but 'it is enough that gravity really exists and acts according to the laws' discovered through its method, because in this way it can 'explain all the motions of the heavenly bodies and of the sea'.[111] In this Newton follows the pathway opened up by Galileo.

This approach exerted a disruptive and highly creative influence in terms of epistemological thinking: according to Newton, it is possible to make epistemic statements only in terms of *how* natures works and not regarding *why* nature is. Scientific knowledge concerns mainly the relations and interactions of phenomena in experience, and not their causes in the ontological sense of this word. This results in a radical change in the field of metaphysics, which will become fully comprehensible only at the end of the eighteenth century.

Newton was tempted to introduce a Rule 5 in support of Rule 4, but in the end he never included it in the *Principles*. Rule 5 survives in two different versions in the same folio,[112] but it is the version of the *recto* that is the most disruptive:

> Rule 5. Whatever things are not derived from objects themselves, whether by the external senses or by the sensation of internal thoughts, are to be taken for hypotheses. Thus I sense that I am thinking, which could not happen unless at the same time I were to sense that I am. But I do not sense that some idea is innate. And I do not take for phenomena only things which are made known to us by the five external senses, but also those which we contemplate in our minds when thinking: such as, I am, I believe, I understand, I remember, I think, I wish, I am unwilling, I am thirsty, I am hungry, I am happy, I am sad, etc. And those things which follow from the phenomena neither by demonstration nor by the argument of induction, I hold as hypotheses.[113]

This rule shows the influence of Descartes and Locke,[114] especially in the definition of what phenomena are, on which Rule 4 is based. As phenomena, Newton included not only what may be known through sensation, but also psychological states. This introduction of the inner states of the mind into experimental philosophy may well be the reason why Newton discarded Rule 5: there was no obvious way for him to subject these to experiments or infer any generalizations. It would have been like introducing into the third stage of regressus a hypothesis from which everything derives, whereupon the synthesis would have been unreliable.

In conclusion, Newton's experimental philosophy is not a naïve epistemology based only on experience,[115] but one also based on advanced logical and mathematical techniques, which must be employed in order to reach the strongest conclusions, especially in establishing the laws of nature. If so, Newton's experimental philosophy rises from the ashes of Aristotelian logic, making Newton's own novel version of the theory of regressus. The popularity of Newton's method shapes the entire eighteenth century and the greater part of the nineteenth, going very well beyond physics. Famously, George Cheyne aimed to find a law of attraction for spiritual beings, and David Hume aspired to be the Newton of moral philosophy.[116] If Rule 5 had been known, it would have aroused – if this were possible – a great deal more interest in the Newtonian project. In any case, Newton's epistemology left its mark on the collective imagination. In the mid-eighteenth century, in his posthumous work titled *Account of Newton's Discoveries* (1748), the mathematician Colin Maclaurin identifies Newton's major achievement with regressus. Indeed, Newton,

> in order to proceed with perfect security, and to put an end for ever to disputes, proposed that, in our inquires into nature, the methods of analysis and synthesis should be both employed in a proper order.[117]

Immediately after this, Denis Diderot and Jean Le Rond d'Alembert in their *Encyclopédie, ou Dictionnaire raisonné des sciences, des arts et métiers* celebrated Newton's experimental philosophy for its capacity to combine the empirical and the theoretical moments of knowledge, through the application of mathematics to experience. Newton's merit is to have shown to the world what his predecessors had merely glimpsed – that is, 'how to apply geometry to physics and to shape, by joining experiment and calculation, an exact, profound, luminous and new science'.[118] In a recent article, George E. Smith has outlined how Newton's *Principles* changed physics and they certainly did.[119] Newton's *Principles*, however, also forever revolutionized epistemology and metaphysics, both of which took totally new directions and faced entirely new challenges. The gauntlet to reform these disciplines on the path established by Newton was picked up by a university professor coming from a remote city of Prussia, Königsberg. This philosopher was Immanuel Kant, whose firm conviction was that if 'Newton had lived longer, he alone would have discovered more than all human beings together would have discovered in a thousand years'.[120]

Conclusion

Königsberg. 1756. One of the most esteemed professors of the university, Martin Knutzen, had died five years previously. His chair in logic and metaphysics was vacant. A new generation of philosophers were standing ready to apply for the position. With this in mind, in accordance with the law established by Frederick II, his former student Immanuel Kant – full of hope, soon to be disappointed – was preparing a new work titled *The Employment in Natural Philosophy of Metaphysics Combined with Geometry ... the Physical Monadology*. The topic would have pleased his mentor, a well-known and eclectic philosopher – one of the first to introduce Locke and Newton in Germany. The main idea of Kant's academic writing – as the title suggests – was to reconcile mathematics with metaphysics. The conciliation presupposes a division, having its origin in the controversy between Newton and Newtonians on the one side, and Cartesians and Leibnizians on the other. The main battlefields were (1) whether Newton's approach to physics were merely a mathematical description of the world or something more, (2) how eventually to substantiate it within a metaphysical system and (3) how this metaphysics might be envisaged if it did not endorse mechanical philosophy.[1]

Kant had already broached this problem in his first public work, *Thoughts on the True Estimation of Living Forces* (1747), written under the direction of Knutzen. In *A New Elucidation of the First Principles of Metaphysical Cognition* (1755), the question remained hidden, a predominantly metaphysical position being adopted, according to the Leibnizian and Wolffian tradition. In 1755, Kant published anonymously *The Universal Natural History and Theory of the Heavens or Essay on the Constitution and the Mechanical Origin of the Whole Universe According to Newtonian Principles*. The booklet embodies a piece of research on the origin of the universe, written according to Newtonian laws of nature. It was Kant's belief that Newton had achieved a perfect system capable of describing the system of the world. But Newton's strength was also his weakness. Indeed, Newton's main contribution was exclusively restricted to the mathematical field and the physical part of his cosmology had yet to be written. According to Kant, Newton had worked only on the mathematical half of natural philosophy, intentionally neglecting physical and metaphysical investigations.[2] In the *Physical Monadology*, Kant's reflections take another direction, which will mark his pathway in future philosophical enquiry, and lead ultimately to his *Critique of Pure Reason* (1781).

In the *Physical Monadology*, the conspicuous influence of Newton's methodology on Kant's early philosophy emerges for the first time, so much so that scholars speak of a 'Newtonian conversion' or of a 'Newtonian revolution'.[3] The work opens with a clear, though reserved, endorsement of Newtonian epistemology:

> Clear-headed philosophers, who are seriously engaged in the investigations of nature, unanimously agree, indeed, that punctilious care must be taken lest anything concocted with rashness or with a certain arbitrariness of conjecture should insinuate itself into natural science, or lest anything be vainly undertaken in it without the support of experience and without the mediation of geometry. Certainly, nothing can be thought more useful to philosophy, or more beneficial to it, than this rule ... if we follow this sound path, we can exhibit the laws of nature though not the origin and causes of these laws. For those who only hunt out the phenomena of nature are always that far removed from deeper understanding of the first causes. Nor will they ever attain knowledge of nature itself of bodies.[4]

Clear-headed philosophers are Newtonians, who are against the introduction of hypotheses in natural science without a yardstick in experience. Kant is alluding to the fact that Cartesians or Leibnizians of any kind proceed with a wrong methodology in natural science, producing castles in air in order to explain natural effects. He will come to call such approaches 'dogmatic' or 'rationalist'. In complete contrast are Newtonians, who are clearly identified as empiricists, whose empiricism, however, prevents knowledge of the causes of things. At this stage in Kant's early philosophical development, he had a strong metaphysical commitment to the possibility of knowing the ultimate substratum of reality, even if he was perfectly aware of the potentialities of Newton's mathematical approach, albeit with limitations. He thought that metaphysics alone – which Newtonians said should be removed from physics – could provide illumination for intricate issues such as the infinite divisibility of space, free motion in the void and universal attraction without mechanical cause. There was no easy solution to these questions – indeed, the marriage between mathematics and metaphysics is more difficult for Kant than the mating of griffins with horses. For our present purpose, not so important is how Kant tried to solve these problems as that he thought a solution was possible through an application of mathematics to experience.

While Kant was writing these words he had on the desk at his side John Keill's *An Introduction to Natural Philosophy* (1702).[5] Although Keill's work was conceived and published before both *Opticks* (1704) and the second edition of the *Principles* (1713), Newton's methodological lesson emerges overwhelmingly in the 'Preface' and in the first lecture, titled *Of the Method of Philosophizing*. The book is written mainly to refute mechanical philosophy and to show how this metaphysics is incompatible with mathematics. The Cartesians' main error is epistemological and lies in the fact that 'ignorant men of geometry presume to philosophize, and to give the causes of natural things'.[6] In so doing, mechanical philosophers 'have embraced the shadows of philosophy'.[7] They would have abandoned the secure path opened by Galileo, who was the first, according to Keill, to penetrate the secrets of nature by means of mathematics, framing a new science of motion and showing a new epistemology 'whereby the mechanical causes of things might be discovered'.[8] Mechanical philosophers pursue another methodology. They appeal to hypotheses like miracles – contrary to the laws of nature in order to explain the works of nature. The main problem of using hypotheses in the field of natural investigation is that they offer no means of explaining other phenomena besides those for which they were introduced or designed for. An important and crucial step in scientific method is therefore missing – that is, that of anticipating and providing new knowledge about possible future phenomena. The sure path of science, according to Keill, is that outlined by 'Sir Isaac Newton, whose prodigious Genius has laid open more and abstruser mysteries of nature, than men could ever have hoped for'.[9] In natural science – for Keill – what all his predecessors in mechanical philosophy have done amounts to no more than one-tenth of what he has discovered with his mathematical method.

Keill provides Kant with a very clear picture of the advancement of Newtonian philosophy by distinguishing four methods for investigating the causes of natural things. None of them is perfect; none of them must be completely endorsed; all of them have weaknesses and strengths. Their advocates comprise: (1) Platonists, (2) Aristotelians, (3) experimental philosophers and (4) mechanical philosophers. While Platonic positions are praised for their use of mathematics, mechanical philosophy is criticized for its erroneous combining of metaphysics with mathematics. Keill does not fully endorse experimental philosophy, which no doubt has led to new discoveries, but many times experiments go astray through fancy theories. He is more interested in Aristotelian philosophy. He is extremely critical of Aristotelians' explanation of nature by means of matter and form, since they do not really reveal the causes of things. They mainly provide essences and definitions of the things themselves. Yet, says Keill, this should not prevent the use of terms like quality, faculty or attraction. These terms – in Newtonian fashion – do not define true and physical causes and modes of action that pertain essentially to things, but actions that 'may be intended and remitted', through which it is possible 'to express the ratios of the forces or their augmentation and diminution'.[10] Keill speaks of qualities that are intended and remitted before any of Newton's public statements in this respect, and before Rule 3 was introduced into the 1713 edition of the *Principles*. There was no doubt a fruitful scientific conversation between Keill and Newton in defining these kinds of qualities, which are useful for explaining nature without providing any metaphysical foundation.

This is clear in the case of gravity, about which – according to Keill – we may say, that it is a quality 'whereby all bodies are carried downwards, whether its cause arises from the virtue of the central body, or is innate to matter itself, or whether it proceeds from the action of the ether by a centrifugal force and so tending upwards, or finally, whether it is produced after any other manner whatsoever.'[11] The metaphysical investigation of the causes is not relevant; it is secondary, as if it were impossible for the limits of human mind to grasp. This does not mean in Keill's view that these qualities are occult. They can be understood as occult, in the same sense as mathematical equations where unknown quantities are designated by letters like x, but not in the sense that they cannot be investigated in their intensions and remissions. For, 'however ignorant we are of the nature of qualities, and how much whatsoever the modus of operations (*operandi modus*) is hidden from us',[12] it is always possible to demonstrate the theorem concerning their intension and remission, according to which every quality that 'is propagated every way in right lines from a center, is diminished in a duplicate proportion of distance from that center'.[13] This theorem is mathematically demonstrable in a universal way, 'whatever is the nature of the quality'.[14] The investigation of the nature of the quality is useless, because this law would work in any case, whatever the object is. But what is more important, according to Keill, is that

> after the ratios of the forces are discovered in given conditions or suppositions (*conditionibus seu suppositionibus*), then those ratios are to be compared with the phenomena of nature, that there may be found out, what conditions of the forces belong to every sort of bodies.[15]

In other words, the discovery of the laws is fruitful because then, in a second phase – corresponding to the synthetic stage of regressus – it is possible to apply them for the description of other phenomena. Before doing this, according to Keill, many experiments should be carried out until what has been discovered becomes an undoubted principle or axiom. There should be an intermediate stage between the discovery – that is the analytic stage – and the application of the conclusions as universal laws. This intermediate stage of performing experiments to establish the principles is call '*negotio*' and should observe three necessary rules in order to minimize errors as much as possible, and to proceed on the secure path of science.[16] Keill is redefining the Aristotelian theory of regressus in his own terms, and he suggests that the rules of philosophizing represent an essential link between the two stages. This process sheds light on what Newton will do eleven years later in the 1713 edition of the *Principles*, but Keill's rules are very different from what Newton elaborates.

The first rule prescribes proceeding according to the method of geometers – that is, to set definitions – which are necessary in order to acquire knowledge of things. To distinguish this position from that of the Aristotelians, Keill writes that these definitions are not logical. Indeed, these definitions do not consist of genus and difference, and nor do they show the intimate essence or the ultimate cause of things. This kind of research does not lead to scientific knowledge and does not concern physics. Kant is very critical of this point of Keill's repudiation of metaphysical research. But Keill, as with the majority of early Newtonians, insists on confessing his own profound ignorance of the intimate natures and causes of things (*intimae rerum naturae & causae*).[17] The only possible way of acquiring reliable knowledge is via sensation, or by means of another property; but in this case too one that is always known by sensation. This kind of knowledge does not give an account of the causes of things, nor provide definitions in the way that logicians would. Rather, it furnishes a description, 'whereby the thing described may be clearly and distinctly conceived and likewise be distinguished from everything else.'[18] The Newtonian philosopher resorts to a terminology strictly related to the tradition of regressus theory. Indeed, the purpose of the whole process is to identify the simplest properties that experience shows belong to the things themselves. This would be the intermediate stage of regressus. Then, having once determined these simple properties, by means of a 'geometrical method' it is possible to deduce other properties of the same things.[19] This is the third stage, which is synthesis. Against the

Cartesians, Keill states that these properties belong and pertain to things, but they are not constituents of the essences and natures. That they are essential elements of things is a mere conjecture or hypothesis. Keill emphasizes the distinction between properties that certainly pertain to things themselves (*proprietates rebus ipsis certo competentes*) and properties that are really in things (*rebus ipsis revera insunt*). These are two very different epistemological approaches, which will influence the genesis of *Critique of Pure Reason*, and also the very distinction between *noumena* and *phaenomena*. From the Newtonian standpoint, there is merely description of how things are experienced, while from the Cartesian and Leibnizian perspective, there is an attempt to say something about the very nature of things – that is, how things are independently of experience. For instance, Keill says, 'the Cartesians say that a fluid has its parts in continual motion, but it does not appear from our senses, or experience, or reason, that such is the nature of a fluid'.[20] This is confirmed only by the hypothesis that they have introduced, but it is contrary to true mathematical demonstrations.

The second rule of philosophizing for investigating natural truth asserts the need 'to consider only the conditions that were supposed at first, abstracting for a time from all other considerations whatsoever'.[21] This corresponds to what Newton will conceive as Rule 1 – that is, to describe a phenomenon by examining the least number of conditions possible because the human mind is limited and can be misled in considering too many conditions at once. Keill's third and final rules follow on from the second one, and prescribe beginning with 'the most simple cases first', and then adding conditions and causes as a way of investigating more complex phenomena.

Keill helped to popularize Newton's idea in very simple terms, and Kant was struck by his epistemological assumptions, making it possible to say something true about reality by means of mathematics, without defining the nature and essence of things. This seemed to be a perfect compromise between experience and theory, missing in Cartesian or Leibnizian metaphysics. Kant submitted the *Physical Monadology* on 23 March 1756, and it was discussed on 10 April. The very same day there appeared in a weekly periodical of Königsberg the first part of Kant's third and final essay on earthquakes, bearing the title *Continued Observations on the Earthquakes That Have Been Experienced for Some Time*.[22] At the very beginning of the text, Kant reiterates Newton's greatest achievement as having purged physics of foolish ideas, banishing miracles and hypotheses from the investigation of natural phenomena.[23] The same adhesion to Newton's epistemology is emphasized in *The Only Possible Argument in Support of a Demonstration of the Existence of God* (1763), in which Kant states that it is in vain for any metaphysical attempt to demolish what has been established on the basis of empirical observations and mathematical deduction by introducing fancy definitions.[24]

Kant's confidence in Newton's epistemology is not without limitations. Indeed as Keil pointed out, Newton's approach is perfect for mathematical descriptions, but fails to answer genuine metaphysical questions, which in Kant's view had extreme relevance for the advancement of scientific and philosophical thought. Already in *The Universal Natural History and Theory of the Heavens*, he aimed to integrate metaphysics with mathematics. In the 1760s, Kant emphasizes the necessity of overcoming the hiatus between the two disciplines, and tries to find a solution. However, his attempt was destined to fail, in its acknowledgement that metaphysics as it was conceived and mathematics had two very different methods and epistemologies. This is the result of the *Inquiry Concerning the Distinctness of the Principles of Natural Theology and Morality* (1763).

Kant's work originated in the 1761 announcement by the Berlin Academy of a prize for an essay on whether metaphysical truths in general – and in particular the first principles of natural theology and morality – were capable of evident proofs like mathematics, and, in the case of a negative answer, what the nature of their certainty was. In other words – reversing the perspective – the Berlin Academy was actualizing the old vexed question posed by Alessandro Piccolomini in the Renaissance – that is, whether demonstrations in mathematics have the same degree of certainty as those in philosophy. The solution to this problem, according to Kant, is pivotal because it could put an end to infinite philosophical opinions, and place metaphysics on the sure and unique path of science, just as in natural science 'Newton's method transformed the chaos of physical hypotheses into a secure procedure based on experience and geometry'.[25] The solution cannot be metaphysical because metaphysics is not a certain science and therefore the result could not be definitive. Therefore, Kant proceeds only from the 'certain data of experience and deduction drawn immediately from them'.[26] At this stage, Kantian method reflects Newton's epistemology from phenomena and from deduction. Kant reiterates his adoption of Newtonian epistemology in a long passage:

> The true method of metaphysics is basically the same as introduced by Newton into natural science and which has been of such benefit to it. Newton's method maintains that one ought, on the basis of certain experience and, if need be, with the help of geometry, to seek out the rules in accordance with which certain phenomena of nature occur. Even if one does not discover the fundamental principle of these occurrences in the bodies themselves, it is nonetheless certain that they operate in accordance with this law. Complex natural events are explained once it has been clearly shown how they are governed by these well-established rules. Likewise in metaphysics: by means of certain inner experience, that is to say, by means of immediate and self-evident inner consciousness, seek out those characteristic marks which are certainly to be found in the concept of any general property. And even if you are not acquainted with the complete essence of the thing, you can still safely employ those characteristic marks to infer a great deal from them about the thing in question.[27]

Kant is clearly referring to the case of gravity, which, even though its cause and nature are unknown, makes possible the description of natural phenomena. In the same way, metaphysics should renounce any attempt to know the essence of things, and its main business should be finding those characteristic marks that can be attributed with certainty to a thing, and using them for its representation.

The search for inherent characteristics was typical of Renaissance Aristotelians, who focused on 'essential attributes', but it received new vigour with the study of combinatorics in the eighteenth century.[28] What Kant is claiming here, however, is not the understanding of essential attributes, but attributes which can be known with certainty, in a very Newtonian fashion. What is important is epistemic certainty, not its foundationalism or essentialism. Though Kant believes in the possibility of characterizing the foundation and basis of reality, he is already sceptical as to the possibility of securing an absolute grasp, or an intellectual intuition, of it. Kant's investigation may already be characterized as epistemological, rather than ontological or – better – he seems to be in the process of transforming metaphysics into epistemology. By endorsing Newton's method for metaphysics, Kant blurs the distinction between phenomena and noumena, which will become clear in the *Critique of Pure Reason*. The former is that aspect of a thing which the mind can know with certainty through the cognition of characteristic marks, and which is functional for a scientific description of reality. The latter is the unknowable nature of a thing, lying behind the phenomenon. Newtonian gravity represents this distinction eloquently at this stage of Kant's philosophical development. Kant's emphasis on inner consciousness in the *Inquiry* shows that he has already understood that the problem needs to be solved epistemologically within the workings of the mind. It is also clear that Kant was missing his famous 'Deduction' that would make this knowledge certain and universal.

Kant's view is that Newton's epistemology is such that no metaphysical argument is sufficient to threaten its conclusions. Going much further than Newton, for Kant Newtonians had established that will directly attract each other at a distance. This is unchallengeable from a metaphysical standpoint, the reason being that metaphysicians start in the wrong place – namely, with the idea that bodies act upon each other only through contact.[29] Metaphysicians start with definitions like geometricians, but this is not the right way for investigating natural things. This is what Kant calls in *The Only Possible Argument*, the 'mania for method':

> The imitation of the mathematician, who advances with a sure step along a well-surfaced road, has occasioned a large number of such mishaps on the slippery ground of metaphysics.[30]

But Kant here is not dealing with the combination or reconciliation of mathematics with metaphysics, but rather with the application of the geometrical order of exposition and explanation to the philosophical field.[31] The main error for Kant in this kind of metaphysics is its rationalism – that is, starting with something produced by definition in the mind and then deducing consequences without any yardstick in experience. This is the wrong way of proceeding in metaphysics, because its object is always something given by experience and never constructed by the mind. Kant was thinking of Cartesian and Wolffian metaphysics. If metaphysics follows this path, it can never achieve the certainty of mathematics. Metaphysics should proceed from analysis, which makes possible a clear and distinct understanding of the concept, to synthesis, which 'subsumes compound cognitions under the simplest cognitions'.[32] Kant proposes, therefore, a twofold method of analysis and synthesis, which starts from experience, discovers simple concepts and explains other natural phenomena through these.

The main source of errors relating to metaphysical knowledge lies in the intermediate phase between analysis and synthesis, which corresponds to the process whereby characteristic marks are identified that can be attributed to the concept of a thing. Indeed, in the continuous attentive effort of this process of identification of 'pure understanding',[33] it is easy 'for the characteristic mark of an abstracted concept to escape our attention without our noticing, for there is nothing sensible which can reveal to us the fact that the characteristic mark has been overlooked'.[34] This was the main problem – as we have seen – with the intermediate stage of regressus in determining the specific attribute of a thing. It led, according to Kant, to the misattribution of a characteristic mark, causing one thing to be confused with another, and producing thus erroneous knowledge – a problem evident to Renaissance logicians.

Metaphysics was far from being established as a true and certain science. As conceived by Cartesians, Wolffians – but also in the very recent attempt of Christian August Crusius – Kant viewed metaphysics as inadequate for achieving integration with mathematics. In his *Attempt to Introduce the Concept of Negative Magnitudes into Philosophy* (1763), Kant explains that metaphysics has frequently armed itself against mathematics and, 'where it might, perhaps, have been able to gain secure foundations on which to base its reflections', it has turned its attention to 'subtle fictions' or 'obscure and untestable abstractions', rather than to distinct, certain, 'intelligible and evident cognitions'.[35] To reconcile metaphysics with mathematics, Kant understands that he will need to reform the traditional conception of metaphysics, and he spends the silent 1770s trying to elaborate a philosophy in agreement with Newton's physics.[36] This philosophy resulted in the transcendental idealism of the *Critique of Pure Reason*.

*

Kant's transcendental idealism constitutes the final development of the epistemology in early modern philosophy. His programmatic intent is to overcome dogmatism and empiricism.[37] In doing this, Kant aspires to a scientific knowledge with a level of certainty equal to that of mathematics, even if mathematics is not the foundation of his cognitive theory. Famously, Kant professes a peculiar kind of criticism that limits scientific knowledge to a specific field – that of phenomena – but in doing so he denies the possibility of scientific knowledge in relation to what Aristotelians called essence. The field of scientific knowledge, therefore, becomes narrower and does not concern itself with essences. Newton must have taught him something, without diminishing the possibility of constructing an epistemology.

Kant is aware of his detachment from the previous epistemological tradition, whence he makes clear to his contemporaries the revolutionary reach of his *Critique of Pure Reason*, which in its first edition aroused both consenting and dissenting voices, in particular regarding the lack of originality and clarity in his philosophical system. Kant articulates his system by assessing the relationship between his own transcendental philosophy and the mathematical sciences, discovering thus their common epistemological roots. Kant's philosophical system takes shape in a lucid and definitive way in the *Prolegomena to Any Future Metaphysics*, where we find an escalation of epistemological questions, starting with (1) how is pure mathematics possible? and continuing with (2) how is pure physics possible? Next comes (3) how is metaphysics in general possible? and then the final question (4) how is metaphysics as science possible? This can be reframed as the more general question (5) how is scientific knowledge possible from pure reason?

Before answering this last essential question, it is important to understand why and how Kant moves away from past philosophies of empiricism and rationalism, closing the circle of early modern philosophy initiated by Francis Bacon. It is not by chance that Bacon's exergue from the preface of the *Great Instauration* opens the second edition of the *Critique of Pure Reason*.

Kant's favourite target is without doubt rationalism or dogmatism, not least because it represented the most popular philosophical movement in the universities at that time. This dogmatism is represented by Christian Wolff and his school. Kant sees Wolffian philosophy as based on 'the presumption of getting on solely with pure cognition from (philosophical) concepts according to principles, which reason has been using for a long time, without first inquiring in what way and by what right it has obtained them'.[38] In particular, the problem with dogmatism is the absence of any yardstick in experience, abstracting completely from the empirical dimension, and working only on the basis of universal principles that are generated in the mind and ineffective for the elaboration of a theory of knowledge that might provide a scientific description of reality. A pure, formal and rationalist epistemology would have been an empty vessel without any real content – an a priori epistemology which does not lead to any new knowledge, explaining only what has been already given in terms of thought.

At the opposite end of the scale, the epistemology of empiricism would lead to a relative and subjective type of knowledge, lacking universal validity and the proper character of science in being based only on the gathering of empirical data. Indeed, Lockean epistemology for Kant derived its concepts from experience, but this derivation is almost accidental and not universal and necessary, and therefore – even if providing a description of experience – such concepts could not pave the way to scientific knowledge.[39] It was David Hume, according to Kant, who understood that for universal and objective knowledge it was necessary that the origins of concepts should be a priori and not in experience. However, Hume did not develop an adequate epistemology because his solution also derived concepts from a subjective necessity – erroneously considered objective – namely, habit. The epistemology of empiricism cannot lead to scientific knowledge because the necessity of its conclusions is subjective and a posteriori.

Kant's aim is to overcome these two epistemological positions – that is, rationalism and empiricism – putting to rest at the same time various forms of pessimism and scepticism. He saw the right pathway for science as having been suggested and stimulated by Bacon and Galileo's epistemologies, which shone a light on all those who study nature. Their merit was that

> they comprehended that reason has insight only into what it itself produces according to its own design; that it must take the lead with principles for its judgments according to constant laws and compel nature to answer its questions … for otherwise accidental observations, made according to no previously designed plan, can never connect up into a necessary law, which is yet what reason seeks and requires. Reason … must approach nature with its principles in one

hand, according to which alone the agreement among appearances can count as laws, and in the other hand, the experiments thought out in accordance with these principles.[40]

Reason and experiment should work together in metaphysics, as in the science of nature, in order to develop a reliable epistemology.[41] Kant aims to assess the possibility of acquiring universal and necessary knowledge which might acquire its content from experience. In other words, within the context of his transcendental idealism, this amounts to answering the question: How are a priori synthetic judgements possible?

Synthetic a priori judgements are that particular kind of cognitive judgement that in adding something new from experience to the concept of the subject has an objective, necessary and universal validity. Objectivity and necessary validity represent for Kant two reciprocal characteristics on which scientific knowledge of the object of experience should be based. Such an epistemological attitude is possible only through changing the existing relation between the subject and the object of knowledge, and redefining the concept of truth. It is Kant's famous Copernican revolution, on which his transcendental idealism is based.

According to Kant, knowledge is scientific – that is, universal and necessary – if and only if grounded on universal and necessary laws. Therefore, he states that

> either these laws are taken from nature by means of experience, or, conversely, nature is derived from the laws of the possibility of experience in general, and is fully identical with the mere universal lawfulness of experience. The first contradicts itself, for the universal laws of nature can and must be cognized *a priori* (i.e., independently of all experience) and set at the foundation of all empirical use of the understanding; so only the second remains.[42]

Kant makes clear that these laws cannot be found in the object – that is, in the experience – otherwise one would fall into the error of searching for something necessary and universal in what is accidental and subjective, as empiricism did. In developing this idea of universal laws of nature, Kant has in mind Newton. In *Reflection 5414*, he opposes Newton's scientific method to Kepler's empiricism:

> One can certainly discover rules empirically, but not laws – as Kepler did in comparison to Newton – for to the latter belongs necessity; and therefore, the fact that they can be cognized a priori.[43]

According to Kant, what Newton had done in the field of physics, he was now doing for metaphysics. In metaphysics, Kant's solution was that the universality and necessity of laws must derive from the subject – that is, from the mind in its capacity to know things. All rational beings have these universal and necessary elements that allow for the picking up of what is accidental coming from experience, and to make of this a form of scientific, universal and necessary knowledge. In this way, the highest legislation for nature must lie in the understanding, which means 'we must not seek the universal laws of nature from nature by means of experience, but, conversely, must seek nature, as regards its universal conformity to law, solely in the conditions of the possibility of experience that lie in our sensibility and understanding'.[44]

This is not the place to explain in detail Kant's theory of knowledge, and how sensation and understanding interact: rivers of ink have been written on these topics.[45] Rather, it is important to emphasize that the subjective conditions of the mind guarantee the objectivity of knowledge – that is, that knowledge is scientific. On these subjective conditions, which are common to all mankind, rest the concept of truth. Truth cannot be any longer understood as the agreement of knowledge with its object:

> If truth consists in the agreement of a cognition with its object, then this object must thereby be distinguished from others; for a cognition is false if it does not agree with the object to which it is related even if it contains something that could well be valid. Now a general criterion of truth would be that which was valid of all cognitions without any distinction among their objects. But it is clear that since with such a criterion one abstracts from all content of cognition (relation to its object), yet truth concerns precisely this content, it would be completely impossible and absurd to ask for a mark of the truth of this content of cognition, and thus it is clear that a sufficient and yet at the same time general sign of truth cannot possibly be provided. Since ... we have called the content of a cognition its matter, one must therefore say that no general sign of the truth of the matter of cognition can be demanded, because it is self-contradictory.[46]

Thus truth is characterized by the agreement of a 'cognition with the general and formal laws of understanding'.[47] The formal laws of the understanding represent those subjective conditions also called transcendental, because without them it would not be possible to acquire scientific knowledge. These laws are the pure principles of the understanding – that is, those rules in relation to which the subject relates necessarily and universally with the object in order to acquire scientific knowledge.

These structures of the subject are the pure, a priori forms of space and time in relation to sensibility, which have the role of providing the object of experience to the subject, and the pure concepts of the understanding, which gives determinate meaning to the object. In particular, it is the synthesis of the pure concepts of the understanding with the manifold nature of possible experience coming from space and time that produces knowledge. Such synthesis characterizes a priori synthetic judgement – that is, those universal and necessary judgements which extend knowledge. For Kant, however, mathematical judgements are also for the most part synthetic a priori judgements, and therefore the certainty of scientific knowledge of experience is equal to mathematical knowledge. Scholars have maintained that Kant modelled his synthetic a priori judgements of knowledge on mathematics, yet this is not completely correct. In fact, in the first edition of the *Critique of Pure Reason*, synthetic a priori judgements were developed independently of the synthetic a priori character of mathematics, and only after the *Prolegomena*, with the clarification of his philosophical system, did Kant come to understand that metaphysics through these judgements has the same degree of universality and necessity as mathematics, at least at the epistemological level.

Kant's so-called revolution and his transcendental idealism seem to write the final words of the epistemology developed in the previous two centuries. He does this at a high price. Indeed, if universality and necessity of knowledge are granted for what the subject provides a priori, this means that the object of knowledge could be known only through these filters – in other words, through a partial and never emendable perspective. For Kant, scientific knowledge is possible only with respect to the phenomenon, that which appears, and never the thing in itself, the substance – that is, the noumenon. Scientific knowledge is always knowledge of phenomena, and this is grounded on the pure concepts of the understanding.

A number of scholars, then, have recognized that Kant himself, even if always implicitly, consistently admitted, writing also in the *Metaphysical Foundations of Natural Science*, that these concepts foresee a kind of knowledge of the phenomenon, which is compatible with Newtonian physics.[48] Kantian's theory of knowledge, therefore, shares with mathematics not only the fact that both are grounded in synthetic a priori judgements but the conviction that the knowledge of the world is substantially a physical-mathematical knowledge, making possible the objective knowledge of phenomena, which may be understood under the lens of the Newtonian system. Indeed, if what we have previously said is true – that is, 'the understanding does not draw its (*a priori*) laws from nature, but prescribes them to it',[49] then the laws that the understanding imposes are the laws of Newton's physics.[50] Thus in the *Prolegomena*, Kant writes that

> the possibility of experience in general is thus at the same time the universal law of nature, and the principles of the former are themselves the laws of the latter. For we are not acquainted with nature except as the sum total of appearances, i.e., of the representations in us, and so we cannot get the laws of their connection from anywhere else except the principles of their connection in us, i.e., from the conditions of necessary unification in one consciousness, which unification constitutes the possibility of experience.[51]

Kant's Copernican revolution provides the metaphysical foundation of Newton's physics, a foundation that was missing according to Cartesians and Leibnizians. This is possible only because scientific knowledge is not about essences. Kant further develops this idea of nature and universal law through working on Newton. The law of gravitation, Kant states, appears to be something that resides 'as necessary in the nature of the things themselves', as if 'no other law of attraction save that of the inverse square of the distances can be conceived as suitable for a system of the world'.[52] This is only an illusion, however. Indeed, the universal validity of this law does not rest in things in themselves, which are totally unknown to the mind, but it lies in the understanding, so much so that

> the understanding is the origin of the universal order of nature; in that it comprehends all appearances under its own laws and thereby first brings about experience *a priori* (with respect to its form), in virtue of which everything that is to be cognized only through experience is necessarily subject to its laws. For we are

not concerned with the nature of *the things in themselves*, which is independent of the conditions of both our senses and understanding, but with nature as an object of possible experience, and here the understanding, since it makes experience possible, at the same time makes it that the sensible world is either not an object of experience at all, or else is nature.[53]

Kant solves Newton's problem of how to apply mathematical laws to things in nature, without saying anything about the nature of things. For Kant, mathematical laws are not merely description of things, for which the Newtonian apparatus could remain a hypothesis; they are not essential to things, just as gravity was not for Newton; rather, they are the laws through which the mind knows the world and constitutes nature, and without which the object of knowledge would not be an object at all.[54] Kant is very well aware of Newton's dilemma and since he does not want to charge him of inconsistency – as Cartesians and Leibnizians did – he argues that the founder of the theory of gravitation believed that gravity was an essential attribute of matter, though he could not support it publicly.[55] The truth was that Newton in his epistemology could solve only provisionally what Kant believed to be an absolute certainty. As we have seen in the previous chapter, Newton is clear in the *Scholium Generale* that he has not yet found the cause of gravity, and his theory should be considered valid and scientific unless proven by some new phenomena to be incorrect. Newton could not give the answer that the philosopher of Königsberg was expecting, because it would have been either incompatible with Newtonian methodology (i.e. admitting gravity as essential to matter), or with Kant's transcendental idealism itself. Indeed, as Michael Friedman has correctly pointed out, Kant's claim must be 'understood in the context of his transformation and reinterpretation of metaphysical questions generally'.[56] Kant solves Newton's problem, we have said, but from his own perspective, which, like that of Newton, rejected in this field the scientific knowledge of essences.

The connection between transcendental idealism and mathematics is clear in the *Metaphysical Foundations of Natural Science*. Kant establishes that we can properly call science only that knowledge whose certainty is necessary. Indeed, any other kind of cognition, such as histories of natural things in the Baconian sense, consist of mere description, and only improperly of knowledge. A proper natural science presupposes a metaphysics of nature: 'for laws, that is, principles of the necessity of that which belongs to the *existence* of a thing, are concerned with a concept that cannot be constructed, since existence cannot be presented *a priori* in any intuition'.[57] A metaphysics of nature can deal with the 'laws that make possible the concept of nature in general, even without relation to any determinate object of experience',[58] and it is called transcendental. But it can deal also with 'a particular nature of this or that kind of thing, for which an empirical concept is given'.[59] In the latter case, it is a special metaphysics of nature – that is, it is not general.

Regarding this special metaphysics, Kant states that 'there can be only as much proper science as there is mathematics therein'.[60] Kant explains this statement establishing that a priori knowledge 'means to cognize it from its mere possibility'.[61] In this case, however, we have knowledge of a determinate object and – as Kant establishes in the *Critique of Pure Reason* – 'in order to cognize the possibility of determinate natural things, and thus to cognize them a priori, it is still required that the intuition corresponding to the concept be given a priori'.[62] He calls this operation 'construction of concepts' and points out that the 'rational cognition through construction of concepts is mathematical'.[63] A pure natural science of determinate objects is therefore possible only through mathematics and, 'since in any doctrine of nature there is only as much proper science as there is a priori knowledge therein, a doctrine of nature will contain only as much proper science as there is mathematics capable of application there'.[64] In this way, for Kant it is possible to achieve scientific knowledge of an object that is epistemologically equivalent to mathematical knowledge.

Mathematics is no longer an epistemic ideal to which one should aspire and which was hardly achievable starting from the knowledge of objects of experience because of their accidental character. Knowledge of experience is mathematical knowledge. In contrast to Locke and Leibniz, Kant is able to develop an epistemology – which profitably combines the empirical and the rational characters in order to acquire universal and necessary knowledge – because he shifts the problem of knowledge from objects to subjects. The necessary connection between cause and effect, between substance and accident, which Aristotelians were looking for by means of the regressus and to which early modern philosophers aspired, is no longer in the object, but in the subject itself. Being in the subject and being a priori, there is no longer need for a work of the intellect which would find the universal laws in objects. It is rather a law of the understanding itself, which is given by nature or – better – nature is the same law. In outlining his epistemology, Kant breaks definitively with the Aristotelian tradition and with all the attempts to reform it.

The axis shift from the object to the subject also changes the opaque zone of the epistemology. If before, what remains opaque was an operation of the intellect – that is, the subject – which guarantees scientific knowledge, now what is opaque is the object, which is no longer known in its totality, or at least the claim to its complete knowledge has collapsed, narrowing cognition to only what the understanding can know scientifically of it, excluding thus essences. Essences are no longer a subject of scientific knowledge. The fact that the laws of the understanding applied to nature are the mathematical laws of Newton's natural philosophy remains a Kantian presupposition, based on the conviction of the validity of the Newtonian system more than on any genuine epistemological consideration. Kant could not conceive of another form of universal and necessary knowledge which might avoid these laws, and these laws, therefore, were those that the understanding prescribed to nature. What is interesting in this Kantian revolution is that mathematics is no longer regarded as the ideal for epistemology – that is, for philosophy – but it is a restricted case. The foundation of mathematics lies in transcendental philosophy, or in what Kant calls metaphysics, and therefore mathematics cannot be taken as an epistemic model of certainty and universality, but only as a particular case of the application of the laws of the understanding, which provides a scientific knowledge, albeit partial, of the world.

Kant resolves the conflict that Bacon felt at the end of the Renaissance, originating in the contrast between empiricism and rationalism, but its epistemological tension remains. Indeed, scientific knowledge narrows its field by establishing specific boundaries that define what the human mind can scientifically know. There no longer exists a clear and distinct knowledge on the one side, and confused and obscure knowledge on the other; rather, there is scientific knowledge and another kind of cognition entirely, which refers to other fields of human experience such as the emotions, beauty and so on.[65] The relevant epistemological question is no longer how the mind can know the world scientifically, but rather – to borrow Bacon's image – how bees digest flower pollen. With Kant the false dichotomy between empiricism and rationalism is definitively overturned, and with it also came the collapse of the myth of the given and the dualism between scheme and content. In this collapse, we are catapulted directly into the twentieth-century philosophy of mind, in which the digestion of empirical data is resolved in the synchronic relation between mind and world, and not in the diachronic logical process applied to experience, as had been the mode for the Aristotelians and the early modern philosophers. But this is another history of epistemology, still to be written.

Notes

Introduction

1. However, as with any label, sometimes their use is unavoidable in order to show their limited validity. As far as possible, I have avoided the use of cognates like 'inductivism', 'eliminativism', 'inferentialism', 'deductivism' or other labels used in an unhistorical way, even if useful for understanding the various positions. I use the expression 'logic of scientific discovery' to refer to the inventive part of method in epistemology, but with no reference to Karl Popper's ideas.
2. See Anstey (2005). For an assessment of the importance of Anstey's approach, see Levitin (2019). See also Engfer (1996); Bodenmann and Rey (2018); De Boer and Prunea-Bretonnet (2021).
3. Laudan (1981: 23).
4. Martin (2014).
5. Buchdahl (1969: 24).
6. Galison (1997).
7. Bonk (2008).
8. Buchdahl (1969: 8).
9. Just to state a famous example, see Putnam (1994); McDowell (1994).
10. See Thorndike (1951: 584–98).
11. See Panofsky (1972: 1–41).
12. Morhof (1688: 108).
13. Roger Ariew has demonstrated the importance of Scholasticism in René Descartes's thought, and Cees Leijnehorst contextualized the appropriation of the Aristotelian tradition in Thomas Hobbes's natural philosophy. Christia Mercer has revealed the Aristotelian legacy in Gottfried Wilhelm Leibniz, Alfredo Ferrarin exposed Aristotle's influence on Hegel, and Riccardo Pozzo traced the overall impact of Aristotelianism on modern philosophy (Ariew, 1999; Mercer, 2001; Leijenhorst, 2002; Ferrarin, 2005; Pozzo, 2004; Ariew, 2011). I myself made a modest contribution by examining the role of Aristotelianism in the rise of the so-called British empiricism and Kant's philosophy (Sgarbi, 2013, 2016). In the field of history of science, William A. Wallace provided a fundamental reassessment of Galileo Galilei's Aristotelian and Scholastic sources, while André Goddu has demonstrated how Aristotelianism shaped Copernicus's philosophy (Wallace, 1984a; Goddu, 2010).
14. Garber (2009: 9–18); Garber (2016: 133–48).
15. Garber (2009: 142).
16. Annas and Barnes (1985: 4).
17. Indeed, the very first occurrence of this word seems to derive from the translation of the German *Wissenschaftslehre*. It became popular only thanks to James Frederick Ferrier's *Institutes of Metaphysic* (1854) where it designates the particular philosophical field that deals with the questions of knowing and the known.
18. Pasnau (2017: 140).

19. Pasnau (2017: 2).
20. For contextual readings in history of philosophy, see Mercer (2019) and Gaukroger (2019).
21. Sgarbi (2018).
22. On the problems of the principles in early modern philosophy, see Anstey (2017).
23. There are two factions, one in favour of and one against continuity between Aristotelian regressus and the modern scientific method. For continuity, see Cassirer (1922); Randall (1940); Crombie (1953); Edwards (1967); Randall (1976); Olivieri (1978); Wallace (1981); Carugo and Crombie (1983); Edwards (1983); Wallace (1983); Crescini (1983); Wallace (1984a, 1992a, b, 1995). For discontinuity, see Gilbert (1963); Schmitt (1969); Poppi (1972); Jardine (1976); Schmitt (1983); Poppi (2004); Palmieri (2007).
24. Osler (1970: 3); Pasnau (2017: 13).
25. Jesseph (2013) shows that after the first half of the century, the influence of Aristotelianism waned.
26. Aristotle, *Posterior Analytics* I.2 71 b 9–13.
27. For a recent discussion on the topic, see Sorell, Kraye and Rogers (2010).
28. We hope that in future, as further research is done, it will be possible to take account of any newly relevant women epistemologists that come to light.
29. On the Alexandristic position in logic, see Risse (1964).
30. Against Bacon's empiricism and experimentalism, see Vanzo (2014).
31. Biener (2018).

1 Averroists

1. Martin (2006); Sgarbi (2021).
2. Paris, Bibliothéque Nationale de France, Ms. Lat. 6535, f. 228v.
3. Craig Martin shows how meteorological investigations based on experience in particular tended to refute Aristotelian explanations from within, reaching the conclusion that Aristotle was not demonstrating, but arguing, for probable knowledge. See Martin (2009).
4. I say 'in general' because there are many variants of regressus. Sometimes authors do not mention the word, but they employ its idea. Others used a variety of terminology like analysis and synthesis, solution and composition, from effects to causes and from causes to effects. There are many ways to refer to regressus, some of which will be examined in this book.
5. Some logicians, like Zabarella, identify the demonstration *propter quid* with the Aristotelian most perfect demonstration (κύριον ἀποδείξιν, *potissima*), while others, like Bernardino Tomitano, identify the demonstration *potissima*, not with the demonstration *propter quid*, but with the entire process of regressus.
6. Randall (1940); Gilbert (1960); Papuli (1983); Wallace (1995).
7. Zabarella (1597: 486f–487a).
8. Pomponazzi (1970: 155).
9. Pomponazzi (1970: 164).
10. Papuli (1983: 232).
11. Randall (1940: 189–92); Papuli (1983: 230–1); Wallace (1995: 82–3).
12. Pomponazzi (1970: 154). Partially translated by Nicholas Jardine (1976).

13. Pomponazzi (1970: 154–5).
14. Aristotle, *Prior Analytics* 46 a 33; Aristotle, *Posterior Analytics* 91 b 15.
15. Aristotle, *Prior Analytics* 46 a 33–46 b 13.
16. Aristotle, *Posterior Analytics* 91 b 35.
17. Pomponazzi (1970: 155).
18. Pomponazzi (1970: 155).
19. Vernia, Lat. Misc. 506, p. 239v.
20. Vernia, Lat. Misc. 506, p. 239v.
21. Trombetta (1502: 19ar).
22. Vernia, Lat. Misc. 506, p. 239v. In another passage of the *Exposition*, Vernia uses the example of the lunar eclipse to emphasize the weakness of inductive knowledge:

    > To know scientifically (*scire*) … is not to know by accident, is not to know from the accident and nor to know from something accidental, as in the case of the demonstration *quia* and of knowledge by accident; instead I mean to know what is from the essential principles of things. Knowing by accident is twofold: one is through something accidental, as I argue of the lunar eclipse: the Moon is eclipsed, something is placed in front of what is eclipsed, then something is placed in front of the Moon; [the second is] the knowledge that we acquire, but which we do not define, such as the knowledge acquired by means of some experiments (*experimentis*).

23. Agrimi and Crisciani (1990); Baroncini (1992).
24. Vernia, Lat. Misc. 506, p. 239v.
25. Nifo (1522: 20vb).
26. Nifo (1508: 7vb). Translation by Randall. I have not been able to find the 1506 edition used by Randall (1940: 192), and Crescini (1965: 141).
27. Nifo (1508: 7vb–8ra). Translation by Randall.
28. Nifo (1549: 6rab). Translation by Randall.
29. Poppi (1970).
30. Pomponazzi (1970: 160).
31. Pomponazzi (1970: 168).
32. Papuli (1983: 231).
33. Pomponazzi (1970: 170).
34. Poppi (1970).
35. Nifo (1549: 6rb). On this topic, see Martin (2009, 2011). René Descartes, Walter Charleton and Robert Boyle, as we shall see, will use this passage of Aristotle to justify their different epistemologies. See Laudan (1981).
36. Averroes, *de coelo* II, comm. 35; Aquinas, In *Aristot. de coelo* II 17, 451.
37. Berti (2009: 435–57).
38. Zimara (1537: 108v).
39. Papuli (1983: 248).
40. Balduino (1557: 11rb).
41. Balduino (1557: 11rb).
42. Biblioteca Universitaria Padova, Ms. 810, ff. 114v–120r. On the manuscript see Simionato (1973); Davi (1994: 103–18).
43. Palmieri (2007: 421).
44. Palmieri (2007: 421).
45. Piccolomini (1551: 203).
46. Piccolomini (1551: 204).

47. Angelo Crescini emphasized the importance of this passage, Crescini (1972: 158–259).
48. Piccolomini (1551: 203–8). Kant expresses a similar concept in *The Only Possible Argument in Support of a Demonstration of the Existence of God*; see *KGS*, II, p. 292.
49. Piccolomini (1551: 209).
50. Piccolomini (1551: 210). See Sgarbi (2021).
51. Capivacci (1562: 78v).
52. Capivacci (1562: 120v).

2 Alexandrists

1. Risse (1964: 278).
2. Zabarella (1597: 481d, 489c–d). Translations from Zabarella's work are from Zabarella (2013), unless otherwise noted.
3. Zabarella (1597: 485b).
4. Zabarella (1597: 485c).
5. Zabarella (1597: 485d–f). For a formalization of 'demonstrative induction' and its role in modern physics, see Dorling (1973).
6. In his *Commentary to Posterior Analytics*, Zabarella writes that

    when we make an induction in necessary matter we do not enumerate all the singulars, because knowing a few the intellect begins to see the essential connection between the two terms. Therefore the intellect, breaking off the enumeration of the remaining individuals, immediately gathers the universal from those few, for the illation from an essential predication to a universal predication is necessary. Thus in cases such as this the essential connection between the terms is so manifest that the universal can be grasped from only a few individuals, and perhaps even from only one.

    See Zabarella (1597: 1281d–e).
7. Zabarella (1597: 486c).
8. Zabarella (1597: 486d).
9. Zabarella (1597: 486f).
10. Zabarella (1597: 487a).
11. Knowable *per se* seem to refer to a kind of analytic knowledge.
12. Zabarella (1597: 238f).
13. Zabarella (1597: 255d–e).
14. Zabarella (1597: 255f).
15. Zabarella (1597: 255f–256a); translation slightly modified.
16. Zabarella (1597: 256b). In Latin, the sentence reads 'est processus a cognitione eiusdem rei imperfecta ad perfectam per eiusmet diligentem inspectionem'. The words Descartes employs for characterizing his mental inspection, which makes possible the transition from obscure and confused to clear and distinct knowledge are similar: 'ejus perceptio non visio, non tactio, non imaginatio est, nec unquam fuit, quamvis prius ita videretur, sed solius mentis inspectio, quae vel imperfecta esse potest & confusa, ut prius erat, vel clara & distincta'. *AT*, VII, p. 31. Peter Dear suggests a possible influence of Zabarella's 'meditation' or 'mental consideration' on the title of Descartes's *Meditations*. See Dear (1995a: 60–2). Timothy J. Reiss points out that the lack of evidence for a personal connection does not diminish the likelihood of Zabarella's

influence on Descartes, 'first because personal connection, if it exists (as it does not here), is likely to be uninterestingly trivial, and second because possible lines of influence are always overdetermined and anyway manifold.' See Reiss (2000: 195). Many scholars have attempted to reconstruct Zabarella's influence on Descartes's methodology, basing this on his three years of training in philosophy at the Jesuit college of La Flèche, or on his stay in Germany (Edwards 1983: 215; Reiss 2000: 198; Mehl 2001; McMullin 2008: 94–5), but none have recognized the similarities between the two philosophers on the notion of mental inspection.

17. Zabarella (1607: 1017).
18. Zabarella (1607: 1000).
19. Zabarella (1607: 256b).
20. Zabarella (1597: 487d–488b).
21. Zabarella (1597: 490d–e).
22. Zabarella (1597: 490e–f).
23. Zabarella (1597: 492b–c).
24. Zabarella (1597: 493d–e).
25. Zabarella (1597: 493e).
26. Zabarella (1597: 266f–267a).
27. Berti (2009: 435–57).
28. Even Zabarella's enemy Bernardino Petrella shares his vision of the necessity of negotiation: 'the true effect does not lead to proper and distinct cause, rather indistinct; only after the negotiation does the intellect know that that cause is proper to the thing', Petrella (1571: 41v).
29. Fabri (1606: 44).
30. Smith (1613: 116).
31. Flavell (1619: 138).
32. Crankanthorpe (1622: 356).
33. Airay (1628: 160).
34. Burgersdijk (1637: 272).
35. Burgundo (1651: 189).
36. Carleton (1649: 189).
37. Martini (1652: d. XI, §5).
38. Jungius (1652: part 2, sect. 1, ch. 2).
39. Newton (1671: 126).
40. Marsh (1679: 222–3).
41. Toletus (1580: 167va).
42. Brutscher (1602: 2).
43. Wallace (1988: 143–5). Crombie and Carugo suggest Ludovicus Carbo as Galilei's source (Crombie and Carugo 1996: 169–72). There are undeniable correspondences and similarities between the *Additamenta* published in 1597 by Carbo and Galileo's Ms. 27, but only on *praecognita*. Carbo is silent on regressus.
44. Lorinus (1620).
45. Now preserved in the manuscript Urb. Lat. 1471.
46. Lorinus (1620: 555).
47. There is no strict relationship between Lorinus's text and Vallius's book, and Wallace's hypothesis seems on the weak side (Wallace 1988).
48. One might argue that it was not an act of full-blown plagiarism, but rather that it is more likely that Vallius introduced the section on regressus into the final printed version. The author seems to confess as much in the introduction to his work, saying

that 'we have so expanded our entire composition ... so in those works [Carbo's works] you have what he took from me, in this that I have prepared more fully and at length' (Vallius 1622: 1). This indicates that Vallius may have elaborated at some point his conception of regressus and of mental examination. This means that it is impossible to confirm any real influence of Vallius on Galileo, who could have read the same ideas directly in Zabarella. In any case, his research is a reworking of Zabarella's text.
49. Vallius (1622: 345).
50. Vallius (1622: 345).
51. Vallius (1622: 346). Wallace has unjustifiably emphasized Vallius's conceptual pairing of *'formaliter/materialiter'* (Wallace 1995: 94–5), which, in fact, already features in Zabarella.
52. Vallius (1622: 346). Galileo used the same terminology in his manuscripts.
53. Baldi (1641: 269).
54. Baldi (1641: 270).
55. Baldi (1641: 272).
56. Baldi (1641: 278).
57. Baldi (1641: 274).
58. Baldi (1641: 281).
59. Cremonini (1596: 4v–5r).
60. Martin (2009: 287).

3 Francis Bacon

1. Mill (1843: vol. 1, 378); Malherbe (1996: 97).
2. Cf. Crescini (1972: 101). See also Cohen (1949: 99–106); Hattaway (1978: 183–97); Cohen (1980: 219–31); Malherbe (1984: 179–200); Milton (1987: 49–74); Malherbe (1990: 69–87); Gaukroger (2001: 132–64). On Bacon's alleged Aristotelianism cf. Larsen (1962: 435–50); Kosman (1964).
3. Malherbe (1996: 76). On natural history in Bacon see Gaukroger (2006: 367–96).
4. Sgarbi (2013).
5. My thesis is substantially different from that proposed by Alistair C. Crombie, according to which 'Bacon's method was a combination of the inductive and deductive processes already seen by his medieval predecessor', namely, Robert Grosseteste. It also differs from that elaborated by John H. Randall on the continuity between Paduan regressus and Bacon's methodology. I accept Antonio Pérez-Ramos's critique of the previous two positions, and I see a clear distance between Bacon's own and traditional logical projects (Pérez-Ramos 1988). However, in contrast to Pérez-Ramos, I do not assess Bacon's shift away from Aristotelian doctrines by considering only Aristotle's texts, but I focus on the Aristotelian tradition closer to the English philosopher (see Pérez-Ramos 1988: 120, 234–8).
6. *OFB*, IV, p. 109.
7. *OFB*, IV, p. 109.
8. *OFB*, IV, p. 109.
9. *OFB*, IV, p. 111.
10. *OFB*, XI, p. 3.
11. *OFB*, XI, p. 31.

12. *OFB*, XI, p. 17.
13. *OFB*, XI, p. 33.
14. *OFB*, XI, pp. 33–4. Translation is slightly modified.
15. *OFB*, XI, p. 71.
16. *OFB*, XI, pp. 74–5.
17. *OFB*, XI, p. 75.
18. *OFB*, XI, p. 159.
19. *OFB*, XI, p. 253.
20. See *OFB*, XI, p. 253.
21. *OFB*, XI, pp. 109–10.
22. *OFB*, XI, p. 110.
23. *OFB*, XI, p. 110. Translation is slightly modified.
24. *OFB*, XI, p. 110.
25. Digby (1580: 4v–5r).
26. Case (1584: 134).
27. Powell (1594: 338–40).
28. Powell (1594: 349).
29. Balfour (1618: 713–14).
30. Smith (1613: 120).
31. Smith (1613: 120–1).
32. Sanderson (1618: 226–7).
33. Sanderson (1618: 152).
34. Flavell (1619: b. 2, 48).
35. Flavell (1619: b. 2, 51).
36. Keckermann (1613: 255–6).
37. Scheibler (1654–8: 45).
38. *OFB*, VI, pp. 97–8.
39. *OFB*, XI, p. 33. Translation is slightly modified.
40. *OFB*, XI, p. 37.
41. *OFB*, XI, p. 33.
42. *OFB*, XI, p. 161.
43. *OFB*, XI, p. 161.
44. *OFB*, XI, p. 175. Translation is slightly modified.
45. *OFB*, XI, pp. 33–5.
46. *OFB*, XI, p. 87.
47. *OFB*, XI, p. 161. Translation is slightly modified.
48. *OFB*, XI, p. 215. Translation is slightly modified.
49. *OFB*, XI, p. 153.
50. *SEH*, III, p. 616; *SEH*, III, p. 385.
51. Celsus (1935: 7).
52. Celsus (1935: 17).
53. Celsus (1935: 17).
54. Celsus (1935: 9).
55. Celsus (1935: 9).
56. Celsus (1935: 11).
57. Celsus (1935: 11).
58. Celsus (1935: 17).
59. Celsus (1935: 17).
60. Celsus (1935: 19).

61. Celsus (1935: 19).
62. Indeed, Celsus introduced a third sect, the so-called Methodists, but as Owsei Temkin has pointed out, they 'have no right to consider themselves a new sect, for they are either dogmatists or merely bad Empiricists'. See Temkin (1935: 261).
63. In opposition to the dogmatics, Celsus supported the idea that experience came first and reasoning later, while for the former experience was merely a test of what had been previously conceived (Celsus 1935: 11). That Bacon endorsed Celsus's epistemology is clear from § 73 of *Novum Organon*, where he writes, 'Celsus admits it frankly and sensibly: first the experiences of medicine were discovered, and then men philosophised about them and sought and assigned causes; it did not happen the other way round, that the experiences were discovered or suggested by philosophy and a knowledge of causes.' A similar quote is in *The Refutation of the Philosophies*, *SEH*, III, p. 578. Misinterpreting *De medicina*, Bacon attributes an empirical perspective to Celsus, as is clear from the fifth book of *De dignitate et augmentis scientiarum*, where there appears an almost direct quotation (*SEH*, I, p. 617). However, the substance remains the same: both the Roman writer and the Lord Chancellor believed in the collaboration of experience and mind.
64. *SEH*, III, p. 232. The translation has been revised.
65. Pérez-Ramos (1988: 236–7).
66. *OFB*, XI, p. 215.
67. *OFB*, XI, p. 215.
68. *OFB*, VI, p. 99.
69. Wear (1985: 118–45; 312–17).
70. *OFB*, XI, p. 221.
71. *OFB*, XI, p. 237.
72. The legal terminology is particularly interesting since the appearance in the court is always after the collection of evidence. The collection of proof is the empirical stage, whereas the judgement of the court is the work of the intellect.
73. *OFB*, XI, p. 253. Translation is slightly modified. In the 1713 edition of *Principles*, Newton speaks of qualities that are intended and remitted, as we shall see in Chapter 10.
74. The reference is to alchemical separation through fire.
75. *OFB*, XI, p. 255. Translation is slightly modified.
76. *OFB*, XI, p. 255.
77. *OFB*, XI, p. 255. Translation is slightly modified.
78. Crescini (1972: 6–7).
79. *OFB*, XI, p. 163.
80. *OFB*, XI, p. 163.
81. *WFB*, III, p. 241.
82. *WFB*, III, p. 241.
83. *WFB*, III, pp. 569.
84. *OFB*, XI, p. 289.
85. A similar account is in Plato, *Politicus* 262 a 4–263 b 9.
86. *Phaedrus* 265 e 1–2.
87. *Phaedrus* 265 d 3–7.
88. Wendin (1987: 213).
89. *Phaedrus* 265 d6; *Politicus* 284 d2.
90. *OFB*, IV, p. 111.

91. *OFB*, IV, p. 389.
92. Kuhn (1977: 44).
93. *OFB*, IV, pp. 111–12.
94. *OFB*, IV, pp. 111–12.
95. *OFB*, IV, p. 112.
96. Yates (2010: 370–3). See Cohen (1980).
97. *WFB*, IV, p. 270.
98. *WFB*, IV, p. 271.
99. *WFB*, IV, p. 271.
100. *WFB*, II, p. 368.
101. *OFB*, XI, p. 59.
102. *OFB*, XI, p. 75.
103. *OFB*, IV, p. 113.
104. *OFB*, IV, p. 113.
105. *OFB*, IV, p. 111:

    This part of the invention, concerning the invention of sciences, I purpose (if God give mee leave) hereafter to propound, having digested it into two partes; whereof the one I tearme 'Experientia Literata', and the other 'Interpretatio Naturæ:' the former being but a degree and rudiment of the latter. But I will not dwell too long, nor speake too great upon a promise.

106. *OFB*, XI, p. 321.
107. Lohne (1968); Dumitru (2013); Jalobeanu (2015).
108. Schwartz (2017).
109. *OFB*, XI, p. 89. Translation is slightly modified.
110. *OFB*, XI, p. 119.
111. *OFB*, XI, p. 201.
112. *OFB*, XI, p. 201.
113. *OFB*, IV, p. 83.
114. *OFB*, IV, p. 83.
115. *OFB*, XI, p. 201.
116. *OFB*, XI, p. 203. Translation is slightly modified.
117. *OFB*, XI, p. 203.
118. *OFB*, XI, p. 203.
119. *OFB*, XI, p. 205.
120. *OFB*, XI, p. 263.
121. Fletcher (2005). See also Anstey (2015) for a better understanding of the impact of Ramus on Bacon.
122. Ramus (1572: 63).
123. *OFB*, III, p. 239.
124. *OFB*, III, p. 239.
125. *OFB*, III, pp. 240–1.
126. *OFB*, XI, p. 209.
127. *OFB*, XI, p. 65.
128. Rees (1986); Mori (2017). For a partial re-evaluation of the role of mathematics for practical knowledge in Bacon, see Domski (2013); Jalobeanu (2016).
129. Dumitru (2013: 45–61).

4 Galileo Galilei

1. Laird (1997: 255).
2. Laird (1997: 255).
3. For a recent discussion, see Dawes (2016).
4. Randall (1940).
5. Gilbert (1963); Schmitt (1969); Poppi (1969, 1972); Jardine (1976); Schmitt (1983).
6. Edwards (1967); Wallace (1974, 1981, 1983, 1984a, b, 1991, 1992a, b, 1997, 1998, 2003, 2004, 2007).
7. Wallace (2000: 102).
8. Drake (1985).
9. Carugo and Crombie (1983: 32–49); Garin (1983: 29).
10. Palmieri (2007: 420–1).
11. Jardine (1976: 304).
12. Scholars in favour of the continuity thesis between the Aristotelian tradition and Galileo start their investigation with the famous passage in which Galileo professed himself a Peripatetic in his letter to the Aristotelian Fortunio Liceti on 17 October 1640. *OG*, XVIII, p. 248. Charles B. Schmitt suggested that Galileo's declaration of Aristotelianism was a 'rhetorical exaggeration' (Schmitt 1983: 28). If we are to believe Galileo's words, Aristotle's main contribution to the history of philosophy and science was in the field of logic and, in particular, epistemology. According to Wallace, this note makes it possible for us to 'appreciate the project in which Galileo was involved throughout his life' (Wallace 1992a: 295) – that is, in Crombie's words, 'his lifelong adherence to the conception of true demonstration set out by Aristotle in the *Posterior Analytics*' (Crombie 1996: 488).
13. *OG*, VIII, p. 197.
14. Wallace (1992b: xv–xvi).
15. For a different interpretation, see Wisan (1978: 43).
16. Wallace (1992a: 89).
17. Wallace (1992a: 88).
18. Wallace (1992a: 90).
19. For a detailed bibliography on this passage, see Mignucci (1975: 178–84).
20. Enrico Berti has shown the uniqueness of Zabarella's interpretation of metaphysics as a universal science, and its peculiar character in anticipating some of the key issues of early modern philosophy. See Berti (2009: 239–54).
21. Zabarella (1597: 781c–d).
22. Poppi (1970: 162–3).
23. Zabarella (1597: 528b–529c).
24. Zabarella (1597: 810d–e). For Zabarella's conception of dialectic and its opposition to metaphysics as a universal science, see Berti (2009: 245–53).
25. Berti (2009: 252).
26. Proclus (1560: 5).
27. Proclus (1560: 2).
28. Proclus (1560: 26–7); Friedlein (1873: 46).
29. Proclus (1560: 4); Friedlein (1873: 8).
30. Zabarella (1597: 829c).
31. Wallace (1992a: 128, 132).
32. Wallace (1992a: 138).

33. Wallace (1992a: 138).
34. Wallace (1992a: 155).
35. Wallace (1992a: 155).
36. Wallace (1992a: 155).
37. Wallace (1992a: 155).
38. Wallace (1992a: 155–6).
39. Wallace (1992a: 157).
40. Wallace (1992a: 158).
41. Wallace (1992a: 158).
42. Wallace (1992a: 160).
43. Wallace (1992a: 162).
44. Wallace (1992a: 163).
45. Wallace (1992a: 160).
46. Wallace (1992a: 182).
47. Wallace (1992a: 182).
48. Wallace (1992a: 182).
49. Wallace (1992a: 163).
50. Wallace (1992a: 183).
51. Wallace (1992a: 183).
52. Wallace (1992a: 183).
53. Wallace (1992a: 183).
54. This condition will be required also by Barrow as we shall see in Chapter 8.
55. Wallace (1988b, 1992b: 187).
56. McMullin (1990: 62–3).
57. Wallace (1992a: 183–4).
58. Wisan (1978: 35). Galileo adds that there is a difference between a mathematical demonstration and the most perfect demonstration. Mathematical demonstrations can be said to be perfect only 'because of their highest degree of certitude', for 'they abstract from matter', which is the first cause of uncertainty, or 'because of the preeminent method they follow'. See Wallace (1992a: 171).
59. Berti (2009: 339).
60. Fredette (2017); Salvia (2017). All translations of *The Older Works on Motion* are taken from Raymond Fredette, http://echo.mpiwg-berlin.mpg.de/MPIWG:ZR79G89F.
61. Camerota and Helbing (2000).
62. *OG*, I, pp. 260, 263. See Fredette (2017) for a contextualization of this passage.
63. *OG*, I, pp. 264–5.
64. *OG*, I, p. 266.
65. *OG*, I, p. 266.
66. Wallace (1992b: 248–9).
67. Giusti (1993, 1994).
68. *OG*, I, p. 273.
69. *OG*, I, pp. 296–8.
70. *OG*, I, pp. 318–19.
71. Berti (2009: 334).
72. Heath (1908: vol. 3, 442). In Latin, the text sounds: 'resolutio est assumptio quaesti tanquam concessi per ea quae sequuntur in verum aliquod concessum. Compositio vero est assumptio concessi per ea quae sequuntur in quaesti terminationem sive occupationem' (Euclid 1517: 230v).

73. Viète (1591: 4r). The passage was considered so spurious that both Nicolò Trataglia and Federico Commandino exclude it from their editions.
74. *OG*, I, pp. 322–3.
75. *OG*, VIII, p. 202.
76. Galileo had at least some of Piccolomini's astronomical works in his own library; his editions of *On the Fixed Stars* (1570) and *Of the Sphere of the World* (1573) contain his marginal notes.
77. Piccolomini (1558: 22r–v).
78. Piccolomini (1558: 22v), translated by Suter (1969: 213).
79. Piccolomini (1558: 22v), translated by Suter (1969: 212).
80. Piccolomini (1558: 22v), translated by Suter (1969: 212).
81. Piccolomini (1558: 22v), translated by Suter (1969: 212–13), slightly modified.
82. Piccolomini (1558: 23r), translated by Suter (1969: 213).
83. Piccolomini (1558: 23r), translated by Suter (1969: 213).
84. In the *Proem* of the second edition of *Of the Sphere*, Piccolomini states that the mathematical sciences have the highest degree of certainty, not for their subject but for their way of going about proving. See Piccolomini (1566: 2).
85. *OG*, II, p. 211.
86. *OG*, II, pp. 211–12.
87. *OG*, II, pp. 211–12.
88. See Berti (2009: 336).
89. *OG*, II, p. 251.
90. This is Wallace's opinion: see Wallace (1992b: 198–207). For a better explanation of the phenomenon, see Miller (2018: 306).
91. *OG*, III.1, p. 72.
92. This could seem like the first stage of regressus, but no induction, no demonstration, only arguments from other authors are considered.
93. Here Galileo seems to apply the intermediate stage of regressus.
94. *OG*, III.1, p. 75. This would constitute the third stage of regressus, which is not fully carried through here. See Miller (2018: 310). The other two cases mentioned by Wallace do not fit regressus. Indeed, for the discovery of Jupiter's satellites, Galileo gives only a hypothesis but he does not proceed to the consequences because – by his own admission – he has not calculated the periods of their motions for the purpose of making a prediction. For the mountains on the circumference on the moon, Galileo does not provide a proof, but engages only in reasoning by analogy in support of his hypothesis. A genuine attempt at a proof would only have become possible in 1664, thanks to Gian Domenico Cassini.
95. *OG*, IV, pp. 49–50.
96. *OG*, IV, p. 67.
97. *OG*, IV, p. 67.
98. Pappus (1877: II, p. 634).
99. Pappus (1589: 157r): 'in resolutione enim id quod quaeritur tamquam factum ponentes, quid ex hoc contingat, consideramus: & rursum illius antecedens, quousque ita progredientes incidamus in aliquod iam cognitum, vel quod sit è numero principiorum'.
100. *OG*, IV, p. 108.
101. *OG*, IV, pp. 108–9.
102. *OG*, V, p. 188.
103. *OG*, IV, p. 521.

104. Jardine (1976: 304, 307).
105. *OG*, IV, p. 521. Alastair C. Crombie translates 'gradazione' with 'progression' (Crombie 1996: 209).
106. Carugo and Crombie (1983: 49).
107. Aristotle, *Topics*, 114 b 37–115 a 6. Cf. Carugo and Crombie (1983: 49).
108. See Koertge (1977: 390); Wisan (1978: 35).
109. Crombie and Carugo traced the application of this rule back in a rather uncontextualized way to the practical mathematical arts like music, mechanics and engineering. Cf. Carugo and Crombie (1983: 49–50). Noretta Koertge finds an application of scientific reasoning, which is for her 'the most reliable and mature statement of his views on science', in the *Two New Sciences*:

> If we find in fact that moveables of different weight differ less and less in speed as they are situated in more and more yielding mediums: and that finally, despite extreme differences in weight, their diversity of speed in the most tenuous medium of all (though not void) is found to be very small and almost unobservable, then it seems to me that we may believe, by a highly probable guess, that in the void all speeds would be entirely equal.

OG, VIII, p. 117. See Koertge (1977: 407). Showing the decreasing effects starting from the decrease of a property leads Galileo to show the true nature of an event, removing all accidental aspects. That Galileo was relating this particular kind of process to what was the negotiation of the intellect for the Aristotelians is clear in the *Dialogue on the Two Chief Systems*, where he states that 'if it is true that one effect can have only one basic cause, and if between the cause and the effect there is a fixed and constant connection, then whenever a fixed and constant alteration is seen in the effect, there must be a fixed and constant variation in the cause'. The framework is evidently that of finding an essential connection between cause and effect, removing all possible accidents which do not characterize the nature of the phenomenon. Finding a constant and essential connection, as we have seen, was the primary target of the negotiation of intellect with the aim of transforming a mere probable and conjectural form of cognition into certain and true foundations for scientific knowledge.

110. *OG*, VII, p. 217.
111. *OG*, VII, p. 75. Translated by Stillman Drake (Galilei 1967). See Berti (2009: 337–8).
112. *OG*, VII, p. 432. Translated by Stillman Drake (1967).
113. *OG*, VII, p. 432. Translated by Stillman Drake (1967). Translation is slightly modified.
114. *OG*, VII, p. 229. Translated by Stillman Drake (1967).
115. *OG*, VII, p. 233. Translated by Stillman Drake (1967).
116. *OG*, VII, p. 229. Translated by Stillman Drake (1967).
117. *OG*, VII, p. 234.
118. Palmerino (2016).
119. *OG*, VIII, p. 50. Translated by Stillman Drake (Galilei 1974).
120. *OG*, VIII, p. 51.
121. *OG*, VIII, p. 175. Translated by Stillman Drake (Galilei 1974).
122. *OG*, VIII, p. 175. Translated by Stillman Drake (Galilei 1974).
123. *OG*, VIII, p. 197.
124. *OG*, VIII, p. 197. Translated by Stillman Drake (Galilei 1974).
125. *OG*, VIII, p. 202.

126. *OG*, VIII, pp. 202–3.
127. Mertz (1980).
128. *OG*, VIII, 212.
129. *OG*, XVII, pp. 90–1. Translated by Stillman Drake (Drake 1978). Emphasis is mine. The same passage appears almost verbatim in the letter to Giovanni Battista Baliani written on 7 January 1639.

5 Thomas Hobbes

1. Ward (1654: 58).
2. Hungerland and Vick (1981: 24); Edwards (1983: 205–20).
3. Hanson (1990: 587–626); Schuhmann (1990: 331–49); Pécharman (1995: 105–24); Pécharman (2016: 21–58).
4. Talaska (1988).
5. *OL*, IV, p. 390.
6. See Prins (1990). The most insightful reflections are to be found in Dear (1998) and Hattab (2014). Peter Dear has brilliantly identified that for Zabarella mental examination resolves the problem of convertibility, that is, of necessary concomitants. In Dear's reconstruction, the possibility of mental examination 'relied on the tenet that the mind could grasp universals (as causes necessarily were) corresponding to something metaphysically real' (Dear 1998: 152) – an unacceptable conception to Hobbes and this is the reason why there would be no evidence of mental examination in his thought. For Hobbes, 'universals could only refer to concepts in the mind' and therefore true physical causes remained unknowable (Dear 1998: 152). Since Zabarella was a realist and Hobbes was, in contrast, a nominalist, the two methods were destined to be different in essence. In addition, Helen Hattab has maintained that 'unlike Hobbes, for Zabarella knowledge from the effects is eminently demonstrative', and that 'Hobbes's philosophical method, in contrast to the method proper of Zabarella and other Aristotelians, yields not knowledge of the causes of things themselves, but rather of the causes of our conceptions of things' (Hattab 2014: 474). She correctly suggests that certain differences between Zabarella's and Hobbes's methods come from the latter's appropriations from Protestant logicians like Bartholomeus Keckermann and Franco Burgersdijk, whose works circulated widely in England in the first half of the seventeenth century.
7. Duncan (2003: 106–7). Stewart Duncan supported the link from Zabarella to Galileo and on to Hobbes through the discussions taking place in Mersenne's circle.
8. *OL*, I, p. 2.
9. *OL*, I, p. 58.
10. Hattab (2014: 471).
11. Prins (1990: 40).
12. *OL*, IV, p. 39.
13. Baroncini (1992: 39–62).
14. *OL*, I, p. 3.
15. *OL*, I, p. 59.
16. *OL*, I, pp. 107–8.
17. *OL*, I, p. 68.

18. *OL*, I, p. 70.
19. *OL*, I, p. 68.
20. *OL*, I, pp. 68–9.
21. *OL*, I, pp. 77–9.
22. *OL*, I, p. 108.
23. *OL*, I, pp. 108–9.
24. *OL*, IV, p. 23.
25. *OL*, IV, p. 23.
26. Zabarella (1597: 829c).
27. *OL*, I, p. 72.
28. *OL*, I, pp. 72–3.
29. Sacksteder (1980: 131–46).
30. *OL*, I, p. 252.
31. This is the reason why Kant thought mathematics and philosophy had two distinct methods, as we shall see in the conclusion.
32. *OL*, IV, p. 156.
33. *OL*, I, p. 62
34. Leijenhorst (2002: 171–218).
35. *OL*, I, p. 65.
36. *OL*, I, p. 65.
37. Prins (1990: 37).
38. Prins (1990: 31).
39. Zabarella (1597: 6a–c).
40. Zabarella (1597: 6b).
41. Zabarella (1597: 21f–22a).
42. Zabarella (1597: 48e–f).
43. Campanini (1986: 53).
44. Campanini (1986: 53).
45. Gargani (1971: 32–96).
46. Prins (1990: 41–2).
47. Against strongly conventionalist interpretations of Hobbes's epistemology, see the lucid account of Jesseph (2010: 123).
48. It is hard to know Hobbes's source for his knowledge of the Zabarellean method. As I have shown in this book, and as recent studies have likewise pointed out, Zabarella's doctrines were well-known in seventeenth-century England (Sgarbi 2013). We can now discount the mediation of Harvey or Galilei, as many scholars have suggested. Harvey has not such an evolved methodology nor so profound an understanding of Zabarella as we have seen in Hobbes (Watkins 1968: 41–2). As for Galileo, the Pisan scientist deals with the theory of mental examination particularly in his manuscripts, and Hobbes could hardly have had access to them. In his printed works, Galileo makes use of regressus, but he never goes into any detail. Both Hobbes and Galileo tried to mathematize, or geometrize, the process that clarifies the cause of effects. As we have seen, Galileo substitutes for mental examination mathematical analysis, and Hobbes, instead, conceives as the ground of every effect the motion generated by the cause. Besides this, however, their perspectives are profoundly dissimilar. As Jesseph has correctly pointed out, 'the evidence for any direct connection between Hobbes and Galileo' in the field of methodology is 'at best inconclusive'. See Jesseph (1999: 239, 2004).

49. Pécharman (2016: 47–8). Pécharman has now provided the most convincing reconstruction of Hobbes's method.
50. Charles B. Schmitt has splendidly pointed out that, even within the Aristotelian tradition, in authors like Zabarella and Cremonini we can find extreme divergences in the interpretation of Aristotle, leading to the endorsement of different philosophies. See Schmitt (1983: 10–12); Sgarbi (2017).

6 René Descartes

1. Ariew (1999).
2. *AT*, III, p. 185.
3. *AT*, VI, pp. 4–5.
4. See Ariew (1999, 2011).
5. Garber (1988).
6. On Descartes's criticism towards Aristotelian syllogistic, see Gaukroger (1989: 6–24); Douglas (2017); Gaukroger (2018).
7. See Sorell (2010).
8. Leibniz (*GP*, IV, 329). For a lucid examination of the problem, see Gaukroger (2002: 79–84). This topic has been discussed at length by the scholarship: see Gewirth (1943); Humber (1981); Curley (1986); Smith (2001).
9. Among the numerous publications on the topics for their contextualization deserving mention are Cozzoli (2008) and Dubouclez (2013). On the Aristotelian traces in Descartes, see Reiss (2000).
10. *AT*, X, p. 371.
11. *AT*, X, p. 381. In this logical context, 'absolute' and 'absolutely' are synonymous with simpliciter (απλώς).
12. *AT*, X, p. 383.
13. *AT*, X, p. 394.
14. Zabarella (1597: 238a–238b).
15. *AT*, X, p. 418.
16. *AT*, X, p. 379.
17. See Beck (1952: 155–70); Hintikka (1978: 75–88); Timmermans (1999); Florka (2001: 69–118); Palkoska (2017).
18. Here I consider reduction as a logical process and not as a reductionist approach like Grosholz (1991).
19. See *AT*, X, p. 430.
20. Brissey (2014, 2018).
21. *AT*, X, p. 395.
22. *AT*, X, p. 388.
23. *AT*, X, p. 389.
24. *AT*, X, p. 431.
25. *AT*, X, p. 440.
26. See Savini (2008). But also Galison (1984) for an application of comparison in the study of light.
27. *AT*, X, p. 435.
28. See *AT*, X, p. 435.
29. *AT*, X, p. 440.

30. *AT*, X, p. 427.
31. *AT*, X, p. 439. See Savini (2008: 154).
32. A number of scholars have anticipated my conclusions in their thinking. However, while Edwin M. Curley had recognized that 'having a clear and distinct idea of a thing ... is a matter of recognizing that there are certain properties we cannot but ascribe to a thing of that kind (clarity) and others which we are not at all compelled to ascribe to it (distinctness)', he did not attribute this process to enumeration as a divisive process (Curley 1986: 170). Kurt Smith, on the other hand, while establishing that – where the thing was complex – something was clear when it exhibited through enumeration the elements that constituted its nature and the relation that unified them (Smith 2001: 294), he reduced the entire Cartesian analysis to enumeration and related this process not to Aristotle but to the Platonic method of collection and division, like Bacon. This process, however, is insufficient for Descartes, as we shall see (Smith 2010, especially chs 4, 5, 6, 7). Alan Gewirth clearly identified this process and its similarities with Aristotle's thought, but he failed to recognize that it was traceable back to what the Aristotelians called negotiation of the intellect (see Gewirth 1943: 17–36).
33. Aristotle, *Posterior Analytics* 91 b 35.
34. See Larmore (1980).
35. *AT*, X, p. 368.
36. *AT*, X, pp. 400–1.
37. *AT*, X, pp. 400, 425, 454.
38. *AT*, X, p. 362.
39. *AT*, X p. 365.
40. *AT*, X p. 365.
41. *AT*, X p. 365.
42. *AT*, X, p. 369. This is the common definition of demonstration in the Aristotelian tradition.
43. *AT*, X, p. 366.
44. *AT*, X, p. 365.
45. See Hatfield (1988).
46. See Recker (1993).
47. *AT*, X, p. 369.
48. *AT*, X, p. 370.
49. Schouls (1980: 36).
50. For a different interpretation, see Raftopoulos (2004: 282).
51. *AT*, X, p. 392.
52. *AT*, X, pp. 394–5. For a detailed examination of the example of the anaclastic lines, see Garber (2001: 36–8).
53. *AT*, X, p. 394.
54. See *AT*, X, p. 405; *AT*, X, pp. 408–9. Zabarella also believes that division and induction 'may exert the mind's perspicacity in considering' simple things and definitions, see Zabarella (1597: 256c).
55. *AT*, X, pp. 401–2.
56. Aristotle, *Nicomachean Ethics*, 1143 a 35–1143 b 1.
57. *AT*, X, p. 407.
58. See Schouls (1980: 38).
59. Della Rocca (2005).
60. See Morris (1969: 161–9).

61. *AT*, X, p. 408.
62. Schouls (1980: 40).
63. *AT*, X, p. 409.
64. As we shall see, Descartes wrote in a letter to Jean-Baptist Morin that all the causes that explain effects fall into the category where the relation between cause and effect is univocal (*AT*, II, p. 199).
65. *AT*, X, p. 421.
66. *AT*, VIII.1, p. 22.
67. *AT*, X, p. 427.
68. *AT*, X, p. 428.
69. Desmond M. Clarke characterizes Descartes as an 'innovative Aristotelian' in this sense. See Clarke (1982: 197–205).
70. Renaissance Aristotelian philosophers called νοῦς an internal eye of the mind, with a clear reference to intuition. See Duodo (1577: 41): 'νοῦς sive mens est oculus internus animae'.
71. See De Koninck 1997. According to De Koninck, 'simple' for Descartes means something different from the Aristotelian 'indivisible', for the latter is the ultimate notion at which division stops. But as outlined already, my own view posits a correspondence between the Cartesian 'simple' and the Aristotelian 'indivisible' on the grounds that for Descartes what is simple is absolute – that is, not further solvable or analysable, and ultimately indivisible.
72. *AT*, X, p. 420.
73. As Remì Brague correctly argues, this verb represents 'the privileged metaphor of the activity of the intellect' – that is, the intellection. See Brague (1988: 356).
74. The verb is employed also in *Metaphysics*, 1072 b 21.
75. Aristotle, *Metaphysics*, 1051 b 23–30.
76. See Marion (1992: 115–39).
77. Aristotle, *Metaphysics*, 1041 b 10–11.
78. Aristotle, *Metaphysics*, 1025 b 14–18.
79. Aristotle, *Metaphysics*, 1025 b 17–18.
80. See Marion (1992: 132–9).
81. Some scholars believe that the *inspectio* is a more elaborate version of the *intuitus* first outlined in the *Rules for the Direction of the Mind* (see Clarke 1982: 58). Others believe that it is an openly anti-Aristotelian doctrine (see Carriero 2008: 122–7; De Pierris 2015: 25), through which Descartes developed a new model of mind in opposition to the Peripatetic and Scholastic one. Commenting brilliantly on the role of the mental inspection, John Carriero in particular maintains that 'Descartes's conception of mind can be understood as a reaction to Aquinas's' (see Carriero 2008: 68) or more generally to 'the Aristotelian thesis that the body is the proper object of the human intellect' (Carriero 2008: 122). His thesis is articulated by means of four assumptions which distinguish the novelty of the Cartesian conception of mind from its predecessors. *First*, in the Aristotelian tradition human understanding takes place in relation to something coming from experience as a form of abstraction deriving from sensation: 'on Aquinas's theory, the human intellect is unable to function according to its natural mode of cognition, without its lower faculties … without them the intellect would have no cognitive access to intelligible nature' (Carriero 2008: 125). In contrast, for Descartes there is a genuine faculty, namely, the intellect, which by means of the mental inspection, can understand without any recourse to sensation or imagination – that is, it 'does not require material from the

lower faculties for its operation, rather our lower cognitive faculties depend on and presuppose it' (Carriero 2008: 69). *Second*, according to the Aristotelian framework – and in Aquinas in particular – understanding is built on sensation and imagination, while in Descartes sensation and imagination are dependent on understanding. *Third*, for the Aristotelians, mind moves through a series of objects – that is, 'proper sensibles, common sensibles, universal natures' (Carriero 2008: 124) – while, for Descartes, 'the same structure that is given to us (perhaps confusedly) when we sense is what we also understand' (Carriero 2008: 124). *Fourth*, for Descartes, understanding 'is a matter of using judgment to penetrate from outward forms into inner constitution, rather than a matter of abstracting the commonalities shared by members of a species' (Carriero 2008: 124). Mental inspection thus conceived would constitute a radical departure from the Aristotelian philosophy of mind. Finally, *inspectio mentis* is usually associated with Cartesian innatism and intellectualism (see Schmaltz 1997; Williams 2005: 222; Boyle 2009; Ben-Yami 2015: 218–19; De Pierris 2015: 25–34). For instance, Bernard Williams, who has influenced a number of Cartesian studies advocates the idea according to which 'our conception of material things cannot be derived from experience but ... is innate, the product of our own faculty of thinking', the mental inspection (Williams 2005: 236).
82. *AT*, VII, p. 31.
83. Williams (2005: 208–10).
84. *AT*, VII, p. 34.
85. Garber (2001: 245).
86. *AT*, VII, p. 30.
87. Pasnau (2017: 22–30).
88. This passage is relevant to understand Newton's Rule 3 on intended and remitted qualities.
89. *AT*, VII, pp. 30–1.
90. *AT*, VII, p. 32.
91. *AT*, VII, p. 31.
92. See *AT*, X, p. 32.
93. See Garber (1992).
94. Fabrizio Baldassarri clearly shows an epistemological continuity from the *Rules* to his later correspondence to Marin Mersenne and Isaac Beeckman. See Baldassarri (2017).
95. *AT*, X, p. 430.
96. *AT*, X, p. 389.
97. *AT*, X, p. 389.
98. See *AT*, I, p. 559.
99. *AT*, I, p. 559.
100. *AT*, I, 195. See Baldassarri (2014, 2017).
101. Garber (2001: 101).
102. Garber (2001: 103).
103. Garber (2001: 103).
104. *AT*, VI, p. 77.
105. McMullin (2008: 94–5).
106. *AT*, IX, p. 327.
107. See Martin (2011: 145).
108. *AT*, II, p. 199.
109. *AT*, II, p. 199.
110. *AT*, II, p. 199.

111. *AT*, II, p. 200.
112. Baldassari (2018).
113. Garber (2001: 94). See also Garber (1992).
114. Brissey (2018: 93). Patrick Brissey shows how this method works in the *Meteors* in the case of parhelia. On the limits of experience see Sakellariadis (1982).
115. *AT*, VII, pp. 496–7.
116. *AT*, X, pp. 515–7.
117. Blake (1966); Clarke (1991); Dobre and Nyden (2013).
118. See Williams (2005: 208–10).
119. In Rule 12, Descartes writes that 'when we consider things in the order that corresponds to our knowledge of them, our view of them must be different from what it would be if we were speaking of them in accordance with how they exist in reality' (*AT*, X, p. 418). Descartes suspends ontological considerations in favour of purely epistemological ones. This could seem in contradiction with the overtly metaphysical project in *Meditations*, but in the *Second Meditation*, when Descartes discusses mental inspection, he is dealing with how minds know.
120. Clarke (1982: 58).
121. Van de Pitte (1998: 455).
122. Scarapelli Cory (2014: 69–114).
123. Williams (2005: 236).
124. *AT*, VIII, p. 358.
125. *AT*, X, pp. 358–9.
126. For a clear understanding of the problem, see Nolan (1997); Nelson (2008); De Rosa (2010).
127. Clarke (1982: 50).
128. Clarke (1982: 50).
129. Buickerood (1985).
130. The best reconstruction is Rabouin (2009).
131. *AT*, X, pp. 156–7.
132. *AT*, X, p. 179.
133. *AT*, X, p. 331. See Sasaki (2003).
134. *AT*, X, pp. 333–4.
135. *AT*, X, pp. 373–8.
136. Berti (1972).
137. *AT*, I, p. 339.
138. *AT*, VI, p. 19.
139. *AT*, pp. 155–6.
140. For the problematic and original nature of this inversion, see Timmermans (1999). On the notion of a priori, see Pickavé (2005).
141. The passage was: 'The method of proceeding from things better known to things to know and the reverse path from the latter to the former, that is analysis and synthesis'. In Latin, the same sentence sounds '& a notioribus ad ea, quae quaeruntur via, & ab his ad ea transitus, quae sane Resolutiones & Compositiones appellantur' (Proclus 1560: 4). While in Greek: 'ἡ ἀπὸ τῶν γνωριμωτέρων ὁδὸς ἐπὶ τὰ ζητούμενα καὶ ἡ ἐχ τούτων ἐπ' ἐκεῖνα μετάβασις, ἃς δὴ καλοῦσιν ἀναλύσεις καὶ συνθέσις' (Friedlein 1873: 8).
142. Friedlein (1873: 18–19).
143. This marks a substantially different position from the conception of analysis and synthesis in the *Meditations* and that of reduction and ascension in the *Rules*. Even if

the processes seem similar, in the *Rules* the propositions – that is, principles – were obscure and convoluted, while in the *Meditations* they were the very first principles.
144. The best characterization of Descartes's *mathesis universalis* in relation to ancient mathematics is Napolitano Valditara (1988). This research, however, does not focus on analysis and synthesis.
145. Friedlein (1873: 45–6).
146. *AT*, IXB, p. 14.
147. Hooke (1665: preface).

7 Isaac Barrow

1. On Piccolomini and certainty, see Giacobbe (1972a), De Pace (1993).
2. Piccolomini (1565: 102r). I quote from the 1565 edition, which has numbered pages.
3. Piccolomini (1565: 102r).
4. Piccolomini (1565: 102v).
5. Piccolomini (1547: 24r–v).
6. Piccolomini (1547: 24v).
7. Perera (1576: 75b). On Perera, see Giacobbe (1977); De Pace (1993).
8. Perera (1576: 26b).
9. Perera (1576: 78b–79a).
10. On Catena and certainty, see Giacobbe (1973); De Pace (1993).
11. Catena (1556: 66).
12. Catena (1556: 55).
13. Zabarella (1597: 732a–733e).
14. On Barozzi and the problem of certainty, see Giacobbe (1972b); De Pace (1993).
15. Barozzi (1559: 24r). Giuseppe Biancani uses this argument: see Dear (1995a: 40). On Biancani, see Giacobbe (1976); De Pace (1993).
16. Barozzi (1559: 22v–25v).
17. Rose (1977: 156).
18. For a detailed discussion of these positions, see Mancosu (1992).
19. For Barrow's knowledge of Aristotle, see Feingold (1990a).
20. Feingold (1990b).
21. Feingold (1990b: 354, 369).
22. Feingold (1990a: 68).
23. Barrow (1683: 2). I quote from the Latin edition because it is more accurate in using Greek terminology and in referring to Aristotle. However, I use whenever possible, unless noted otherwise, the English translation made by John Kirby in 1734 under the title *The Usefulness of Mathematical Learning Explained and Demonstrated*.
24. Barrow (1683: 4).
25. Barrow (1683: 31–2).
26. Barrow could not have known either Beeckman's *Journal* or Descartes's *Rules*.
27. Barrow (1683: 60).
28. Barrow (1683: 59).
29. Barrow (1683: 60).
30. Barrow (1683: 60–1).
31. Barrow (1683: 61).
32. Barrow (1683: 64).

33. Barrow (1683: 65).
34. Barrow (1683: 74).
35. Barrow (1683: 74).
36. The nature and the origin of these principles are discussed in Lecture 7.
37. Barrow (1683: 78).
38. Barrow (1683: 80).
39. Barrow (1683: 80).
40. Barrow (1683: 81). In the English version, the reference to Aristotle is not as evident as in the Latin version, where, instead of intellection and intellect, the terms νοεῖν and νοῦς are used.
41. Barrow (1683: 82).
42. Barrow (1683: 82). See Dear (1995a: 30).
43. Barrow (1683: 82).
44. Barrow (1683: 83).
45. Barrow (1683: 83). Dear correctly suggests that Barrow could be the source of Newton's usage of induction in mathematics. See Dear (1995a: 25, 241).
46. In the Port Royal logic, there is a similar idea: see Dear (1995a: 29).
47. Barrow (1683: 84).
48. Barrow (1683: 85).
49. Barrow (1683: 86).
50. John Wallis takes a different position and considers arithmetic the queen of the mathematical sciences: see Wallis (1657: 15, 53, 183).
51. Barrow (1684: 115).
52. Barrow (1684: 117).
53. Barrow (1684: 122).
54. Barrow (1684: 125–6).
55. Barrow (1684: 126).
56. Barrow (1684: 130).
57. Barrow (1683: 86).
58. Barrow (1683: 86).
59. Barrow (1683: 86).
60. Barrow (1683: 88).
61. Barrow (1683: 90).
62. Barrow (1683: 90).
63. Barrow (1683: 90).
64. Barrow (1683: 92).
65. Barrow (1683: 95).
66. Barrow (1683: 97).
67. Barrow (1683: 98).
68. For the role of voluntarism in seventeenth-century science, and on God's significance in Barrow's philosophy of mathematics, see Malet (1997).
69. Barrow (1683: 98–9).
70. Barrow (1683: 99).
71. Barrow (1683: 99).
72. Barrow (1683: 100).
73. Barrow (1683: 114).
74. Barrow (1683: 117).
75. Friedlein (1873: 75).
76. Piccolomini (1565: 95v).

77. Giovanni Crapulli is in favour of this interpretation of a common science being mathematics, containing arithmetic and geometry. See Crapulli (1969: 39).
78. Proclus (1560: 5).
79. Crapulli (1969: 71).
80. Burtt (1932: 150–62).

8 Robert Hooke

1. Sprat (1667: 31).
2. The best study on Sprat's epistemology and on its limits is still Wood (1980). Now see also Lynch (2001); Hunter (2016).
3. Sprat (1667: 37). On the Royal Society as the house of the experiment, see Shapin (2011).
4. Hunter (1981, 1982); Schaffer and Shapin (1985); Hunter and Wood (1986: 50); Hunter (1988, 1989); Hunter and Schaffer (1989); Sargent (1994); Hunter (1994); Sargent (1995); Hunter (2000); Feingold (2001); Lynch (2002); Hunter (2003); Feingold (2005); Feingold (2006); Hunter (2007); Anstey and Hunter (2008), Hunter (2009, 2011, 2015).
5. Looked at in detail, a lack of homogeneity is common among the fellows (Hunter 2007). Hunter has shown the interplay between the genesis of some of Robert Boyle's epistemological ideas and the general endeavour of the Royal Society.
6. 'Espinasse (1956); Hesse (1964, 1966); Oldroy (1972, 1980); Bennett (1980); Pugliese (1982); Oldroy (1987); Hunter and Schaffer (1989); Hunter (2003); Cooper and Hunter (2017). Hesse and Oldroy emphasize the Baconian aspects of Hooke's methodology, while Hunter offers a more balanced view in showing Cartesian influences.
7. Hooke (1661: 6).
8. Hooke (1661: 6).
9. Hooke (1661: 9–10).
10. Hooke (1661: 24–6).
11. Hooke (1661: 26).
12. Hooke (1665: 21).
13. Singer (1976); Hintzman (2003).
14. Hooke (1665: preface).
15. Hooke (1665: preface).
16. Hooke (1665: preface).
17. Hooke (1665: preface).
18. Hooke (1665: preface).
19. Hooke (1665: preface).
20. Hooke (1665: preface).
21. Hooke (1665: preface).
22. Hooke (1665: preface).
23. Hooke (1705: 3).
24. Hooke (1705: 3).
25. Hooke (1705: 3).
26. Hooke (1705: 3).
27. The date being 1687 is concluded from the allusion to the forthcoming publication of Newton's *Principles.*
28. Hooke (1705: 338).

29. Hooke (1705: 329).
30. Hooke (1705: 5).
31. Hooke (1705: 5).
32. Hooke (1705: 6).
33. Hooke (1705: 6).
34. Hooke (1705: 7).
35. *AT*, IXB, p. 14.
36. Hooke (1705: 19).
37. Classified Papers of the Royal Society 20/39, 1r.
38. Classified Papers of the Royal Society 20/39, 1r.
39. Classified Papers of the Royal Society 20/39, 1r.
40. Classified Papers of the Royal Society 20/39, 1r.
41. Classified Papers of the Royal Society 20/39, 1r.
42. Classified Papers of the Royal Society 20/39, 1r.
43. Classified Papers of the Royal Society 20/39, 1r.
44. Classified Papers of the Royal Society 20/39, 1r.
45. Classified Papers of the Royal Society 20/39, 1r.
46. Classified Papers of the Royal Society 20/39, 1r.
47. Classified Papers of the Royal Society 20/39, 1r.
48. Classified Papers of the Royal Society 20/39, 1r.
49. Classified Papers of the Royal Society 20/39, 1v.
50. Classified Papers of the Royal Society 20/39, 1v.
51. Classified Papers of the Royal Society 20/39, 1v.
52. Classified Papers of the Royal Society 20/39, 1v.
53. Classified Papers of the Royal Society 20/39, 1v.
54. Classified Papers of the Royal Society 20/39, 2r.
55. Viète (1591: 4r).
56. Viète (1591: 4r).
57. Viète (1591: 4r).
58. Viète (1591: 4r).
59. Viète (1591: 4r).
60. Jeake (1696: 334).
61. Classified Papers of the Royal Society 20/72, pp. 1r–2v. See in particular, p. 2r where Hooke lists the Aristotelian categories and tries to include them in his algebra.
62. Hooke (1705: 7).
63. In other documents, Hooke proposes similar methods for making experiments and preparing natural histories. See Derham (1726: 25–7); Oldroy (1987: 151–9). On the ideas of social accreditation in doing experiments, see Dear (1995a: 23); Shapin (2011: 65–124).
64. Hooke (1705: 35).
65. Hooke (1705: 34).
66. Hooke (1705: 44).
67. Trinity College Library, Ms. O.11a.1/14, see Oldroy (1980: 30).
68. The list is in Hooke (1705: 43–4). David R. Oldroy has provided a useful scheme, which I reproduce here in a somewhat different fashion for the parts that serve my argument. See Oldroy (1987: 148–9).
69. Hooke (1705: 28).
70. Aristotle, *Topics*, 114 b 37–115 a 6.
71. Hooke (1705: 61).

72. Hooke (1705: 61).
73. For a similar methodology in Newton, see Cambridge University Library, Add. Ms. 3790, f. 243r, 280v.
74. Hooke (1705: 61).
75. Hooke (1705: 61). The mathematician and fellow of the Royal Society William Neile shares with Hooke this concern about the necessity of finding true causes being the main task of philosophy. He explicitly writes that 'experiments themselves are but a dry entertainment without the indagation of causes'. See Hunter and Wood (1986: 79). However, as far as we know, Neile never elaborated a methodology for finding causes.
76. The date 1687 is extrapolated from the allusion to the forthcoming publication of Newton's *Principles*.
77. See Lynch (2001: 74).
78. Hooke (1705: 330).
79. Hooke (1705: 330).
80. Harriot (1631: 1).
81. Harriot (1631: 1).
82. Harriot (1631: 1).
83. Harriot (1631: 1).
84. Harriot (1631: 1).
85. Harriot (1631: 2).
86. Harriot (1631: 2).
87. Harriot (1631: 2).
88. Hooke (1705: 61).
89. Hooke (1705: 331).
90. Hooke (1705: 331).
91. Hooke (1705: 330).
92. Hooke (1705: 331).
93. Hesse emphasizes this aspect in Hesse (1966: 81).
94. Hooke (1705: 330).
95. The manuscript was published by Michael Hunter in 2003.
96. I quote from Hunter (2003: 107). Classified Papers of the Royal Society 20/77, 1r.
97. I quote from Hunter (2003: 107). Classified Papers of the Royal Society 20/77, 1r.
98. In all these stages, Hooke's process does not differ significantly from Galileo's epistemology. For the Pisan scientists knowledge starts from sensate experiences acquired through the use of the telescope. These experiences were reduced to mathematical terms in order to provide a mathematical description, which led to the general definition of a law. This law was tested and, once confirmed, it was applied for the discovery of phenomena.
99. On Galileo's early reception in England, see Feingold (1983).

9 John Locke

1. Woolhouse (2007: 18).
2. For a revision of Locke's empiricism, see Anstey (2003).
3. Locke (1975: I, 1, §8, 47).
4. *AT*, VII, p. 181.
5. Locke (1975: I, 1, §8, 47).

6. Locke (1990: 2).
7. Locke (1990: 15–16).
8. Locke (1990: 16–17).
9. Locke (1990: 22–5).
10. Locke (1990: 28).
11. Locke (1990: 26).
12. This conception will be fundamental, as Anstey has pointed out, in the elaboration of Locke's conception of knowledge in the *Essay*. Anstey (2011: 113–4).
13. Locke (1990: 22).
14. Anstey (2011: 112).
15. Locke (1990: 27).
16. Locke (1990: 27).
17. Locke (1990: 28).
18. Locke (1990: 30).
19. Locke (1990: 30).
20. Locke (1990: 31).
21. Locke (1990: 30).
22. Garber (1982); Downing (2001: 515–35).
23. Downing (1992).
24. Locke (1975: II, 23, §11, 301).
25. Locke (1990: 142).
26. Locke (1990: 176).
27. Locke (1975: 7).
28. Locke (1975: I, 1, §7, 25).
29. Locke (1975: I, 2, §9, 51).
30. Locke (1975: IV, 17, §1, 668).
31. Locke (1975: IV, 17, §2, 668).
32. Locke (1975: IV, 17, §3, 669).
33. Locke (1975: IV, 2, §3, 532).
34. Locke (1975: IV, 2, §3, 532).
35. Owen (2000: 45).
36. See Winkler (2003: 167–73).
37. Locke (1975: I, 2, §12, 53).
38. Locke (1975: I, 2, §15, 55).
39. Locke (1975: I, 2, §15, 55).
40. Locke (1975: I, 2, §20, 58).
41. Locke (1975: II, 1, §2, 104).
42. Locke (1975: II, 1, §4, 105).
43. Locke (1975: II, 9, §1, 143).
44. Locke (1975: II, 10, §1, 149).
45. Locke (1975: II, 10, §7, 152).
46. Locke (1975: II, 11, §1, 155).
47. Locke (1975: II, 11, §1, 155).
48. Locke (1975: II, 11, §3, 157).
49. Locke (1975: II, 2, §1, 157).
50. Locke (1975: II, 8, §8, 134).
51. Locke (1975: II, 11, §2, 134).
52. Locke (1975: II, 11, §9, 156).
53. Locke (1975: II, 29, §4, 363).

54. Locke (1975: II, 29, §6, 364).
55. Locke (1975: II, 29, §11, 367).
56. Locke (1975: IV, 14, §4, 653).
57. Locke (1975, IV, 1, §2, 525).
58. Locke (1975, IV, 1, §4, 526).
59. Locke (1975, IV, 1, §4, 526).
60. Owen (2000, 33, 35).
61. Pappas (1998, 288).
62. Locke (1975, IV, 3, §9, 544).
63. Locke (1975, IV, 3, §16, 548. Cf. Anstey 2011, 31–45).
64. Locke (1975, IV, 1, §8, 527).
65. Locke (1975, IV, 2, §1, 531).
66. Locke (1975, IV, 2, §2, 532).
67. Locke (1975, IV, 2, §4, 532).
68. Locke (1975, IV, 2, §7, 533).
69. Locke (1975, IV, 7, §11, 599).
70. Locke (1975, IV, 2, §10, 535).
71. Locke (1975, IV, 2, §11, 535).
72. Locke (1975, IV, 12, §9, 644).
73. Locke (1975, IV, 12, §10, 645).
74. Locke (1975, IV, 12, §10, 646).
75. Anstey (2011, 46–69).
76. On the utility of hypothesis in Locke see Anstey (2011, 70–89) and Ducheyne (2013). See also Laudan (1981, 59–71). On Locke's use of Baconian 'history' see Anstey (2002).
77. Romanell (1984, 71).
78. Locke (1975, IV, 16, §12, 666).
79. Locke (1975, IV, 16, §12, 665–6).
80. Locke (1975, IV, 16, §12, 665).
81. Locke (1975, IV, 16, §12, 666).
82. Anstey (2011, 76).
83. Locke (1823, 275).
84. Locke (1975, IV, 16, §12, 666).

10 Gottfried Wilhelm Leibniz

1. *GP*, VII, pp. 514–27.
2. The time span of thirty years does not substantially affect the epistemological approach, which remains more or less the same. I have therefore permitted myself to range between texts written in different periods to render Leibnizian theory clear and coherent.
3. *GP*, VI, p. 608.
4. *GP*, VI, p. 617.
5. *GP*, VI, p. 617.
6. *GP*, VI, p. 610.
7. *GP*, VI, p. 611.
8. *GP*, VI, pp. 608–9.

9. *GP*, VI, p. 610.
10. *GP*, VI, p. 611.
11. *GP*, VI, p. 611.
12. *GP*, I, p. 370, translated by Leroy E. Loemker, slightly modified.
13. Parkinson (1982: 5).
14. *GP*, VI, pp. 490–1.
15. *GP*, VI, p. 506.
16. *GP*, II, p. 452. In the *New Essays*, Leibniz writes that the truth or true criterion of sensible things consists in the connection of phenomena (*GP*, V, p. 355).
17. *A*, VI, 4, p. 1501.
18. *GP*, VI, p. 599.
19. *A*, VI, 4, pp. 1501–2.
20. Nadler (1989: 79–100).
21. *A*, II, 1, p. 542.
22. *GP*, IV, p. 422.
23. *GP*, IV, p. 422.
24. *GP*, IV, p. 422.
25. *GP*, IV, p. 423.
26. *GP*, IV, p. 423.
27. *GP*, IV, p. 423.
28. *GP*, IV, p. 423.
29. *GP*, IV, p. 425.
30. *GP*, IV, p. 425.
31. *GP*, VII, p. 296.
32. *GP*, VII, p. 296.
33. *GP*, VII, p. 296.
34. *GP*, VII, p. 296.
35. *GP*, VII, pp. 296–7.
36. *GP*, VII, p. 297.
37. *GP*, VII, p. 297.
38. *CL*, p. 190.
39. *CL*, p. 190.
40. Anne-Lise Rey (2013).
41. *A*, VI, 4, p. 1984.
42. In a letter sent in 1671 to Magnus Hesenthaler (*A*, II, 1, p. 200), Leibniz wrote that there are mixed propositions, which are deducted from theorems and observations combined together.
43. *A*, VI, 4, p. 1993.
44. *A*, VI, 4, pp. 1993–4.
45. *A*, VI, 4, p. 1996.
46. *A*, VI, 4, p. 1997.
47. *A*, VI, 4, p. 1997.
48. *A*, VI, 4, p. 1997. On Leibniz's calculus, see Mugnai 2015.
49. *A*, VI, 4, p. 1998.
50. Arthur (2004).
51. *A*, VI, 4, p. 1998.
52. *A*, VI, 4, p. 1999.
53. *A*, VI, 4, p. 1999.
54. *A*, VI, 4, p. 1999.

55. *A*, VI, 4, p. 2000.
56. *A*, VI, 4, p. 2000.
57. *A*, VI, 4, p. 2001.
58. *A*, VI, 4, p. 2001.
59. *A*, VI, 4, p. 2001.
60. *A*, VI, 4, p. 2001.
61. *A*, VI, 4, p. 2002.
62. *A*, VI, 4, p. 2002.
63. *A*, VI, 4, p. 2002.
64. *A*, VI, 4, p. 2003.
65. *A*, VI, 4, p. 2005.
66. *A*, VI, 4, p. 2005.
67. *A*, VI, 4, pp. 2005–6.
68. *A*, VI, 4, p. 2006.
69. *A*, VI, 4, p. 2006.
70. *A*, VI, 4, p. 2006.
71. *A*, VI, 4, p. 2006.
72. *A*, VI, 4, pp. 2006–7.
73. *A*, VI, 4, p. 2007.
74. *A*, VI, 1, p. 199.
75. *A*, VI, 1, p. 199.
76. *A*, VI, 6, p. 446.
77. *A*, VI, 6, p. 484.
78. *A*, VI, 6, p. 450.
79. *A*, VI, 6, p. 450.
80. Cambridge University Library, Add. Ms. 3968.39 f. 586v.

11 Isaac Newton

1. I quote the English text of Newton's *Principles* from the translation of Cohen and Whitman's edition (Cohen and Whitman 1999), unless otherwise stated. Where necessary, I quote from the various original editions. Quotations from Newton's manuscripts are from *The Newton Project*, edited by Rob Iliffe and Scott Mandelbrote, available at http://www.newtonproject.ox.ac.uk. I have maintained additions and spelling, but not deletions. In [] I indicate possible variations or conjectures.
2. Feingold (2004: 29–41, 66–75).
3. Cohen (1992).
4. Régis (1688).
5. I consider 'Leibnizians' all those philosophers and scientists who supported Leibniz in his controversy with Newton; however, there was never a Leibnizian sect as such.
6. *GP*, VII, p. 343.
7. On the preparation of these notes, see Ducheyne and Dhondt (2021).
8. Cambridge University Library, Add. Ms. 3790, f. 285r.
9. Various sources like Robert Sanderson, Samuel Smith and Johannes Magirus have been suggested: McGuire and Tamny (1983); Wallace (1988); Mamiani (2001); Feingold (2004: 9); Ducheyne (2005a, 2012). Ducheyne has provided the most insightful and detailed investigation of Newton's logic of scientific discovery.

10. Sprat (1667: 31). It is well-known that these directives were blatantly disregarded by many members, who focused mainly on experiments.
11. See Buchwald and Feingold (2013: 15); Levitin (2016: 55). In many Aristotelian textbooks, as we have seen, the treatment of regressus was quite rudimentary; sometimes the word regressus was missing, and many of them did not even mention the intermediate stage of regressus.
12. I consider Newton's manuscript notes essential for understanding the historical development of his epistemological thought, but by no means can they be taken as indicating the final stage of his thinking: they are a series of reflections, attempts and failures which do not always reflect his definitive position.
13. Shapiro (2004).
14. See Levitin (2016: 76).
15. Cambridge University Library, Add. Ms. 3790, f. 244v.
16. Cambridge University Library, Add. Ms. 3790, f. 243r, 244v.
17. Cambridge University Library, Add. Ms. 3790, f. 243r, 280v.
18. On the epistemological relevance of the *Quaestiones*/Queries, see Anstey (2004).
19. See Anstey (2005).
20. Newton (1706: 314).
21. Newton (1706: 342).
22. Newton (1718: 376).
23. For Newton's analysis and synthesis in mathematics, see Guicciardini (2002, 2009).
24. Guicciardini (2009: 321–3).
25. Feingold (2001). For the relation with Locke, see Domski (2012). For the relation between mathematics and natural philosophy, see Domski (2013).
26. This, of course, goes beyond the fact that mathematics according to Piccolomini, as we have seen in Chapter 2, does not capture causal relations between things, and therefore cannot be a science in the strict Aristotelian sense.
27. For a different perspective on the relation between natural philosophy and mathematics, see Garrison (1987).
28. Newton (1706: 347).
29. Guicciardini (2009: 322). See also Stein (1991); Belkind (2012).
30. Cambridge University Library, Add. Ms. 3790, f. 243r, 280v.
31. Cambridge University Library, Add. Ms. 3790, f. 243r, 280v. On this manuscript note, see McGuire (1970).
32. Janiak (2008: 50–86). However, Newton uses the term quality rather than quantity.
33. Cambridge University Library, Add. Ms. 3984.14, f 1r.
34. Cambridge University Library, Add. Ms. 3984.14, f 1v.
35. Newton (1672c: 4004).
36. Belkind characterizes three different kinds of induction employed by Newton: Baconian induction, demonstrative induction and universal induction. Belkind (2020).
37. Newton (1959: I, 96–7).
38. According to Alan E. Shapiro, Newton accepted the probabilism of his contemporaries only in the last decades of his life. See Shapiro (1993: 14). I prefer to speak of provisionalism, rather than probabilism.
39. Guicciardini (2009: 12, 23). Members of the Royal Academy had different epistemological positions, see Shapiro (1983); Hunter and Wood (1986); Feingold (2001).
40. Newton (1672a: 3075). See Schaffer (1989).

41. Newton (1672a: 3077).
42. Newton (1672a: 3079).
43. Hooke (1665: 73).
44. Newton (1672b: 5014).
45. Cambridge University Library, Add. Ms. 3984.14, f 1r.
46. *GP*, III, pp. 518–19.
47. Shapiro (2004: 202).
48. Saurin (1709: 148). See Shank (2018: 363).
49. Newton (1718: 344).
50. Newton (1718: 369).
51. Newton (1718: 377).
52. Schliesser (2007); Levitin (2016: 67–75).
53. Cambridge University Library, Add. Ms. 3970, f. 109v.
54. Cambridge University Library, Add. Ms. 3970, f. 109v.
55. John F. McDonald suggests a distinction between 'property hypotheses', which are *principia cognoscendi*, and 'causal hypotheses', which are *principia essendi*. Given the complexity of the term 'hypothesis' in Newton, I prefer to use the Aristotelian distinction used in the regressus, which better contextualizes the Newtonian reappraisal of this epistemological doctrine. See McDonald (1972).
56. Newton (1718: 380).
57. Newton (1718: 380–1).
58. Cohen and Whitman (1999: 382).
59. Cohen and Whitman (1999: 382). Translation is slightly modified.
60. Cohen and Whitman (1999: 382).
61. Cohen and Whitman (1999: 381).
62. Cohen and Whitman (1999: 381).
63. Cohen and Whitman (1999: 385).
64. Cohen and Whitman (1999: 385).
65. Cohen and Whitman (1999: 386).
66. Cohen and Whitman (1999: 386).
67. Cohen and Whitman (1999: 386).
68. Cohen and Whitman (1999: 793). Translation is slightly modified.
69. Cohen and Whitman (1999: 408).
70. Koyré (1965); Cohen (1969); Finocchiaro (1974); McGuire (1995); Smith (2002); Spencer (2004); Harper (2011); Ducheyne (2012, 2015); Belkind (2017); Levitin (2021). I am very much indebted to all these studies in my reconstruction.
71. Biener (2018). In the next few pages, I hope to provide more evidence in favour of his thesis.
72. Biener (2018: 3). He means a third stage beyond the two usually considered – that is, analysis and synthesis.
73. Biener (2018: 4).
74. Ducheyne (2015).
75. Cohen and Whitman (1999: 794). I follow Spencer's suggestion in using 'explicate' instead of 'explain'. See Spencer (2004: 260).
76. Cohen and Whitman (1999: 795).
77. Cohen and Whitman (1999: 795).
78. Cohen and Whitman (1999: 385).
79. Cohen and Whitman (1999: 391). Translation is modified.
80. Cohen and Whitman (1999: 391).

81. Cohen and Whitman (1999: 393).
82. Cohen and Whitman (1999: 392).
83. Cohen and Whitman (1999: 390). Translation is slightly modified.
84. *OFB*, XI, p. 320. On similarities between Bacon and Newton, see Ducheyne (2005b).
85. *OFB*, XI, p. 320.
86. Hume (1751: 61).
87. Cohen and Whitman (1999: 391).
88. Cohen and Whitman (1999: 795).
89. *OFB*, XI, p. 209.
90. *OFB*, XI, p. 209.
91. I am extremely indebted to his investigation.
92. Cohen and Whitman (1999: 795–6). Translation is slightly modified. I provide only the core of the central part of the explanation because the reasoning repeats.
93. I do not discuss all the various possible interpretations; rather, I follow those that are more convincing according to my reconstruction, being conscious that all have strengths and weaknesses.
94. Cambridge University Library, Adv. B. 39. I. Curiously enough, McGuire in his English translation does not include the words in brackets.
95. Finocchiaro (1974).
96. McGuire (1995: 248–50).
97. Jardine (1974: 110).
98. *OFB*, XI, p. 210.
99. After this point, the manuscript is unreadable and any reconstruction is mere conjecture.
100. Cambridge University Library, Add. Ms. 4005.15, 81r–v.
101. On this point, my interpretation differs from McGuire's. See also Okruhlik (1989).
102. Trinity College Library, N.Q. 16.200.
103. Cohen and Whitman (1999: 391).
104. Cambridge University Library, Add. Ms. 3965, f. 266r:

    Hypoth. III The qualities of bodies which cannot be intended and remitted and which apply to all bodies in which it is possible to set up experiments are qualities of all bodies universally. The same is to be understood of qualities of all bodies of the same kind. This evidently is the Foundation of all Philosophy. For otherwise one could not derive the qualities of insensible bodies from the qualities of sensible [ones]; Hypoth. IV Things which agree as to all their known qualities with other kinds of things are not to be considered as new kinds of things.

    See Cohen (1969: 315); McGuire (1995: 98).
105. Belkind (2017).
106. Cambridge University Library, Add. Ms. 3968, f. 436r.
107. Cohen and Whitman (1999: 796).
108. For the role of God in Newton's epistemology, see Janiak (2008: 163).
109. Biener's working hypothesis seems to be correct.
110. Cohen and Whitman (1999: 943).
111. Cohen and Whitman (1999: 943). Translation is slightly modified.
112. The second version, completely different from the first one, can be found in Cambridge University Library, Ms. Add. 3965, f. 419v:

Rule V. In experimental philosophy propositions from phenomena acquired by induction cannot be disputed by hypothesis For if arguments from hypotheses would be admitted against inductions, inductive arguments, on which the whole of experimental philosophy is based, could always be overturned by contrary hypotheses. If a certain proposition collected by induction should be not sufficiently accurate, it ought be corrected, not by hypotheses but by phenomena of nature that are to be more widely and accurately observed.

See Ducheyne (2015).
113. Cambridge University Library, Add. Ms. 3965, f. 419r. See Koyré (1965: 262).
114. Domski (2012).
115. For the complex attitude of Newton towards empiricism, see Biener and Schliesser (2014).
116. Cheyne (1715: I, 50).
117. MacLaurin (1748: 8).
118. Diderot and D'Alembert (1751–72: VI, 299).
119. Smith (2012).
120. *KGS*, XXVIII, p. 294.

Conclusion

1. The problem should not be confused with the adoption of geometrical method into writing in philosophy (*more geometrico*), but rather couched as the use of mathematics for explaining philosophy.
2. *KGS*, I, p. 231.
3. Hahn (1988); Schönfeld (2000: 173). Here I consider only the methodological and epistemological influences of Newton on Kant and not the great impact that the Newtonian conceptions of force, time and space had on the Kantian conception of nature. For this, see Friedman (2013).
4. *KGS*, I, p. 475. Translation has been slightly modified.
5. Kant had the 1739 edition and made use of it in the *Physical Monadology* (*KGS*, I, p. 486). The terminology of contraposition between geometry as the style of Newtonians and metaphysics as the style of Rationalists is borrowed from this text. It will have an influence in Kant's writings for at least a decade.
6. Keill (1702: praefatio). See Guicciardini 1999.
7. Keill (1702: praefatio).
8. Keill (1702: praefatio).
9. Keill (1702: praefatio).
10. Keill (1702: 3).
11. Keill (1702: 3).
12. Keill (1702: 3–4).
13. Keill (1702: 4).
14. Keill (1702: 5).
15. Keill (1702: 5).
16. Keill (1702: 6).
17. Keill (1702: 6).
18. Keill (1702: 6).
19. Keill (1702: 6).

20. Keill (1702: 7).
21. Keill (1702: 7).
22. The journal was *Wöchentliche Königsbergische Frag- und Anzeigungs-Nachrichten*.
23. *KGS*, I, p. 466.
24. *KGS*, II, p. 139.
25. *KGS*, II, p. 275.
26. *KGS*, II, p. 275.
27. *KGS*, II, p. 286.
28. Martin (1985).
29. *KGS*, II, p. 288.
30. *KGS*, II, p. 71.
31. In other works Kant writes that

> the use to which mathematics can be put in philosophy consists either in the imitation of its method or in the genuine application of its propositions to the objects of philosophy. With respect to the first of these two uses: it has not been noticed that it has had only one benefit … by contrast, the second use to which mathematics has been put in philosophy has been all the more beneficial to the parts of philosophy affected. These parts of philosophy, by turning the doctrines of mathematics to their own advantage, have attained to heights, to which they would not otherwise have been able to aspire. (*KGS*, II, p. 167)

32. *KGS*, II, p. 290.
33. Kant's terminology recalls Zabarella's 'fixing the eyes of the mind', and Descartes' 'fixing the mind' on the object.
34. *KGS*, II, p. 292.
35. *KGS*, II, p. 168.
36. In the *Dreams of a Spirit-Seer Elucidated by Dreams of Metaphysics* (1766), Kant articulates the foundations of a metaphysics of morals based on a moral unity conceived like Newton's law of attraction:

> We sense within ourselves a constraining of our will to harmonise with the general will. To call this sensed constraining moral feeling, is to speak of it merely as a manifestation of that which takes place within us, without establishing its causes. Thus it was that Newton called the certain law governing the tendencies inherent in all particles of matter to draw closer to each other the gravitation of matter, not wishing to entangle his mathematical demonstrations in possible vexatious philosophical disputes concerning the cause of those tendencies. Nonetheless, he did not hesitate to treat gravitation as a genuine effect produced by the universal activity of matter operating on itself, for this reason he also gave it the name attraction.' (*KGS*, II, p. 335)

37. Despite strong differences between Bacon's conception of empirics and Kant's notion of empiricism (see Vanzo 2014), a constant tension between two opposite epistemological poles is conspicuous in early modern philosophy in the attempt to elaborate a middle way.
38. *KGS*, III, BXXXV.
39. *KGS*, III, B 127.
40. *KGS*, III, B XIII.
41. Vanzo (2012).
42. *KGS*, IV, p. 319.

43. *KGS*, XVIII, p. 176. See Friedman (1992: 163).
44. *KGS*, IV, p. 4, 319.
45. See Allison (1983); Guyer (1987); Kitcher (1993); Longuenesse (1998); Dicker (2004).
46. *KGS*, III, A 59/B 83.
47. *KGS*, III, A 59/B 84.
48. Brittan (1978: 117–42); Friedman (1992: 136–64).
49. *KGS*, IV, p. 320.
50. See Butts (1974).
51. *KGS*, IV, p. 319.
52. *KGS*, IV, p. 321.
53. *KGS*, IV, p. 322.
54. Friedman (1992: 165–210).
55. *KGS*, IV, pp. 514–5:

> It is commonly supposed that Newton did not at all find it necessary for his system to assume an immediate attraction of matter, but, with the most rigorous abstinence of pure mathematics, allowed the physicists full freedom to explain the possibility of attraction as they might see fit, without mixing his propositions with their play of hypotheses. But how could he ground the proposition that the universal attraction of bodies, which they exert at equal distances around them, is proportional to the quantity of their matter, if he did not assume that all matter, merely as matter, therefore, and through its essential property, exerts this moving force? ... He rightly abstracted from all hypotheses purporting to answer the question as to the cause of the universal attraction of matter, for this question is physical or metaphysical, but not mathematical. And, even though he says in the advertisement to the second edition of his *Optice*, 'to show that I do not take gravity for an essential property of bodies, I have added one question concerning its cause,' it is clear that the offense taken by his contemporaries, and perhaps even by Newton himself, at the concept of an original attraction set him at variance with himself. For he could by no means say that the attractive forces of two planets, those of Jupiter and Saturn for example, manifested at equal distances from their satellites (whose mass is unknown), are proportional to the quantity of matter of these heavenly bodies, if he did not assume that they attracted other matter merely as matter, and thus according to a universal property of matter.

Buchdahl (1970); Friedman (1990); Booth (1996); Janiak (2010).
56. Friedman (1990: 198).
57. *KGS*, IV, p. 469.
58. *KGS*, IV, p. 469.
59. *KGS*, IV, p. 470.
60. *KGS*, IV, p. 470.
61. *KGS*, IV, p. 470.
62. *KGS*, IV, p. 470.
63. *KGS*, IV, p. 470. On the mathematical construction as construal, see Hanna (2006: 331–40).
64. *KGS*, IV, p. 470.
65. For the cognitive value of non-synthetic a priori determining judgements, see Makkreel (1994).

References

Agrimi, Joel, and Crisciani Chiara. 1990. 'Per una ricerca su experimentum-experimenta: riflessione epistemologica e tradizione medica (secoli XIII–XV)'. In *Presenza del lessico greco e latino nelle lingue contemporanee*, edited by Pietro Janni and Innocenzo Mazzini, 9–49. Macerata: Università degli Studi di Macerata.

Airay, Christopher. 1628. *Fasciculus praeceptorum logicorum*. Oxford: Turner.

Allison, Henry E. 1983. *Kant's Transcendental Idealism. An Interpretation and Defense*. New Haven: Yale University Press.

Annas, Julia, and Jonathan Barnes. 1985. *The Modes of Scepticism: Ancient Texts and Modern Interpretations*. Cambridge: Cambridge University Press.

Anstey, R. Peter, and Michael Hunter. 2008. 'Robert Boyle's "Design about Natural History"'. *Early Science and Medicine* 13: 83–126.

Anstey, Peter R. 2002. 'Locke, Bacon and Natural History'. *Early Science and Medicine* 7: 65–92.

Anstey, Peter R. 2003. 'Locke on Method in Natural Philosophy'. In *The Philosophy of John Locke: New Perspectives*, edited by Peter R. Anstey, 26–42. London: Routledge.

Anstey, Peter R. 2004. 'The Methodological Origins of Newton's Queries'. *Studies in History and Philosophy of Science A* 35: 247–69.

Anstey, Peter R. 2005. 'Experimental versus Speculative Natural Philosophy'. In *The Science of Nature in the Seventeenth Century: Patterns of Changes in Early Modern Natural Philosophy*, edited by Peter R. Anstey and John Schuster, 215–42. Dordrecht: Springer.

Anstey, Peter R. 2011. *John Locke and Natural Philosophy*. Oxford: Oxford University Press.

Anstey, Peter R. 2015. 'Francis Bacon and the Laws of Ramus'. *HOPOS: The Journal of the International Society for the History of Philosophy of Science* 5: 1–23.

Anstey, Peter R. 2017. 'Introduction'. In *The Idea of Principles in Early Modern Thought*, edited by Peter R. Anstey, 1–15. London: Routledge.

Ariew, Roger. 1999. *Descartes and the Last Scholastics*. Ithaca: Cornell University Press.

Ariew, Roger. 2011. *Descartes among Scholastics*. Leiden: Brill.

Aristotle. 1984. *Complete Works of Aristotle*. Princeton: Princeton University Press.

Arthur, Richard T. W. 2004. 'The Enigma of Leibniz's Atomism'. *Oxford Studies in Early Modern Philosophy* 1: 183–227.

Baldassarri, Fabrizio. 2014. 'Between Natural History and Experimental Method. Descartes and Botany'. *Society & Politics* 8: 43–60.

Baldassarri, Fabrizio. 2017. 'Per experientiam scilicet, vel deductionem. Descartes' Battle for *Scientia* in the Early 1630s'. *Historia Philosophica* 15: 115–33.

Baldassarri, Fabrizio. 2018. 'Seeking Intellectual Evidence in Sciences: The Role of Botany in Descartes' Therapeutics'. In *Evidence in the Age of the New Sciences*, edited by James A. T. Lancaster and Richard Raiswell, 47–75. Dordrecht: Springer.

Baldi, Gherardo. 1641 *Institutionum logicalium Pars Prior*. Florence: Massa.

Balduino, Girolamo. 1557. *De regressu demonstrativo*. Napoli: Cancer.

Balfour, Robert. 1618. *Commentarius in organum Aristotelicum*. Bordeaux: Milangium.
Baroncini, Gabriele. 1992. *Forme di esperienza e rivoluzione scientifica*. Florence: Olschki.
Barozzi, Francesco. 1559. *Opusculum, in quo una Oratio, et duae Quaestiones*. Padua: Percacino.
Barrow, Isaac. 1683. *Lectiones habitae in Scholis Publicis Academia Cantabrigiensis. An. Dom. MDCLXIV*. London: Wells.
Barrow, Isaac. 1684. *Lectiones habitae in Scholis Publicis Academia Cantabrigiensis. An. Dom. MDCLXV*. London: Wells.
Beck, Leslie J. 1952. *The Method of Descartes. A Study of the Regulae*. Oxford: Oxford University Press.
Belkind, Ori. 2012. 'Newton's Scientific Method and the Universal Law of Gravitation'. In *Interpreting Newton. Critical Essays*, edited by Andrew Janiak and Eric Schliesser, 138–68. Cambridge: Cambridge University Press.
Belkind, Ori. 2017. 'On Newtonian Induction'. *Philosophy of Science* 84: 677–97.
Belkind, Ori. 2020. 'Newton's Methodology'. In *Encyclopedia of Early Modern Philosophy and the Sciences*, https://doi.org/10.1007/978-3-319-20791-9_101-1.
Bennett, Jim. 1980. 'Robert Hooke as Mechanic and Natural Philosopher'. *Notes and Records of the Royal Society*, 35: 33–48.
Ben-Yami, Hanoch. 2015. *Descartes' Philosophical Revolution. A Reassessment*. Dordrecht: Springer.
Berti, Enrico. 1972. 'Le origini del matematismo moderno'. *Giornale critico della filosofia italiana* 51: 337–65.
Berti, Enrico. 2009. *Nuovi studi aristotelici. IV.2 – L'influenza di Aristotele. Antichità, Medioevo e Rinascimento*. Brescia: Morcelliana.
Biener, Zvi. 2018. 'Newton's *Reguale Philosophandi*'. In *Oxford Handbook of Isaac Newton*, edited by Eric Schliesser and Chris Smeenk, 1–23. Oxford: Oxford University Press.
Biener, Zvi and Erich Schliesser. (eds). 2014. *Newton and Empiricism*. Oxford: Oxford University Press.
Blake, Ralph M. 1966. 'The Role of Experience in Descartes' Theory of Method'. In *Theories of Scientific Method: The Renaissance through the Nineteenth Century*, edited by Ralph M. Blake, Curt J. Ducasse and Edward. H. Madden, 75–103. Seattle: University of Washington Press.
Bodenmann, Siegfried, and Anne-Lise Rey (eds). 2018. *What Does It Mean to Be an Empiricist? Empiricisms in Eighteenth-Century Science*. Boston: Springer.
Bonk, Thomas. 2008. *Underdetermination: An Essay on Evidence and Limits of Natural Knowledge*. Dordrecht: Springer.
Booth, Edward. 1996. 'Kant's Critique of Newton'. *Kant-Studien* 87: 149–65.
Boyle, Deborah A. 2009. *Descartes on Innate Ideas*. London: Bloomsbury.
Brague, Rémi. 1988. *Aristote et la question du monde*. Paris: PUF.
Brissey, Patrick. 2014. 'Rule VIII of Descartes' Regulae ad directionem ingenii'. *Journal of Early Modern Studies* 3: 9–31.
Brissey, Patrick. 2018. 'Towards Descartes' Scientific Method: A Posteriori Evidence and the Rhetoric of Les Météores'. In *Evidence in the Age of the New Sciences*, edited by James A. T. Lancaster and Richard Raiswell, 77–99. Dordrecht: Springer.
Brittan, Gordon. 1978. *Kant's Theory of Science*. Princeton: Princeton University Press.
Brutscher, Johannes. 1602. *Disputatio philosophica ex universa logicae*. Ingolstadt: Eder.
Buchdahl, Gerd. 1969. *Metaphysics and the Philosophy of Science. The Classical Origins – Descartes to Kant*. Oxford: Blackwell.

Buchdahl, Gerd. 1970. 'Gravity and Intelligibility: Newton and Kant'. In *The Methodological Heritage of Newton*, edited by Robert E. Butts and John W. Davis, 74–102. Toronto: University of Toronto Press.

Buchwald, Jed, and Feingold, Mordechai. 2013. *Newton and the Origin of Civilization*. Princeton: Princeton University Press.

Buickerood, James G. 1985. 'The Natural History of the Understanding: Locke and the Rise of Facultative Logic in the Eighteenth Century'. *History and Philosophy of Logic* 6: 157–90.

Burgersdijk, Franco. 1637. *Institutionum logicarum libri duo*. Cambridge: Cambridge University Press.

Burgundo, Claudio C. 1651. *Prodromus scientiarum artiumue liberalium*. Venice: Ferretti.

Burtt, Edwin A. 1932. *The Metaphysical Foundations of Modern Physical Science*. London: Kegan.

Butts, R. E. 1974. 'Kant's Mathematization of Scientific Experience'. *Proceedings of the Heraclitean Society* 1: 271–80.

Camerota, Michele and Otto Helbing. 2000. 'Galileo and Pisan Aristotelianism: Galileo's "De motu antiquiora" and the "Quaestiones de motu elementorum" of the Pisan Professors'. *Early Science and Medicine* 5: 319–365.

Campanini, Massimo. 1986. 'Realtà della natura e verità del conoscere in Jacopo Zabarella'. *Annali della Facoltà di Lettere e Filosofia dell'Università degli Studi di Milano* 39: 51–72.

Capivacci, Girolamo. 1562. *De differentiis doctrinarum seu de Methodis, philosophis, theologis, juriscosultis et medicis necessarium*. Padua: Pasquati.

Carleton, Thomas. 1649. *Philosophia universa*. Antwerp: Meur.

Carriero, John. 2008. *Between Two Worlds. A Reading of Descartes's Meditations*. Princeton: Princeton University Press.

Carugo, Adriano, and Alistair C. Crombie. 1983. 'The Jesuits and Galileo's Ideas of Science and of Nature'. *Annali dell'Istituto e Museo di Storia della Scienza di Firenze* 8: 3–68.

Case, John. 1584. *Summa veterum interpretum in universam dialecticam Aristotelis*. London: Vautrollerius.

Cassirer, Ernst. 1922. *Das Erkenntnisproblem in der Philosophie und Wissenschaft der neueren Zeit*. Berlin: Cassirer.

Catena, Pietro. 1556. *Universa loca in logicam Aristotelis in Mathematicas disciplinas hoc novum opus declarat*. Venice: Marcolini.

Celsus, Aulus Cornelius. 1935. *De medicina*. Cambridge, MA: Harvard University Press.

Chyene, George. 1715. *Philosophical Principles of Religion*. London: Strahan.

Clarke, Desmond M. 1982. *Descartes' Philosophy of Science*. Manchester: Manchester University Press.

Clarke, Desmond M. 1991. 'Descartes's Use of "Demonstration" and "Deduction"'. In *René Descartes: Critical Assessments*, vol. 1, edited by George J. D. Moyal, 237–47. London: Routledge.

Cohen, I. Bernard. 1969. 'Hypotheses in Newton's Philosophy'. *Proceedings of the Boston Colloquium for the Philosophy of Science 1966/1968*, edited by Robert S. Cohen and Marx W. Wartofsky, 304–26. Dordrecht: Reidel.

Cohen, I. Bernard. 1992. 'The Review of the First Edition of Newton's *Principia* in the *Acta Eruditorum*, with Notes on the other Reviews'. In *The Investigation of Difficult Things*, edited by Peter M. Harman and Alan E. Shapiro, 323–54. Cambridge: Cambridge University Press.

Cohen, I. Bernard and Anne Whitman. 1999. *The Principia: Mathematical Principles of Natural Philosophy*. Berkeley: University of California Press.

Cohen, Laurence J. 1980. 'Some Historical Remarks on the Baconian Conception of Probability'. *Journal of the History of Ideas* 41: 219–31.
Cohen, Morris R. 1949. *Studies in Philosophy and Science*. New York: Holt.
Cooper, Michael, and Michael Hunter. 2017. *Robert Hooke. Tercentennial Studies*. London: Routledge.
Cozzoli, Daniele. 2008. *Il metodo di Descartes*. Macerata: Quodlibet.
Crankanthorpe, Richard. 1622. *Logicae libri quinque*. London: Legati.
Crapulli, Giovanni. 1969. *Mathesis Universalis. Genesi di una idea nel XVI secolo*. Rome: Ateneo.
Cremonini, Cesare. 1596. *Explanatio proeemii libro rum Aristotelis de physico auditu cum introductione ad naturalem Arist. Philosophiam, continente, tractatum de paedia*. Padua: Novelli.
Crescini, Angelo. 1965. *Le origini del metodo analitico. Il Cinquecento*. Udine: Del Bianco.
Crescini, Angelo. 1972. *Il problema metodologico alle origini della scienza moderna*. Rome: Edizioni dell'Ateneo.
Crescini, Angelo. 1983. 'La teoria del regressus di fronte alla epistemologia moderna'. In *Aristotelismo veneto e scienza moderna*, edited by Luigi Olivieri, 280–90. Padua: Antenore.
Crombie, Alistair C. 1953. *Robert Grosseteste and the Origins of Experimental Science 1100–1700*. Oxford: Clarendon Press.
Crombie, Alistair C. 1996. *Science, Art and Nature in Medieval and Modern Thought*. London: Bloomsbury.
Curley, Edwin M. 1986. 'Analysis in the Meditations: the Quest for Clear and Distinct Ideas'. *Essays on Descartes' Meditations*, edited by Amelie O. Rorty, 153–76. Berkeley: University of California Press.
Davi, Maria R. 1994. *Bernardino Tomitano. Filosofo, medico e letterato (1517–1576). Profilo biografico e critico*. Trieste: LINT.
Dawes, Gregory. 2016. 'Experiment, Speculation, and Galileo's Scientific Reasoning'. *Perspective on Science* 24: 343–60.
De Boer, Karin, and Tinca Prunea-Bretonnet. (eds). 2021. *The Experiential Turn in Eighteenth-Century German Philosophy*. London: Routledge.
De Koninck, Thomas. 1997. 'L'intellection des indivisibles et l'appréhension des natures simples: Aristote et Descartes'. *Laval Théologique et Philosophique* 53: 767–83.
De Pace, Anna. 1993. *Le matematiche e il mondo. Ricerche su un dibattito in Italia nella seconda metà del Cinquecento*. Milan: Franco Angeli.
De Pierris, Gabrielle. 2015. *Ideas, Evidence, & Method. Hume's Skepticism & Naturalism Concerning Knowledge & Causation*. Oxford: Oxford University Press.
De Rosa, Raffaella. 2010. *Descartes and the Puzzle of Sensory Representation*. Oxford: Oxford University Press.
Dear, Peter. 1995a. *Discipline and Experience. The Mathematical Way in the Scientific Revolution*. Chicago: University of Chicago Press.
Dear, Peter. 1995b. 'Mersenne's Suggestion: Cartesian Meditation and the Mathematical Model of Knowledge in the Seventeenth Century'. In *Descartes and His Contemporaries: Meditations, Objections, and Replies*, edited by Gary C. Hatfield, 49–62. Chicago: University of Chicago Press.
Dear, Peter. 1998. 'Method and the Study of Nature'. In *The Cambridge History of Seventeenth-Century Philosophy*, edited by Daniel Garber and Michael Ayers, 147–77. Cambridge: Cambridge University Press.

Della Rocca, Michael. 2005. 'Descartes, the Cartesian Circle, and Epistemology Without God'. *Philosophy and Phenomenological Research* 70: 1–33.

Derham, William. 1726. *Philosophical Experiments and Observations of the Late Eminent Dr. Robert Hooke, S.R.S. and Geom. Prof. Gresh. and Other Eminent Virtuosos in His Time*. London: Royal Society.

Dicker, Georges. 2004. *Kant's Theory of Knowledge. An Analytical Introduction*. Oxford: Oxford University Press.

Diderot, Denis, and Jean le Rond D'Alambert. 1751–1766. *Encyclopédie, ou dictionnaire raisonné des sciences, des arts et des métiers*. Paris: Briasson-David-Le Breton-Durand.

Digby, Everard. 1580. *Admonitioni F. Mildapetti Navareni de unica Rami methodo retinenda, responsio*. London: Bynneman.

Dobre, Mihnea, and Tammy Nyden. 2013. *Cartesian Empiricism*. Dordrecht: Springer.

Domski, Mary. 2012. 'Locke's Qualified Embrace of Newton's Principia'. In *Interpreting Newton. Critical Essays*, edited by Andrew Janiak and Eric Schliesser, 48–68. Cambridge: Cambridge University Press.

Domski, Mary. 2013. 'Observation and Mathematics'. In *The Oxford Companion of British Philosophy in the Seventeenth Century*, edited by Peter R. Anstey, 144–68. Oxford: Oxford University Press.

Dorling, Jon. 1973. 'Demonstrative Induction: Its Significant Role in the History of Physics'. *Philosophy of Science* 40: 360–72.

Douglas, Alexander. 2017. 'Descartes's Critique of the Syllogistic'. *History of Philosophy Quarterly* 4: 331–50.

Downing, Lisa. 1992. 'Are Corpuscles Unobservable in Principle for Locke?' *Journal of the History of Philosophy* 30: 33–52.

Downing, Lisa. 2001. 'The Uses of Mechanism: Corpuscularianism in Drafts A and B of Locke's Essay'. in *Late Medieval and Early Modern Corpuscular Matter Theories*, edited by Christoph Lüthy, John E. Murdoch and William R. Newman, 515–34. Leiden: Brill.

Drake, Stillman. 1978. *Galileo at Work: His Scientific Biography*. Chicago: University of Chicago Press.

Drake, Stillman. 1985. 'Review to William A. Wallace. Galileo and his Sources: The Heritage of the Collegio Romano in Galileo's Science. Princeton: Princeton University Press. 1984. Pp. xiv. 371. $42.50'. *The American Historical Review* 90: 970–1.

Dubouclez, Olivier. 2013. *Descartes et la voie de l'analyse*. Paris: PUF.

Ducheyne, Steffen. 2005a. 'Bacon's Idea and Newton's Practice of Induction'. *Philosophica*, 76: 115–28.

Ducheyne, Steffen. 2005b. 'Newton's Training in the Aristotelian Textbook Tradition: From Effects to Causes and Back'. *History of Science* 43: 217–37.

Ducheyne, Steffen. 2012. *The Main Business of Natural Philosophy: Isaac Newton's Natural-Philosophical Methodology*. Dordrecht: Springer.

Ducheyne, Steffen. 2013. 'The Status of the Theory and Hypotheses'. In *The Oxford Companion of British Philosophy in the Seventeenth Century*, edited by Peter R. Anstey, 169–91. Oxford: Oxford University Press.

Ducheyne, Steffen. 2015. 'An editorial history of Newton's *regulae philosophandi*'. *Estudios de Filosofía* 5: 143–64.

Ducheyne, Steffen, and Frederik Dhondt. 2021. 'Explicating his Natural Philosophical Method. A Study of the Development of the Methodological Statements in the Queries to the *Opticks*'. *Revue belge de Philologie et d'Histoire* 99: 341–88.

Dumitru, Claudia. 2013. 'Crucial Instances and Crucial Experiments in Bacon, Boyle, and Hooke'. *Society and Politics* 13: 45–61.

Duncan, Stewart D. R. 2003. *Thomas Hobbes: Metaphysics and Method*. PhD Dissertation, Rutgers, the State University of New Jersey, New Brunswick.

Duodo, Andrea. 1577. *De habitibus intellectus*. Venice: Nicolini.

Edwards, William F. 1967. 'Randall on the Development of Scientific Method in the School of Padua – A Continuing Reappraisal'. In *Naturalism and Historical Understanding*, edited by John P. Anton, 53–68. Albany: SUNY Press.

Edwards, William F. 1983. 'Paduan Aristotelianism and the Origins of Modern Theories of Method'. *Aristotelismo veneto e scienza moderna*, edited by Luigi Olivieri, 205–20. Padua: Antenore.

Engfer, Hans-Jürgen. 1996. *Empirismus versus Rationalismus? Kritik eines philosophiehistorischen Schemas*. Paderborn: Schöningh.

'Espinasse, Margaret. 1956. *Robert Hooke*. Berkeley: California University Press.

Euclid. 1517. *Elementa*. Paris: Stephani.

Fabri, Philippus. 1606. *Philosophia Naturalis Ioan. Duns Scoti*. Venice: Bertoni.

Feingold, Mordechai. 1983. 'Galileo in England: The First Phase'. In *Novità celesti e crisi del sapere*, edited by paolo. Galluzzi 411–20. Florence: Barbera.

Feingold, Mordechai. 1990a. 'Isaac Barrow: Divine, Scholar, Mathematician'. In *Before Newton. The Life and Times of Isaac Barrow*, edited by Mordechai Feingold, 1–104. Cambridge: Cambridge University Press.

Feingold, Mordechai. 1990b. 'Isaac Barrow's Library'. In *Before Newton. The Life and Times of Isaac Barrow*, edited by Mordechai Feingold, 333–72. Cambridge: Cambridge University Press.

Feingold, Mordechai. 2001. 'Mathematicians and Naturalists: Sir Isaac Newton and the Royal Society'. In *Isaac Newton's Natural Philosophy*, edited by Jed Z. Buchwald and I. Bernard Cohen, 77–102. Cambridge, MA: MIT Press, 2001.

Feingold, Mordechai. 2004. *The Newtonian Moment. Isaac Newton and the Making of Modern Culture*. Oxford: Oxford University Press.

Feingold, Mordechai. 2005. 'The Origins of the Royal Society Revisited'. In *The Practice of Reform in Health, Medicine, and Science, 1500–2000*, edited by Margaret Pelling and Scott Mandelbrote, 167–83. Aldershot: Ashgate.

Feingold, Mordechai. 2006. 'Parallel Lives: The Mathematical Careers of John Pell and John Wallis'. *Huntington Library Quarterly*, 26: 69: 451–68.

Ferrarin, Alfredo. 2005. *Hegel and Aristotle*. Cambridge: Cambridge University Press.

Finocchiaro, Maurice A. (1974). 'Newton's Third Rule of Philosophizing: A Role for Logic in Historiography'. *Isis* 65: 66–73.

Flavell, John. 1619. *Tractatus de demonstratione methodicus & polemicus*. Oxford: Lichfield.

Fletcher, Angus. 2005. 'Francis Bacon's Forms and the Logic of Ramist Conversion'. *Journal of the History of Philosophy* 43: 157–69.

Florka, Roger. 2001. *Descartes's Metaphysical Reasoning*. London: Routledge.

Fredette, Raymond. 2017. 'Les "De motu antiquiora" de Galileo Galilei: le lancement de la carrière du filosofo-geometra'. *Philosophia Scientiae* 21: 55–70.

Friedlein, Gottfried. 1873. *Procli Diadochi in primum Euclidis Elementorum librum commentarii*.

Friedman, Michael. 1990. 'Kant and Newton: Why Gravity Is Essential to Matter'. In *Philosophical Perspectives on Newtonian Science*, edited by Phillip Bricker and R. I. G. Hughes, 185–202. Boston: MIT Press.

Friedman, Michael. 1992. *Kant and the Exact Sciences*. Cambridge, MA: Harvard University Press.

Friedman, Michael. 2013. *Kant's Construction of Nature. A Reading of the Metaphysical Foundations of Natural Science*. Cambridge: Cambridge University Press.
Galilei, Galileo. 1960. *Discourse on Bodies in Water*. Urbana: University of Illinois Press.
Galilei, Galileo. 1967. *Dialogue Concerning the Two Chief World Systems – Ptolemaic & Copernican*. Berkeley: University of California Press.
Galilei, Galileo. 1974. *Two New Sciences*. Madison: Wisconsin University Press.
Galison, Peter. 1984. 'Descartes's Comparisons: From the Invisible to the Visible'. *Isis* 75: 311–26.
Galison, Peter. 1997. *Image and Logic. A Material Culture of Microphysics*. Chicago: University of Chicago Press.
Garber, Daniel. 1982. 'Locke, Berkeley, and Corpuscular Scepticism'. In *Berkeley: Critical and Interpretive Essays*, edited by Colin M. Turbayne, 174–93. Minneapolis: University of Minnesota Press.
Garber, Daniel. 1988. 'Descartes, the Aristotelians, and the Revolution That Didn't Happen in 1637'. *The Monist* 71: 471–86.
Garber, Daniel. 1992. *Descartes' Metaphysical Physics*. Chicago: Chicago University Press.
Garber, Daniel. 2001. *Descartes Embodied: Reading Cartesian Philosophy through Cartesian Science*. Cambridge: Cambridge University Press.
Garber, Daniel. 2009. 'Galileo, Newton and All That: If It Wasn't a Scientific Revolution, What Was It? (A Manifesto)'. *Circumscribere: International Journal for the History of Science* 7: 9–18.
Garber, Daniel. 2016. 'Why the Scientific Revolution Wasn't a Scientific Revolution, and Why It Matters'. In *Kuhn's Structure of Scientific Revolutions at Fifty*, edited by Robert Richards and Lorraine Daston, 133–48. Chicago: University of Chicago Press.
Gargani, Aldo. 1971. *Hobbes e la scienza*. Turin: Einaudi.
Garin, Eugenio. 1983. 'Aristotelismo veneto e scienza moderna'. In *Aristotelismo veneto e scienza moderna*, edited by Luigi Olivieri, 3–33. Padua: Antenore.
Garrison, John W. 1987. 'Newton and the Relation of Mathematics to Natural Philosophy'. *Journal of the History of Ideas*, 48: 609–27.
Gaukroger, Stephen. 1989. *Cartesian Logic, An Essay on Descartes's Conception of Inference*. Oxford: Clarendon Press.
Gaukroger, Stephen. 2001. *Francis Bacon and the Transformation of Early-Modern Philosophy*. Cambridge: Cambridge University Press.
Gaukroger, Stephen. 2002. *Descartes' System of Natural Philosophy*. Cambridge: Cambridge University Press.
Gaukroger, Stephen. 2006. *The Emergence of Scientific Culture. Science and the Shaping of Modernity, 1210–1685*. Oxford: Clarendon Press.
Gaukroger, Stephen. 2018. 'Syllogistic and Formal Reasoning: The Cartesian Critique'. In *The Aftermath of Syllogism. Aristotelian Logical Argument from Avicenna to Hegel*, 59–66. London: Bloomsbury.
Gaukroger, Stephen. 2019. 'Introduction'. In *The Philosophy of Knowledge. A History. Vol. III. Knowledge in Modern Philosophy*, 1–6. London: Bloomsbury.
Gewirth, Alan. 1943. 'Clearness and Distinctness in Descartes'. *Philosophy* 69: 17–36.
Giacobbe, Giulio Cesare. 1972a. 'Il Commentarium de certitudine mathematicarum disciplinarum di Alessandro Piccolomini'. *Physis* 14: 162–93.
Giacobbe, Giulio Cesare. 1972b. 'Francesco Barozzi e la Quaestio de certitudine mathematicarum'. *Physis* 14: 357–74.
Giacobbe, Giulio Cesare. 1973. 'La riflessione matematica di Pietro Catena'. *Physis* 15: 178–96.

Giacobbe, Giulio Cesare. 1976. 'Epigoni nel Seicento della Quaestio de certitudine mathematicarum'. *Physis* 18: 5–40.

Giacobbe, Giulio Cesare. 1977. 'Un gesuita progressista nella Quaestio de certitudine mathematicarum: Benito Pereyra'. *Physis*, 19: 51–86.

Gilbert, Neal W. 1960. *Renaissance Concepts of Method*. New York: Columbia University Press.

Gilbert, Neal W. 1963. 'Galileo and the School of Padua'. *Journal of the History of Ideas* 2: 223–31.

Giusti, Enrico. 1993. *Euclides reformatus. La teoria delle proporzioni nella scuola galileiana*. Torino: Bollati Boringhieri.

Giusti, Enrico. 1994. 'Il filosofo geometra. Matematica e filosofia naturale in Galileo'. *Nuncius* 9: 485–98.

Goddu, André. 2010. *Copernicus and the Aristotelian Tradition: Education, Reading and Philosophy in Copernicus's Path to Heliocentrism*. Leiden-Boston: Brill.

Grosholz, Emily. 1991. *Cartesian Method and the Problem of Reduction*. Oxford: Clarendon Press.

Guicciardini, Niccolò. 1999. *Reading the Principia: The Debate on Newton's Mathematical Methods for Natural Philosophy from 1687 to 1736*. Cambridge: Cambridge University Press, 1999.

Guicciardini, Niccolò. 2002. 'Analysis and Synthesis in Newton's Mathematical Work'. In *The Cambridge Companion to Newton*, edited by I. Bernard Cohen and George E. Smith, 308–38. Cambridge: Cambridge University Press.

Guicciardini, Niccolò. 2009. *Isaac Newton on Mathematical Certainty and Method*. Cambridge, MA: MIT Press.

Guyer, Paul. 1987. *Kant and the Claims of Knowledge*. Cambridge: Cambridge University Press.

Hahn, Robert. 1988. *Kant's Newtonian Revolution in Philosophy*. Carbondale: Southern Illinois University Press.

Hanna, Robert. 2006. *Kant, Science and Human Nature*. Oxford: Oxford University Press.

Hanson, Donald W. 1990. 'The Meaning of Demonstration in Hobbes's Science'. *History of Political Thought* 11: 587–626.

Harper, William. 2011. *Isaac Newton's Scientific Method: Turning Data into Evidence about Gravity and Cosmology*. Oxford: Oxford University Press.

Harriot, Thomas. 1631. *Artis Analyticae Praxis*. London: Barker.

Hatfield, Gary. 1988. 'Science, Certainty, and Descartes'. *Proceedings of the Biennial Meeting of the Philosophy of Science Association* 2: 249–62.

Hattab, Helen. 2014. 'Hobbes's and Zabarella's Methods: A Missing Link'. *Journal of the History of Philosophy* 52: 461–85.

Hattaway, Michael. 1978. 'Bacon and Knowledge Broken. Limits of Scientific Method'. *Journal of the History of Ideas* 39: 183–97.

Heath, Thomas L. 1908. *The Thirteen Books of Euclid's Elements*. Cambridge: Cambridge University Press.

Hesse, Mary B. 1964. 'Hooke's Development of Bacon's method'. In *Proceedings of the 10th International Congress for the History of Science*, 265–8. Paris: Hermann.

Hesse, Mary B. 1966. 'Hooke's Philosophical Algebra'. *Isis* 57: 67–83.

Hintikka, Jaakko. 1978. 'A Discourse on Descartes's Method'. *Descartes*, edited by Michael Hooker, 75–88. Baltimore: Johns Hopkins University Press.

Hintzman, Douglas L. 2003. 'Robert Hooke's Model of Memory'. *Psychonomic Bulletin & Review* 10: 3–14.

Hooke, Robert. 1661. *An Attempt for the Explication of the Phenomena*. London: Thompson.
Hooke, Robert. 1665. *Micrographia*. London: Royal Society.
Hooke, Robert. 1705. *Posthumous Works*. London: Royal Society.
Humber, James M. 1981. 'Recognizing Clear and Distinct Perceptions'. *Philosophy and Phenomenological Research* 41: 487–507.
Hume, David. 1751. *An Enquiry concerning the Principles of Morals*. London: Millar.
Hungerland, Isabel P. C., and Georg R. Vick. 1981. 'Misinterpreations of Hobbes: The Correct View'. In *Computatio, sive logica*, edited by I. C. Hungerland and G. R. Vick (eds.), 15–29. New York: Abaris Book.
Hunter, Michael. 1981. *Science and Society in Restoration England*. Cambridge: Cambridge University Press.
Hunter, Michael. 1982. *The Royal Society and Its Fellows, 1660-1700: The Morphology of an Early Scientific Institution*. Chalfont St. Giles: British Society for the History of Science.
Hunter, Michael. 1988. 'Promoting the New Science: Henry Oldenburg and the Early Royal Society'. *History of Science* 26: 165–81.
Hunter, Michael. 1989. *Establishing the New Science: The Experience of the Early Royal Society*. Woodbridge: Boydell Press.
Hunter, Michael. 1994. *Robert Boyle Reconsidered*. Cambridge: Cambridge University Press.
Hunter, Michael. 2000. *Robert Boyle (1627-91): Scrupulosity and Science*. Woodbridge: Boydell Press.
Hunter, Michael. 2003. 'Hooke the Natural Philosophy'. In *London's Leonardo: The Life and Work of Robert Hooke*, edited by Jim Bennet, Michael Cooper, Michael Hunter and Lisa Jardine, 105–62. Oxford: Oxford University Press.
Hunter, Michael. 2007. 'Robert Boyle and the Early Royal Society: A Reciprocal Exchange in the Making of Baconian Science'. *British Journal for the History of Science* 40: 1–23.
Hunter, Michael. 2009. *Boyle. Between God and Science*. New Haven: Yale University Press.
Hunter, Michael. 2011. 'Robert Boyle and Secrecy'. In *Secrets and Knowledge in Medicine and Science, 1500-1800*, edited by Elaine Leong and Alisha Rankin, 83–104. Aldershot: Ashgate.
Hunter, Michael. 2015. *Boyle Studies. Aspects of the Life and Thought of Robert Boyle (1627-91)*. Aldershot: Ashgate.
Hunter, Michael. 2016. *The Image of Restoration Science: The Frontispiece to Thomas Sprat's History of the Royal Society (1667)*. London: Routledge.
Hunter, Michael, and Paul B. Wood. 1986. 'Towards Solomon's House: Rival Strategies for Reforming the Early Royal Society'. *History of Science* 24: 49–108.
Hunter, Michael, and Simon Schaffer. 1989. *Robert Hooke: New Studies*. Woodbridge: Boydell Press.
Jalobeanu, Dana. 2015. 'Robert Boyle's Experimental Philosophy Revisited'. *Society and Politics* 9: 100–2.
Jalobeanu, Dana. 2016. *The Art of Experimental Natural Philosophy*. Bucharest: Zeta Books.
Janiak, Andrew. 2008. *Newton as Philosopher*. Cambridge: Cambridge University Press.
Janiak, Andrew. 2010. 'Newton's Forces in Kant's Critique'. In *Discourse on a New Method: Reinvigorating the Marriage of History and Philosophy of Science*, 91–110. Chicago: Open Court.
Jardine, Lisa. 1974. *Francis Bacon: Discovery and the Art of Discourse*. Cambridge: Cambridge University Press.

Jardine, Nicholas. 1976. 'Galileo's Road to Truth and the Demonstrative Regress'. *Studies in History and Philosophy of Science* 7: 277–318.
Jeake, Samuel. 1696. Λογιστικηλογια or Arithmetick. London: Kettilby-Mount.
Jesseph, Douglas M. 1999. *Squaring the Circle. The War between Hobbes and Wallis.* Chicago: University of Chicago Press.
Jesseph, Douglas M. 2004. 'Galileo, Hobbes, and the Book of Nature'. *Perspectives on Science* 12: 191–211.
Jesseph, Douglas M. 2010. 'Scientia in Hobbes'. in *Scientia in Early Modern Philosophy*, edited by Tom Sorrel, G. A. J. Rogers and Jill Kraye, 117–28. New York: Springer.
Jesseph, Douglas M. 2013. 'Logic and Demonstrative Knowledge'. In *The Oxford Handbook of British Philosophy in the Seventeenth Century*, edited by Peter R. Anstey, 373–90. Oxford: Oxford University Press.
Jungius, Joachim. 1652. *Doxoscopiae physicae minores.* Hamburg: Naumann.
Keckermann, Bartholomaeus. 1613. *Systema systematum.* Hannover: Anton.
Keill, John. 1702. *Introductio ad Veram Physicam seu Lectiones Physicae.* Oxford: Sheldon.
Kitcher, Patricia. 1993. *Kant's Transcendental Psychology.* Oxford: Oxford University Press.
Koertge, Noretta. 1977. 'Galileo and the Problem of Accidents'. *Journal of the History of Ideas* 38: 389–408.
Kosman, Louis A. 1964. *The Aristotelian Backgrounds of Bacon's Novum Organum.* PhD Dissertation, Harvard University, Cambridge, MA.
Koyré, Alexandre. 1965. *Newtonian Studies.* Cambridge, MA: Harvard University Press.
Kuhn, Thomas. 1977. *The Essential Tension: Selected Studies in Scientific Tradition and Change.* Chicago: University of Chicago Press.
Laird, W. Roy. 1997. 'Galileo and the Mixed Sciences'. In *Method and Order in Renaissance Philosophy of Nature. The Aristotle Commentary Tradition*, edited by Daniel A. Liscia and Eckhard Kessler, 253–70. Aldershot: Ashgate.
Larmore, Charles. 1980. 'Descartes' Empirical Epistemology'. In *Descartes. Philosophy, Mathematics and Physics*, edited by Stephen Gaukroger, 6–22. Sussex: Harvester Press.
Larsen, Robert E., 1962. 'The Aristotelianism of Bacon's Novum Organum'. *Journal of the History of Ideas* 4: 435–50.
Laudan, Larry. 1981. *Science and Hypothesis: Historical Essays on Scientific Methodology.* Dordrecht. Springer.
Leijenhorst, Cees. 2002. *The Mechanisation of Aristotelianism: The Late Aristotelian Setting of Thomas Hobbes' Natural Philosophy.* Leiden: Brill.
Levitin, Dimitri. 2016. 'Newton and Scholastic Philosophy'. *British Journal for the History of Science* 49: 53–77.
Levitin, Dimitri. 2019. 'Early Modern Experimental Philosophy. A Non-Anglocentric Overview'. In *Experiment, Speculation and Religion in Early Modern Philosophy*, edited by Alberto Vanzo and Peter R. Anstey, 229–91. London: Routledge.
Levitin, Dimitri. 2021. 'Newton on the Rules of Philosophizing and Hypotheses: New Evidence, New Conclusions'. *Isis* 2: 242–65.
Locke, John. 1823. *The Works of John Locke.* London: Tegg.
Locke, John. 1975. *The Clarendon Edition of the Works of John Locke: An Essay Concerning Human Understanding*, edited by Peter H. Nidditch. Oxford: Oxford University Press.
Locke, John. 1990. *The Clarendon Edition of the Works of John Locke: Drafts for the Essay Concerning Human Understanding, and Other Philosophical Writings: In Three Volumes, Vol. 1: Drafts A and B*, edited by Peter H. Nidditch and G. A. J. Rogers. Oxford: Oxford University Press.

Lohne, Johannes A. 1968. 'Experimentum Crucis'. *Notes and Records of the Royal Society of London* 23: 169–99.

Longuenesse, Béatrice. 1998. *Kant and the Capacity to Judge. Sensibility and Discursivity in the Transcendental Analytic of the Critique of Pure Reason*. Princeton: Princeton University Press.

Lorinus, Jean. 1620. *In universam Aristotelis logicam. Commentarii cum annexis disputationibus Romae ab eodem olim praelecti*. Köln: Cholinus.

Lynch, William. 2001. *Solomon's Child: Method in the Early Royal Society of London*. Stanford: Stanford University Press.

MacLaurin, Colin. 1748. *An Account of Sir Isaac Newton's Philosophical Discoveries*. London: Printed for the Author's Children.

Makkreel, Rudolf A. 1994. *Imagination and Interpretation in Kant. The Hermeneutical Import of the* Critique of Judgment. Chicago: Chicago University Press.

Malet, Antoni. 1997. 'Isaac Barrow on the Mathematization of Nature: Theological Voluntarism and the Rise of Geometrical Optics'. *Journal of the History of Ideas* 58: 265–87.

Malherbe, Michel. 1984. 'L'induction baconienne: De l'échec métaphysique à l'échec logique'. In *Francis Bacon. Terminologia e fortuna nel XVII*, edited by Marta Fattori, 179–200. Florence: Olschki.

Malherbe, Michel. 1990. 'Bacon's Critique of Logic'. In *Francis Bacon's Legacy of Texts*, edited by William A. Sessions, 69–87. New York: AMS Press.

Malherbe, Michel. 1996. 'Bacon's Method of Science'. In *The Cambridge Companion to Bacon*, edited by Markku Peltonen, 75–98. Cambridge: Cambridge University Press.

Mamiani, Maurizio. 2001. 'To Twist the Meaning: Newton's *Regulae Philosophandi* Revisited'. In *Isaac Newton's Natural Philosophy*, edited by Jed Z. Buchwald and I. B. Cohen, 3–14. Cambridge, MA: MIT Press.

Mancosu, Paolo. 1992. 'Aristotelian Logic and Euclidean Mathematics: Seventeenth-Century Developments of the Quaestio de Certitudine Mathematicarum'. *Studies in History and Philosophy of Science Part A* 23: 241–65.

Marion, Jean-Luc. 1992. 'Cartesian Metaphysics and the Role of the Simple Natures'. In *The Cambridge Companion to Descartes*, edited by John Cottingham, 115–39. Cambridge: Cambridge University Press.

Marsh, Narcissus. 1679. *Institutio logicae*. Dublin: Helsam.

Martin, Craig. 2006. 'Experience of the New World and Aristotelian Revisions of the Earth's Climates during the Renaissance'. *History of Meteorology*, 3: 1–15.

Martin, Craig. 2009. 'Conjecture, Probabilism, and Provisional Knowledge in Renaissance Meteorology'. *Early Modern Science and Medicine* 14: 265–89.

Martin, Craig. 2011. *Renaissance Meteorology: Pomponazzi to Descartes*. Baltimore: Johns Hopkins University Press.

Martin, Craig. 2014. *Subverting Aristotle: Religion, History and Philosophy in Early Modern Science*. Baltimore: Johns Hopkins University Press.

Martin, Gottfried. 1985. *Arithmetic and Combinatorics. Kant and His Contemporaries*. Carbondale: Southern Illinois University Press.

Martini, Cornelius. 1652. *Disputationes logicae*. Helmstedt: Müller.

McDonald, John F. 1972. 'Properties and Causes: An Approach to the Problem of Hypothesis in the Scientific Methodology of Sir Isaac Newton'. *Annals of Science*, 28: 217–33.

McDowell, John H. 1994. *The Mind and the World*. Cambridge, MA: Harvard University Press.

McGuire, James E. 1970. 'Newton's *Principles of Philosophy*: An Intended Preface for the 1704 *Opticks* and a Related Draft Fragment'. *The British Journal for the History of Science*, 5: 178–86.

McGuire, James E. 1995. *Tradition and Innovation: Newton's Metaphysics of Nature*. Boston: Kluwer Academic.

McGuire, James E., and Martin Tamny. 1983. *Certain Philosophical Questions: Newton's Trinity Notebook*. Cambridge: Cambridge University Press. 1983.

McMullin, Ernan. 1990. 'Conceptions of Science in the Scientific Revolution'. In *Reappraisals of the Scientific Revolution*, edited by David C. Lindberg and Robert S. Westman, 27–92. Cambridge: Cambridge University Press.

McMullin, Ernan. 2008. 'Explanation as Confirmation in Descartes's Natural Philosophy'. In *A Companion to Descartes*, edited by Janet Broughton and John Carriero, 84–102. Malden: Blackwell.

Mehl, Eduard. 2001. *Descartes en Allemagne, 1619-1620. Le contexte allemand de l'élaboration de la science cartésienne*. Strasbourg: Presses Universitaires de Strasbourg.

Mercer, Christia. 2001. *Leibniz's Metaphysics: Its Origins and Development*. Cambridge: Cambridge University Press.

Mercer, Christia. 2019. 'The Contextualist Revolution in Early Modern Philosophy'. *Journal of the History of Philosophy* 57: 529–48.

Mertz, Donald W. 1980. 'Galileo's Method of Causal Proportionality'. *Studies in History and Philosophy of Science Part A* 11: 229–42.

Mignucci, Mario. 1975. *L'argomentazione dimostrativa in Aristotele*. Padua: Antenore. Leipzig: Teubner.

Mill, John S. 1843. *A System of Logic, Ratiocinative and Inductive*. London: Parker.

Miller, David M. 2018. 'Regressus and Empiricism in the Controversy about Galileo's Lunar Observations'. *Perspectives on Science* 26: 293–324.

Milton, John R. 1987. 'Induction before Hume'. *British Journal for the Philosophy of Science* 38: 49–74.

Morhof, Daniel H. 1688. *Polyhistor Literarius, Philosophicus et Practicus*. Lübeck: Böckmann.

Mori, Giuliano. 2017. 'Mathematical Subtleties and Scientific Knowledge: Francis Bacon and Mathematics, at the Crossing of Two Traditions'. *British Journal for the History of Science* 50: 1–21.

Morris, John. 1969. 'Cartesian Certainty'. *Australasian Journal of Philosophy*, 47: 161–9.

Mugnai, Massimo. 2015. 'Ars characteristica, Logical Calculus, and Natural Languages'. In *The Oxford Companion of Leibniz*, edited by Maria Rosa Antognazza, 177–207. Oxford: Oxford University Press.

Nadler, Steven. 1989. *Arnauld and the Cartesian Philosophy of Ideas*. Manchester: Manchester University Press.

Napolitano Valditara, Linda. 1988. *Le idee, i numeri, l'ordine. La dottrina della mathesis universalis dall'Accademia antica al neoplatonismo*. Napoli: Bibliopolis.

Nelson, Alan. 2008. 'Cartesian Innateness'. In *A Companion to Descartes*, edited by Janet Broughton and John Carriero, 319–33. Malden: Blackwell.

Newton, Isaac. 1672a. 'A Letter of Mr. Isaac Newton, Mathematick Professor in the University of Cambridge, containing his New Theory about Light and Colors'. *Philosophical Transactions* 80: 4004–5007.

Newton, Isaac. 1672b. 'Mr. Newton's Answer to the Foregoing Letter'. *Philosophical Transactions* 85: 5014–18.

Newton, Isaac. 1672c. 'A Serie's of Quere's Propounded by Mr. Isaac Newton, to be Determin'd by Experiments, Positively and Directly Concluding His New Theory of Light and Colours; and Here Recommended to the Industry of the Lovers of Experimental Philosophy, as They Were Generously Imparted to the Publisher in a Letter of the Said Mr. Newtons of July 8.1672'. *Philosophical Transactions* 85: 5004–7.

Newton, Isaac. 1704. *Opticks*. London: Smith-Walford.

Newton, Isaac. 1706. *Optice*. London: Smith-Walford.

Newton, Isaac. 1718. *Opticks*. London: Innys.

Newton, Isaac. 1959. *Correspondence*. Cambridge: Cambridge University Press.

Newton, John. 1671. *An Introduction to the Art of Logick*. London: Passenger.

Nifo, Agostino. 1508. *Aristotelis Physicarum acroasum hoc est naturalium asuscultationum liber interprete*. Venice: Scoto.

Nifo, Agostino. 1522. *Collectanea ac commentaria in libros De anima*. Venice: Scoto.

Nifo, Agostino. 1549. *Aristotelis Physicarum acroasum hoc est naturalium asuscultationum liber interprete*. Venice: Scoto.

Nolan, Lawerence. 1997. 'The Ontological Status of Cartesian Natures'. *Pacific Philosophical Quarterly* 78: 169–94.

Okruhlik, Kathleen. 1989. 'The Foundation of All Philosophy: Newton's Third Rule'. In *An Intimate Relation: Studies in the History and Philosophy of Science*, edited by James Robert Brown and Jürgen Mittelstrass, 97–114. Dordrecht: Kluwer Academic.

Oldroy, David R. 1972. 'Robert Hooke's Methodology of Science as Exemplified in His Discourse of Earthquakes'. *The British Journal for the History of Science* 6: 109–30.

Oldroy, David R. 1980. 'Some 'Philosophical Scribbles' Attributed to Robert Hooke'. *Notes and Records of the Royal Society* 35: 17–32.

Oldroy, David R. 1987. 'Some Writings of Robert Hooke on Procedures for the Prosecution of Scientific Inquiry, Including His 'Lectures of Things Requisite to a Natural History'. *Notes and Records of the Royal Society* 41: 145–67.

Olivieri, Luigi. 1978. 'Galileo Galilei e la tradizione aristotelica'. *Verifiche* 11: 147–66.

Osler, Margaret. 1970. 'John Locke and the Changing Ideal of Scientific Knowledge'. *Journal of the History of Ideas* 31: 3–16.

Owen, David. 2000. *Hume's Reason*. Oxford: Oxford University Press.

Palkoska, Jan. 2017. *The A Priori in the Thought of Descartes: Cognition, Method and Science*. Newcastle upon Tyne: Cambridge Scholars.

Palmerino, Carla Rita. 2016. 'Reading the Book of Nature: The Ontological and Epistemological Underpinnings of Galileo's Mathematical Realism'. In *The Language of Nature: Reassessing the Mathematization of Natural Philosophy in the Seventeenth Century*, edited by Geoffrey Gorham, Benjamin Hill, Edward Slowik and C. Kenneth Waters, 29–50. Minneapolis: University of Minnesota Press.

Palmieri, Paolo. 2007. 'Science and Authority in Giacomo Zabarella'. *Science History Publications* 45: 404–27.

Panofsky, Erwin. 1972. *Renaissance and Renascences in Western Art*. New York: Harper & Row.

Pappas, George. 1998. 'Epistemology in the Empiricists'. *History of Philosophy Quarterly* 15: 285–302.

Pappus. 1589. Mathematicae Collectiones. Venice: De Franceschi.

Pappus. 1877. *Pappi Alexandrini Collectionis quae supersunt*. Berlin: Weidmann.

Papuli, Giovanni. 1983. 'La teoria del regressus come metodo scientifico negli autori della scuola di Padova'. In *Aristotelismo veneto e scienza moderna*, edited by Luigi Olivieri, 221–77. Padua: Antenore.

Parkinson, G. H. R. 1982. 'The Intellectualization of Appearances: Aspects of Leibniz's Theory of Sensation and Thought'. In *Leibniz. Critical and Interpretative Essays*, edited by Michael Hooker, 3–20. Minneapolis: University of Minnesota Press.

Pasnau, Robert. 2017. *After Certainty: A History of Our Epistemic Ideals and Illusions.* Oxford: Oxford University Press.

Pécharman, Martine. 1995. 'La logique de Hobbes et la tradition aristotélienne'. *Hobbes Studies* 8: 105–24.

Pécharman, Martine. 2016. 'Hobbes on Logic, or How to Deal with Aristotle's Legacy'. In *The Oxford Hanbook of Hobbes*, edited by A. L. Martinich and Kinch Hoekstra, 21–58. Oxford: Oxford University Press.

Perera, Benet. 1576. *De communibus omnium rerum naturalium principijs.* Rome: Zanetti.

Pérez-Ramos, Antonio. 1988. *Francis Bacon's Idea of Science and the Maker's Knowledge Tradition.* Oxford: Oxford University Press.

Petrella, Bernardino. 1571. *Quaestiones logicae.* Padua: Pasquato.

Piccolomini, Alessandro. 1551. *L'instrumento della filosofia*, Roma: Valgrisi.

Piccolomini, Alessandro. 1565. *In Mechanicas Quaestiones Aristotelis.* Venice: Curti.

Piccolomini, Alessandro. 1566. *Della sfera del mondo.* Venice: Varisco.

Piccolomini, Alessandro. 1558. *La prima parte de le theoriche.* Venice: Varisco.

Pickavé, Martin. 2005. 'The Notion of A Priori in Descartes and the Medieval Philosophers'. *Les études philosophiques* 75: 433–54.

Pomponazzi, Pietro. 1970. *Corsi inediti dell'insegnamento padovano.* Padua: Antenore.

Poppi, Antonino. 1969. 'Pietro Pomponazzi tra averroismo e galenismo sul problema del regressus'. *Rivista critica di storia della filosofia* 24: 243–66.

Poppi, Antonino. 1970. *Introduzione all'aristotelismo padovano.* Padua: Antenore.

Poppi, Antonino. 1972. *La dottrina della scienza in Giacomo Zabarella.* Padua: Antenore.

Poppi, Antonino. 2004. 'Zabarella or Aristotelianism as a Rigorous Science'. In *The Impact of Aristotelianism on Modern Philosophy*, edited by Riccardo Pozzo, 35–63. Washington, DC: CUA Press.

Powell, Griffith. 1594. *Analysis analyticorum posteriorum sive librorum Aristotelis de demonstratione.* Oxford: Barnes.

Pozzo, Riccardo. (ed.). 2004. *The Impact of Aristotelianism on Modern Philosophy.* Washington, DC: CUA Press.

Prins, Jan. 1990. 'Hobbes and the School of Padua: Two Incompatible Approaches of Science'. *Archiv für Geschichte der Philosophie* 72: 26–46.

Proclus. 1560. *Procli Diadochi Lycii Philosophi Platonici ac Mathematici probatissimi in Primum Euclidis Elementorum librum Commentariorum ad universam mathematicarum disciplinam principium eruditionis tradentium libri IIII.* Padua: Percacino.

Pugliese, David. 1982. *The Scientific Achievement of Robert Hooke: Method and Mechanics.* PhD Thesis, Harvard University, Cambridge, MA.

Putnam, Hilary. 1994. *Words and Life*, edited by James F. Conant. Cambridge, MA: Harvard University Press.

Rabouin, David. 2009. *Mathesis universalis. L'idée de mathématique universelle d'Aristote à Descartes.* Paris: PUF.

Raftopoulos, Athanassios. 2004. 'Cartesian Analysis and Synthesis'. *Studies in History and Philosophy of Science* 34: 265–308.

Ramus, Petrus. 1572. *Dialectica.* Basel: Episcopium.

Randall, John H. 1940. 'The Development of Scientific Method in the School of Padua'. *Journal of the History of Ideas* 1: 177–206.

Randall, John H. 1976. 'Paduan Aristotelianism Reconsidered'. In *Philosophy and Humanism: Renaissance Essays in Honor of Paul Oskar Kristeller*, edited by Edward P. Mahoney, 275–82. New York: Columbia University Press.

Recker, Doren A. 1993. 'Mathematical Demonstration and Deduction in Descartes's Early Methodological and Scientific Writings'. *Journal of the History of Philosophy* 31: 223–44.

Rees, Graham. 1986. 'Mathematics and Francis Bacon's Natural Philosophy'. *Revue Internationale de Philosophie* 159: 399–426.

Régis, Pierre-Sylvain. 1688. 'Philosophiæ Naturalis Principia Mathematica. Autore J.S. Newton Matheseos Professore Lucasiano et Societatis Regalis Sodali'. *Journal des Sçavans* 10: 153–4.

Reiss, Timothy J. 2000. 'Neo-Aristotle and method: Between Zabarella and Descartes'. In *Descartes's Natural Philosophy*, edited by Stephen Gaukroger, John Schuster and John Sutton, 195–227. London: Routledge.

Rey, Anne-Lise. 2013. 'The Status of Leibniz' Medical Experiments: A Provisional Empiricism?'. *Early Science and Medicine* 18: 360–80.

Risse, Wilhelm. 1964. *Die Logik der Neuzeit. Band I*. Stuttgart: frommann-holzboog.

Romanell, Patrick. 1984. *John Locke and Medicine. A New Key to Locke*. Buffalo: Prometheus Books.

Rose, Paul Lawrence. 1977. 'A Venetian Patron and Mathematician of the Sixteenth Century: Francesco Barozzi (1537–1604)'. *Studi veneziani* 1: 119–77.

Sacksteder, William. 1980. 'Hobbes: The Art of Geometricians'. *Journal of the History of Philosophy* 18: 131–46.

Sakellariadis, Spyros. 1982. 'Descartes's Use of Empirical Data to Test Hypotheses'. *Isis* 73: 68–76.

Sanderson, Robert. 1618. *Logicae artis compendium*. Oxford: Lichfield.

Salvia, Stefano. 2017. 'From Archimedean Hydrostatics to Post-Aristotelian Mechanics: Galileo's Early Manuscripts "De motu antiquiora" (ca. 1590)'. *Perspectives on Science* 19: 105–50.

Sargent, Rose-Mary. 1994. 'Learning from Experience: Boyle's Construction of an Experimental Philosophy'. In *Robert Boyle Reconsidered*, edited by Michael Hunter, 57–78. Cambridge: Cambridge University Press.

Sargent, Rose-Mary. 1995. *The Diffident Naturalist. Robert Boyle and the Philosophy of Experiment*. Chicago: University of Chicago Press.

Sasaki, Chikara. 2003. *Descartes's Mathematical Thought*. Dordrecht: Springer.

Saurin, Joseph. 1709. 'Examen d'une difficulté considerable proposée par M. Hughens contre le Systeme sur la cause de la Pesanteur'. *Histoire de l'Académie royale des sciences. Avec les memoires de mathématique et de physique* 1: 131–48.

Savini, Massimiliano. 2008. 'Comparatio vel ratiocinatio. Statuto e funzione del concetto di compartio/comparison nel pensiero di R. Descartes'. In *Descartes et des Letters. Epistolari e filosofia nell'età cartesiana*, edited by Francesco Marrone, 132–69. Florence: Le Monnier.

Scarapelli Cory, T. 2014. *Aquinas on Human Self-Knowledge*. Cambridge: Cambridge University Press.

Schaffer, Simon. 1989. 'Glass Works: Newton's Prisms and the Uses of Experiment'. In *The Uses of Experiment*, edited by David Gooding, Trevor Pinch and Simon Schaffer, 64–107. Cambridge: Cambridge University Press.

Schaffer, Simon, and Steven Shapin. 1985. *Leviathan and the Air-Pump: Hobbes, Boyle, and the Experimental Life*. Princeton: Princeton University Press.

Scheibler, Christoph. 1654–8. *Opus logicum*. Marburg: Chemlin.
Schliesser, Eric. 2007. 'Two Definitions of Cause, Newton, and the Significance of the Humean Distinction between Natural and Philosophical Relations'. *Journal of Scottish Philosophy* 5: 83–101.
Schmaltz, Tad. 1997. 'Descartes on Innate Ideas, Sensation, and Scholasticism: The Response to Regius'. In *Studies in Seventeenth-Century European Philosophy*, edited by M. A. Stewart, 33–73. Oxford: Clarendon Press.
Schmitt, Charles B. 1969. 'Experience and Experiment: A Comparison of Zabarella's View with Galileo's in De Motu'. *Studies in the Renaissance* 16: 80–138.
Schmitt, Charles B. 1983a. *Aristotle and the Renaissance*. Cambridge, MA: Harvard University Press.
Schmitt, Charles B. 1983b. 'L'aristotelismo nel veneto e le origini della scienza moderna: Alcune considerazioni sul problema della continuità'. In *Aristotelismo veneto e scienza moderna*, edited by Luigi Olivieri, 79–103. Padua: Antenore.
Schönfeld, Martin. 2000. *The Philosophy of the Young Kant. The Precritical Project*. Oxford: Oxford University Press.
Schouls, Peter A. 1980. *The Imposition of Method. A Study of Descartes and Locke*. Oxford: Clarendon Press.
Schuhmann, Karl. 1990. 'Hobbes and Renaissance Philosophy'. In *Hobbes Oggi*, edited by Andrea Napoli, 331–49. Milan: Franco Angeli.
Schwartz, Daniel. 2017. 'Crucial Instances and Francis Bacon's Quest for Certainty'. *HOPOS: The Journal of the International Society for the History of Philosophy of Science* 7: 130–50.
Sgarbi, Marco. 2013. *The Aristotelian Tradition and the Rise of British Empiricism. Logic and Epistemology in the British Isles (1570–1689)*. Dordrecht: Springer.
Sgarbi, Marco. 2016. *Kant and Aristotle. Epistemology, Logic, and Method*. New York: SUNY Press.
Sgarbi, Marco. 2017. 'What Does a Renaissance Aristotelian Look Like? From Petrarch to Galilei'. *HOPOS: The Journal of the International Society for the History of Philosophy of Science* 7: 226–45.
Sgarbi, Marco. 2018. 'Renaissance Facultative Logic and the Workings of the Mind: The Cognitive Turn'. In *Philosophy of Mind in the Late Middle Ages and Renaissance*, edited by Stephan Schmid, 270–90. London: Routledge.
Sgarbi, Marco. 2021. 'The Epistemology of Vernacular Aristotelianism in Renaissance Italy: The Case of Alessandro Piccolomini'. *Aither* 8: 190–215.
Shank, John B. 2018. *Before Voltaire: The French Origins of 'Newtonian' Mechanics, 1680–1715*. Chicago: University of Chicago Press.
Shapin, Steven. 2011. *A Social History of Truth: Civility and Science in Seventeenth Century England*. Chicago: University of Chicago Press.
Shapiro, Alan E. 1993. *Fits, Passions, and Paroxysms: Physics, Method, and Chemistry and Newton's Theories of Colored Bodies and Fits of Easy Reflection*. Cambridge: Cambridge University Press.
Shapiro, Alan E. 2004. 'Newton's "Experimental Philosophy"'. *Early Science and Medicine* 9: 168–217.
Shapiro, Barbara J. 1983. *Probability and Certainty in Seventeenth-Century England: A Study of the Relationships between Natural Science, Religion, History, Law and Literature*. Princeton: Princeton University Press.
Simionato, Giustina. 1973. 'Significato e contenuto delle Lectiones inedite di logica di Bernardino Tomitano'. *Quaderni per la storia dell'Università* 6: 111–24.

Singer, B. R. 1976. 'Robert Hooke on Memory, Association and Time Perception'. *Notes and Records of the Royal Society of London* 31: 114–31.

Smith, George E. 2002. 'The Methodology of the Principia'. In *The Cambridge Companion to Newton*, edited by I. Bernard Cohen and George E. Smith, 138–73. Cambridge: Cambridge University Press.

Smith, George E. 2012. 'How Newton's *Principia* Changed Physics'. In *Interpreting Newton. Critical Essays*, edited by Andrew Janiak and Erich Schliesser, 362–97. Cambridge: Cambridge University Press.

Smith, Kurt. 2001. 'A General Theory of Cartesian Clarity and Distinctness Based on the Theory of Enumeration in the Rules'. *Dialogue* 40: 279–309.

Smith, Kurt. 2010. *Matter Matters. Metaphysics and Methodology in the Early Modern Period*. Oxford: Oxford University Press.

Smith, Samuel. 1613. *Aditus ad logicam*. London: Stansby.

Sorell, Tom. 2010. 'Scientia and the Sciences in Descartes'. In *Scientia in Early Modern Philosophy. Seventeenth-Century Thinkers on Demonstrative Knowledge from First Principles*, edited by Tom Sorell, G. A. J. Rogers and Jill Kraye, 71–82. Dordrecht: Springer.

Sorell, Tom, G. A. J. Rogers and Jill Kraye (eds). 2010. *Scientia in Early Modern Philosophy*. New York: Springer.

Spencer, Quayshawn. 2004. 'Do Newton's Rules of Reasoning Guarantee Truth ... Must They?'. *Studies in History and Philosophy of Science, Part A* 35: 759–82.

Sprat, Thomas. 1667. *History of the Royal Society*. London: Martyn.

Stein, Howard. 1991. 'From the Phenomena of Motions to the Forces of Nature: Hypothesis or Deduction?'. *PSA: Proceedings of the Biennial Meeting of the Philosophy of Science Association (1990)* 2: 209–22.

Suter, Rufus. 1969. 'The Scientific Work of Allesandro [sic] Piccolomini'. *Isis* 60: 210–22.

Talaska, Richard A. 1998. 'Analytic and Synthetic Method According to Hobbes'. *Journal of the History of Philosophy* 26: 207–37.

Temkin, Owsei. 1935. 'Celsus' 'On medicine' and the Ancient Medical Sects'. *Bulletin of the History of Medicine* 3: 239–64.

Thorndike, Lynn. 1951. 'Newness and Craving for Novelty in Seventeenth-Century Science and Medicine'. *Journal of the History of Ideas* 4: 584–98.

Timmermans, Benoît. 1999. 'The Originality of Descartes's Conception of Analysis as Discovery'. *Journal of the History of Ideas* 60: 433–47.

Toletus, Franciscus. 1580. *Commentaria in universam Aristotelis logicam*, Venice. Giunti.

Trombetta, Antonio. 1502. *Opus in Metaphysicam Arist. Padue in thomistas discussum: cum questionibus perutilissimis antiquioribus adiectis in optimam seriem redactis: & formalitates eiusdem cum additionibus & dilucidatione diligenti exculte*. Venice: Scoto.

Vallius, Paulus. 1622. *Logica*. Lyon: Rouille.

Van de Pitte, Frederick P. 1998. 'Intuition and Judgment in Descartes' Theory of Truth'. *Journal of the History of Philosophy* 26: 543–70.

Vanzo, Alberto. 2012. 'Kant on Experiment'. In *Rationis Defensor*, edited by James Maclaurin, 75–96. Dordrecht: Springer.

Vanzo, Alberto. 2014. 'From Empirics to Empiricists'. *Intellectual History Review* 24: 517–38.

Viète, François. 1591. *In Artem Analyticen Isagoge*. Tours: Mettayer.

Wallace, William A. 1974. 'Galileo and the Reasoning Ex Suppositione: The Methodology of the Two New Sciences'. In *Proceedings of the 1974 Biennial Meeting of the Philosophy of Science Association*, 79–104. Dordrecht: Springer.

Wallace, William A. 1981. *Prelude to Galileo. Essays on Medieval and Sixteenth-Century Sources of Galileo's Thought*. Dordrecht: Reidel.

Wallace, William A. 1983. 'Aristotelian Influences on Galileo's Thought'. In *Aristotelismo veneto e scienza moderna*, edited by Luigi Olivieri, 349–78. Padua: Antenore.

Wallace, William A. 1984a. *Galileo and His Sources Sources: Heritage of the Collegio Romeno in Galileo's Science*. Princeton: Princeton University Press.

Wallace, William A. 1984b. 'Galileo and the Continuity Thesis'. *Philosophy of Science* 51: 504–10.

Wallace, William A. 1988a. 'Newton's Early Writings: Beginning of a New Direction'. In *Newton and the New Direction in Science*, 23–44. Vatican City: Specola Vaticana.

Wallace, William A. 1988b. 'Randall Redivivus: Galileo and the Paduan Aristotelians'. *Journal of the History of Ideas* 49: 133–49.

Wallace, William A. 1991. *Galileo, the Jesuits and the Medieval Aristotle*. Aldershot: Ashgate.

Wallace, William A. 1992a. *Galileo's Logical Treatises. A Translation, with Notes and Commentary, of His Appropriated Latin Questions on Aristotle's* Posterior Analytics. Dordrecht: Springer.

Wallace, William A. 1992b. *Galileo's Logic of Discovery and Proof: The Background, Content, and Use of His Appropriated Treatises on Aristotle's* Posterior Analytics. Dordrecht: Springer.

Wallace, William A. 1995. 'Circularity and the Paduan Regressus: From Pietro d'Abano to Galileo Galilei'. *Vivarium* 33: 76–97.

Wallace, William A. 1997. 'Domingo de Soto and the Iberian Roots of Galileo's Science'. In *Hispanic Philosophy in the Age of Discovery*, edited by Kevin White, 113–29. Washington, DC.: CUA Press.

Wallace, William A. 1998. 'Galileo's Pisan Studies in Science and Philosophy'. In *The Cambridge Companion to Galileo*, edited by Peter Machamer, 27–52. Cambridge: Cambridge University Press.

Wallace, William A. 2000. 'Dialectics, Experiments, and Mathematics in Galileo'. In *Scientific Controversies: Philosophical and Historical Perspectives*, edited by Peter Machamer, Marcello Pera and Aristides Baltas, 100–24. Oxford: Oxford University Press.

Wallace, William A. 2003. 'Galileo's Jesuits Connections and their Influence on His Science'. In *Jesuit Science and the Republic of Letters*, edited by Mordechai Feingold, 99–126. Cambridge, MA: Harvard University Press.

Wallace, William A. 2004. 'The Influence of Aristotle on Galileo's Logic and its Use in His Science'. In *The Impact of Aristotelianism on Modern Philosophy*, edited by Riccardo Pozzo, 64–83. Washington, DC: CUA Press.

Wallace, William A. 2007. 'Jesuits Influence on Galileo's Science'. In *The Jesuits. Vol. 2. Culture, Science and the Arts, 1540–1773*, 314–35. Toronto: University of Toronto Press.

Wallis, John. 1657. *Operum Mathmaticorum Pars Prima*. Oxford: Lichfield.

Ward, Seth. 1654. *Vindiciae Academiarum*. Oxford: Lichfield.

Watkins, John W. N. 1968. *Hobbes's System of Ideas: A Study in the Political Significance of Philosophical Theories*. New York: Barnes & Noble.

Wear, Andrew. 1985. 'Explorations in Renaissance Writings on the Practice of Medicine'. In *The Medical Renaissance of the Sixteenth Century*, edited by Andrew Wear, Roger K. French and Ian M. Lonie, 118–45, 312–17. Cambridge: Cambridge University Press.

Wendin, Michael V. 1987. 'Collection and Division in the Phaedrus and Statesman'. *Revue de philosophie ancienne* 5: 207–33.

Williams, Bernard. 2005. *Descartes. The Project of Pure Enquiry*. London: Routledge.

Winkler, Kenneth P. 2003. 'Lockean Logic'. In *The Philosophy of John Locke: New Perspectives*, edited by Peter R. Anstey, 154–78. London: Routledge.

Wisan, Winifried L. 1978. 'Galileo's Scientific Method: A Reexamination'. In *New Perspectives on Galileo*, edited by Robert E. Butts and Joseph C. Pitt, 1–57. Dordrecht: Reidel.

Wood, Paul. 1980. 'Methodology and Apologetics: Thomas Sprat's History of the Royal Society,' *The British Journal for the History of Science*, 13: 1–26.

Woolhouse, Roger. 2007. Locke: A Biography. Cambridge: Cambridge University Press.

Yates, Frances. 2010. *Selected Works. Volume III. The Art of Memory*. London: Routledge.

Zabarella, Jacopo. 1597. *Opera logica*. Frankfurt: Zetzner.

Zabarella, Jacopo. 1607. *De rebus naturalibus libri XXX*. Frankfurt: Zetzner.

Zabarella, Jacopo. 2013. *On Methods, on Regressus*, edited by. J. P. McCaskey, Cambridge, MA: Harvard University Press.

Zimara, Marco Antonio. 1537. *Tabula dilucidationum in dictis Aristotelis et Averrois*. Venice: Scoto.

Index

'Espinasse, Margaret 261, 280
Agrimi, Joel 241, 275
Airay, Christopher 39, 243, 275
Aldrovandi, Ulisse 153
Allison, Henry E. 273, 275
Annas, Julia 6, 239, 275
Anstey, Peter R. 1, 179, 239–40, 247, 261, 263–5, 268, 275, 279, 284, 293
Antognazza, Maria Rosa 286
Anton, John P. 280
Ariew, Roger 239, 254, 275
Aristotle 2–4, 6–9, 13–18, 22–5, 27, 31–2, 36–9, 54, 56, 60–2, 66–9, 75–6, 83, 86–7, 99–101, 111, 116–17, 122, 134–40, 143–7, 163, 173, 202, 205, 209, 212, 239–41, 244, 248, 251, 254–6, 259–60, 262, 275, 280, 284–5, 288–90, 292
Arnauld, Antoine 184, 286
Arthur, Richard T. W. 266, 275
Ayers, Michael 278

Baldassarri, Fabrizio 257–8, 275
Baldi, Gherardo 41–2, 244, 275
Balduino, Girolamo 14, 24–6, 37–8, 40, 74, 212–13, 241, 275
Balfour, Robert 49, 245, 276
Baltas, Aristides 292
Barnes, Jonathan 6, 239, 275
Baroncini, Gabriele 245, 252, 276
Barozzi, Francesco 69, 128–9, 131, 135, 138–9, 146, 259, 276, 281, 289
Barozzi, Pietro 13
Barrow, Isaac 10, 85, 131, 136–47, 154–5, 249, 259–60, 276, 280, 285
Beck, Leslie J. 254, 276
Beeckman, Isaac 128
Belkind, Ori 268–70, 276
Ben-Yami, Hanoch 257, 276
Bennett, Jim 261, 276, 283
Berkeley, George 3, 8

Berti, Enrico 79, 82–3, 241, 243, 248–51, 258, 276
Biancani, Giuseppe 88, 136, 140, 259
Biener, Zvi 211, 240, 269, 270–1, 276
Blake, Ralph M. 258, 276
Bodenmann, Siegfried 239, 276
Bonk, Thomas 239, 276
Booth, Edward 273, 276
Bourdin, Pierre 123
Boyle, Deborah A. 257, 276
Boyle, Robert 4–5, 59, 150, 202, 241, 257, 261, 275, 279, 283, 289
Brague, Remì 256, 276
Bricker, Phillip 280
Brissey, Patrick 254, 258, 276
Brittan, Gordon 273, 276
Broughton, Janet 286
Brown, James Robert 285
Brutscher, Johannes 40, 243, 276
Buchdahl, Gerd 3, 239, 273, 276–7
Buchwald, Jed Z. 268, 277, 280, 285
Buickerood, James G. 258, 277
Buonarroti, Michelangelo 140
Burgersdijk, Franco 39, 243, 252, 277
Burgundo, Claudio C. 39, 243, 277
Burtt, Edwin A. 261, 277
Butts, Robert E. 273, 277, 293

Camerota, Michele 249, 277
Campanini, Massimo 253, 277
Campanus of Novara 133
Capivacci, Girolamo 27–8, 36, 42, 51, 54, 98, 242, 277
Carbo, Ludovico 2, 40, 67, 243–4
Carleton, Thomas 39, 243, 277
Carriero, John 256–7, 277, 286
Carugo, Adriano 240, 243, 248, 251, 277
Case, John 48, 277
Cassirer, Ernst 240, 277
Catena, Pietro 134–5, 138, 143, 150, 259, 277, 281

Index

Celsus, Aulus Cornelius 53–4, 245–6, 277, 291
Cheyne, George 222, 271, 277
Clarke, Desmond M. 125, 127, 256, 258, 277
Clavius, Christoph 107
Cohen, I. Bernard 244, 267, 269–70, 277, 280, 282, 285, 291
Cohen, Laurence J. 244, 247, 278
Cohen, Morris R. 244, 278
Cohen, Robert S. 277
Conant, James F. 288
Cooper, Michael 261, 278, 283
Cotes, Roger 201, 203, 208–9, 213, 214–15, 219
Cottingham, John 285
Cozzoli, Daniele 254, 278
Crankanthorpe, Richard 39, 278
Crapulli, Giovanni 261, 278
Cremonini, Cesare 42, 244, 254, 278
Crescini, Angelo 240–2, 244, 246, 278
Crisciani, Chiara 241, 275
Crombie, Alistair C. 240, 243–4, 248, 251, 277–8
Crusius, Christian August 230
Curley, Edwin M. 254–5

D'Alambert, Jean le Rond 279
Daston, Lorraine 281
Davi, Maria R. 241, 278
Davis, John W. 277
Dawes, Gregory 248, 278
De Boer, Karin 239, 278
De Koninck, Thomas 256, 278
De Pace, Anna 259, 278
De Pierris, Gabrielle 256–7, 278
De Rosa, Raffaella 258, 278
Dear, Peter 242, 252, 259–62, 278
Della Rocca, Michael 255, 279
Derham, William 262, 279
Des Bosses, Bartholomew 183, 185
Dhondt, Frederik 267, 279
Dicker, Georges 273, 279
Diderot, Denis 222, 271, 279
Digby, Everard 45, 48, 245, 279
Dobre, Mihnea, 258, 279
Domski, Mary 247, 268, 271, 279
Dorling, Jon 242, 279
Douglas, Alexander 254, 279

Downing, Lisa 264, 279
Drake, Stillman 248, 251–2, 279
Dubouclez, Olivier 254, 279
Ducasse, Curt J. 276
Ducheyne, Steffen 265, 267, 269–71, 279
Dumitru, Claudia 247,
Duncan, Stewart D. R. 94, 252, 280
Duodo, Andrea 256, 280

Edwards, William F. 260, 243, 248, 252, 280
Engfer, Hans-Jürgen, 239, 280
Euclid 79, 133, 136, 138, 146, 249, 280, 282

Fabri, Philippus 39, 243, 280
Fattori, Marta 285
Feingold, Mordechai 259, 261, 263, 267–8, 277, 280, 292
Ferrarin, Alfredo 239, 280
Finocchiaro, Maurice A. 269, 270, 280
Flavell, John 39, 49, 243, 245
Fletcher, Angus 61, 247, 280
Florka, Roger 254, 280
Fredette, Raymond 249, 280
French, Roger K. 293
Friedlein, Gottfried 248, 258–60, 280
Friedman, Michael 235, 271, 273, 280–1

Galilei, Galileo 2, 4–5, 9–10, 40, 65–79, 81–91, 123, 146–7, 157–8, 160–1, 166, 169, 172, 177, 207, 221, 224, 231, 239, 243–4, 259–63, 277, 278–82, 284, 286–7, 289–90, 292–3
Galison, Peter 239, 254, 281
Galluzzi, Paolo 280
Garber, Daniel 5, 121, 123, 239, 254–5, 257–8, 264, 278, 281
Gargani, Aldo 253, 281
Garin, Eugenio 248, 281
Garrison, John W. 268, 281
Gaukroger, Stephen 240, 244, 254, 281, 284, 289
Gessner, Conrad 153
Gewirth, Alan 254–5, 281
Giacobbe, Giulio Cesare 259, 281–2
Gilbert, Neal W. 240, 248, 282
Gilbert, William 4, 87
Giusti, Enrico 249, 282, 290

Goddu, André 239, 282
Gooding, David 289
Gorham, Geoffrey 287
Grosholz, Emily 254, 282
Guicciardini, Niccolò 199–200, 268, 271, 282
Guyer, Paul 273, 282

Hahn, Robert 271, 282
Hanna, Robert 273, 282
Hanson, Donald W. 252, 282
Harman, Peter M. 277
Harper, William 269, 282, 287
Harriot, Thomas 156, 162–5, 263, 282
Hartsoeker, Nicolaas 203
Hatfield, Gary C. 278, 282
Hattab, Helen 252, 282
Hattaway, Michael 244, 282
Heath, Thomas 249, 282
Helbing, Otto 249
Hérigone, Pierre 156
Hesenthaler, Magnus 266
Hesse, Mary B. 149, 261, 263, 282
Hill, Benjamin 287
Hintikka, Jaakko 254, 282
Hintzman, Douglas L. 261, 282
Hoekstra, Kinch 288
Hooke, Robert 10, 85, 131, 137, 149–66, 199, 202–3, 259, 261–3, 269, 276, 278–80, 282–3, 287–8, 291
Hooker, Michael 282, 288
Hughes, R. I. G. 204, 280, 289
Humber, James M. 254, 283
Hume, David 8, 170, 215, 222, 231, 270, 278, 283, 286–7, 290
Hungerland, Isabel P. C. 252, 283

Iliffe, Robert 267

Jalobeanu, Dana 247, 283
Janiak, Andrew 201, 268, 270, 273, 276, 279, 283, 291
Janni, Pietro 275
Jardine, Lisa 217, 270, 283
Jardine, Nicholas 240, 248, 251, 270, 284
Jeake, Samuel 262, 284
Jesseph, Douglas M. 240, 253, 284
Jungius, Joachim 39, 243, 284

Keckermann, Bartholomaeus 49, 245, 252, 284
Keill, John 224–7, 271–2, 284
Kessler, Eckhard 284
Kitcher, Patricia 273, 284
Koertge, Noretta 251, 284
Kosman, Louis A. 244, 284
Koyré, Alexandre 269, 271, 284
Kraye, Jill
Kuhn, Thomas S. 247, 281

Laird, W. Roy. 248, 284
Lancaster, James A. T. 275–6
Larmore, Charle 255, 284
Larsen, Robert E. 244, 284
Laudan, Larry 1, 239, 241, 265, 284
Leijenhorst, Cees 239, 253, 284
Leong, Elaine 283
Levitin, Dimitri 239, 268–9, 284
Lindberg, David C. 286
Liscia, Daniel A. 284
Locke, John 2–4, 7, 11, 107–8, 167–79, 183–9, 192–5, 199, 208, 221, 223, 231, 246, 263–5, 268, 275, 277, 279, 281, 284, 287, 289–90, 293
Lohne, Johannes A. 247, 285
Longuenesse, Béatrice 273, 285
Lonie, Ian M. 293
Lorinus, Jean 2, 40, 67, 79, 243, 285
Lull, Ramon 128
Lüthy, Christoph 279
Lynch, William 261, 263, 285

Machamer, Peter 292
MacLaurin, Colin 222, 271, 285
Maclaurin, James 291
Madden, Edward. H. 276
Mahoney, Edward P. 289
Makkreel, Rudolf A. 273, 285
Malebranche, Nicolas 184
Malet, Antoni 260, 285
Malherbe, Michel 244, 285
Mamiani, Maurizio 267, 285
Mancosu, Paolo 259, 285
Mandelbrote, Scott 267, 280
Marion, Jean-Luc 117, 256, 285
Marrone, Francesco 289
Marsh, Narcissus 39, 243, 285
Martin, Craig 43, 240, 285

Martin, Gottfried 272, 285
Martini, Cornelius 39, 243, 285
Martinich, A. L. 288
Mazzini, Innocenzo 275
McCaskey, J. P. 293
McDonald, John F. 269, 285
McDowell, John H. 239, 285
McGuire, James E. 216, 267–70, 286
McMullin, Ernan 74, 122, 243, 249, 257, 286
Mehl, Eduard 243, 286
Mercer, Christia 239–40, 286
Mersenne, Marin 129, 252, 257, 278
Mertz, Donald W. 252, 286
Mignucci, Mario 248, 286
Mill, John S. 45, 244, 286
Miller, David M. 250, 286
Milton, John R. 244, 286
Mittelstrass, Jürgen 287
Morhof, Daniel H. 5, 239, 286
Mori, Giuliano 247, 286
Morin, Jean-Baptist 123, 256
Morris, John 255, 286
Moyal, George J. D. 277
Mugnai, Massimo 266, 286
Murdoch, John E. 279

Nadler, Steven 266, 286
Napoli, Andrea 290
Napolitano Valditara, Linda 259, 286
Neile, William 263
Nelson, Alan 258, 286
Newman, William R. 279
Newton, Isaac 1–2, 4, 7, 11, 82, 85, 122, 150, 164, 178–9, 194–236, 246, 257, 260–1, 263, 267–73, 275–7, 279–87, 289–92
Nidditch, Peter H. 284
Nifo, Agostino 14, 19–25, 38, 68, 96, 122, 212–13, 241, 287
Nolan, Lawerence 258, 287
Nyden, Tammy 258, 279

Okruhlik, Kathleen 270, 287
Oldroy, David R. 149, 261–2, 287
Olivieri, Luigi 240, 278–81, 287, 290, 292
Osler, Margaret 240, 287
Oughtred, William 156
Owen, David 264–5, 287

Palkoska, Jan 254, 287
Palmerino, Carla Rita 251, 287
Palmieri, Paolo 25, 240–1, 248, 287
Panofsky, Erwin 4, 239, 287
Pappas, George 265, 287
Pappus 75, 83, 85, 89, 161–2, 164, 193, 250, 287
Papuli, Giovanni 240–1, 287
Pardies, Ignace-Gaston 203
Parkinson, G. H. R. 183, 266, 288
Pasnau, Robert 6, 119, 239–40, 257, 288
Pécharman, Martine 252, 254, 288
Pell, John 280
Pelling, Margaret 280
Peltonen, Markku 285
Pera, Marcello 292
Perera, Benet 134, 136, 142, 259, 288
Pérez-Ramos, Antonio 54, 244, 246, 288
Petrella, Bernardino 243, 288
Piccolomini, Alessandro 25–7, 29, 37, 42, 51, 79–81, 133–5, 142, 145–6, 200, 228, 241–2, 250, 259–60, 268, 281, 288, 290–1
Pickavé, Martin 258, 288
Picot, Abbé 131
Pinch, Trevor 289
Pitt, Joseph C. 293
Plato 2, 6, 16, 55–7, 60–1, 69, 72, 155, 246
Pomponazzi, Pietro 13–17, 21, 22, 33, 37, 61, 68, 125, 127, 240–1, 285, 288
Poppi, Antonino 240–1, 248, 288
Powell, Griffith 48–9, 245, 288
Pozzo, Riccardo 239, 288, 292
Prins, Jan 102, 104, 252–3, 288
Proclus 69, 128–31, 133, 136–7, 141, 146, 157, 248, 258, 261, 288
Prunea-Bretonnet, Tinca 239, 278
Pugliese, David 266, 288
Putnam, Hilary 239, 288

Rabouin, David 258, 288
Raftopoulos, Athanassios 255, 288
Raiswell, Richard 275–6
Ramus, Petrus 61, 147, 247, 250, 275, 288
Randall, John H. 19, 65–6, 240–1, 244, 248, 280, 288–9, 292
Rankin, Alisha 283

Recker, Doren A. 255, 289
Rees, Graham 247, 289
Régis, Pierre-Sylvain 195, 267, 289
Reiss, Timothy J. 242–243, 254, 289
Rey, Anne-Lise 187, 239, 266, 276, 282, 287, 289
Richards, Robert 281
Risse, Wilhelm 240, 242, 254, 258, 276, 289
Rogers, G. A. J. 240, 284, 291
Romanell, Patrick 265, 289
Rook, Lawerence 156
Rorty, Amelie O. 278
Rose, Paul Lawrence 259, 289
Rubius, Antonious 107

Sacksteder, William 253, 289
Sakellariadis, Spyros 258, 289
Salvia, Stefano 86, 89, 249, 289
Sanderson, Robert 2, 49, 52, 202, 245, 267, 289
Sargent, Rose-Mary 261, 289
Sasaki, Chikara 258, 289
Saurin, Joseph 204–5, 269, 289
Savini, Massimiliano 254–5, 289
Scarapelli Cory, T. 258, 289
Schaffer, Simon 283, 289
Scheibler, Christoph 49, 245, 290
Schliesser, Eric 269, 271, 276, 279, 290–1
Schmaltz, Tad 257, 290
Schmid, Stephan 290
Schmitt, Charles B. 240, 248, 254, 290
Schönfeld, Martin 271, 290
Schouls, Peter A. 115, 255–6, 290
Schuhmann, Karl 252, 290
Schuster, John 275, 289
Schwartz, Daniel 59, 247, 290
Sessions, William A. 285
Sgarbi, Marco 239–40, 242, 244, 253–4, 290
Shank, John B. 269, 290
Shapin, Steven 261, 289–90
Shapiro, Alan E. 197, 204, 268–9, 277, 290
Shapiro, Barbara J. 268
Simionato, Giustina 241, 290
Singer, B. R. 261, 291
Slowik, Edward 287
Smeenk, Chris 276
Smiglecki, Martin 136, 138
Smith, George E. 222, 269, 271, 282, 291

Smith, Kurt 254–5, 291
Smith, Samuel 2, 39, 49, 202, 243, 245, 267, 291
Sophie Charlotte 183
Sorell, Tom 240, 254, 291
Spencer, Quayshawn 269, 291
Spinoza, Baruch 8, 107
Sprat, Thomas 149, 152, 261, 268, 283, 291, 293
Stein, Howard 268, 291
Stewart, M. A. 290
Suter, Rufus 250, 291
Sutton, John 289

Talaska, Richard A. 252, 291
Tamny, Martin 267, 286
Temkin, Owsei 246, 291
Theon of Alexandria 133
Theon of Smyrna 79, 155, 161–1, 164
Thorndike, Lynn 239, 291
Timmermans, Benoît 254, 258, 291
Toletus, Franciscus 40, 107, 243, 291
Trombetta, Antonio 14–19, 21, 23, 25, 241, 291
Tschirnaus, Ehrenfried Walhter von 184
Turbayne, Colin M. 281

Vallius, Paulus 2, 40, 41, 67, 243–4, 291
Van de Pitte, Frederick P. 258, 291
Vanzo, Alberto 240. 272, 284, 291
Vick, Georg R. 252, 283
Viète, François 79, 155–6, 160, 162, 164, 250, 262, 291
Vossius, Gerhard Johannes 136, 140

Wallace, William A. 40, 77, 82, 239–40, 243–4, 248–50, 267, 279, 292
Wallis, John 99, 131, 136, 138, 260, 280, 284, 292
Ward, Seth 93, 292
Wartofsky, Marx W. 277
Waters, C. Kenneth 287
Watkins, John W. N. 253, 292
Wear, Andrew 246, 293
Wendin, Michael V. 56, 246, 293
Westman, Robert S. 286
White, Kevin 292
Whitman, Anne 267, 269–70, 277
Williams, Bernard 125–7, 257–8, 293

Winkler, Kenneth P. 264, 293
Wisan, Winifried L. 248–9, 251, 293
Wood, Paul 263, 268, 293
Woolhouse, Roger 263, 293

Yates, Frances 247, 293

Zabarella, Jacopo 2, 9, 14, 25, 31–40, 43, 49, 54, 65, 67–9, 72–4, 93–6, 98, 100–4, 108, 122, 131, 135–6, 138, 140, 147, 163, 167, 212–13, 240, 242–4, 248, 252–5, 259, 272, 277, 282, 287–90, 293
Zimara, Marco Antonio 23, 241, 293

www.ingramcontent.com/pod-product-compliance
Lightning Source LLC
Chambersburg PA
CBHW052151300426
44115CB00011B/1621